In Focus – Special Book Series

Wastewater-based Epidemiology at the Frontier of Global Public Health

Editors

Matthew J. Wade, Joshua T. Bunce, Susan Petterson, Christobel Ferguson, Nohelia Castro del Campo, Erica Gaddis and Panagiotis Karanis

IWA PUBLISHING

Published by **IWA Publishing**
Unit 104-105, Export Building
1 Clove Crescent
London E14 2BA, UK
Telephone: +44 (0)20 7654 5500
Fax: +44 (0)20 7654 5555
Email: publications@iwap.co.uk
Web: www.iwapublishing.com

First published 2023

British Library Cataloguing in Publication Data
A CIP catalogue record for this book is available from the British Library

ISBN: 9781789064322

Contents

doi: 10.2166/wh.2023.001

Editorial: Wastewater-based epidemiology at the frontier of global public health

The COVID-19 pandemic demonstrated the necessity for highly interdisciplinary, multi-stakeholder collaboration to enable effective public health practice and response. The application of wastewater monitoring to support the pandemic response was demonstrated quickly, leveraging existing scientific and engineering capability to rapidly develop methods that enabled scaling of monitoring programmes across the globe (Naughton *et al.* 2021; The Economist 2022). The science and practice of wastewater monitoring, or wastewater-based epidemiology (WBE) as it is often termed in the context of surveillance, for public health threats is well-documented. As a highly socialised tool, national, regional, and international communities of interest have joined efforts to provide evidence of the value of the data derived and its application in practice. While the use of wastewater monitoring to detect and characterise targets of public health concern is not new, COVID-19 and the (re)emergence of rare, novel, or unexpected pathogens (e.g., Poliovirus (Klapsa *et al.* 2022; Ryerson *et al.* 2022; Zuckerman *et al.* 2022), Mpox (de Jonge *et al.* 2022), Enteric viruses (Reyne *et al.* 2023), Respiratory illnesses (Wolfe *et al.* 2022)) stresses the importance of continued research and development into the utility of wastewater monitoring and utility of the data it generates. As wastewater monitoring continues to move from the research community into applied public health agencies, there is a need to develop standard methods and data analytics for specific applications. Establishing wastewater-based epidemiology at the frontier of global public health requires a clearer understanding and acknowledgment of its value by public health policy makers, regulators, and government.

This book presents a broad view of WBE thinking and practice, delineating the current understanding and future potential of the field. The topics covered range from geographically distinct reports on the use of WBE to inform on pathogen circulation in a population, through technical developments and data utility, to commentaries on future challenges and opportunities including for WBE 'beyond the pandemic'.

As a pandemic, COVID-19 has seen a global public health response, with regional, national, and sub-national variation in surveillance and intervention. Although wastewater monitoring of the virus has not been adopted universally, a significant number of countries have or continue to provide complementary data on SARS-CoV-2 load or concentration in sewage (Naughton *et al.* 2021), the majority of which are large, well-resourced countries. However, the democratisation of WBE, to a large extent, has enabled less populated, geographically small, or lower resourced nations, to pursue monitoring campaigns. Markt *et al.* (2022) report on the results of a six month monitoring programme in the Principality of Liechtenstein, highlighting the potential of WBE as an 'early-warning' indicator ahead of clinical results in response to lower testing rates. This, and their observation that changes in phenotypic expression across variants may affect the interpretation of wastewater measurements, is not unique and demonstrates some inherent determinism in the use of WBE. Similarly, Fang *et al.* (2022), report on the wastewater monitoring programme in Scotland, and address the issue of variability in both structural and analytical terms, describing approaches to manage inherent and sporadic variation in the data.

Normalisation of wastewater measurements is discussed by Fang *et al.* (2022) as approach to account for systematic variation in the data (e.g., through extraneous, non-human flow contributions). Sakarovitch *et al.* (2022) focus explicitly on the use of normalisation to avoid misinterpretation of the data. Here, they recommend that multicomponent, or composite, normalisation can account for multiple sources of variability and evidence its improvement over single factor normalisation (e.g., using flow rate).

The management, analysis, and use of WBE data is a broad but important topic and assuring the methods employed for WBE is needed to provide reliable and comparable numeric estimates and predictions of community disease burden. Anneser *et al.* (2022) present an example of its use for modelling the relationship between viral concentrations (wastewater and settled solids) measured across three regions in New England, USA and corresponding individual case rates in those geographies. Several statistical models were employed, with promising results suggesting some high level of correspondence between the datasets.

While studies that employ models to ascribe a relationship between wastewater measurements and other measures of disease burden, may find a relationship, the fact that no metric provides complete confidence in the true state and all suffer from inherent biases and uncertainties, urges caution when interpreting modelled relationships. This is discussed by Faraway *et al.* (2022), who provide a summary of some of the data and methodological issues that complicate development of effective predictive models for disease case rates including temporal and spatial heterogeneity, and the effects of different levels of data aggregation.

While most WBE research and practice in recent years has focused on SARS-CoV-2, a groundswell of diverse applications for wastewater monitoring have arisen in the wake of the pandemic, based often on extant knowledge, and integrating recent advances in methods and data science. In the wake of easing COVID-19 restrictions and interventions, much speculation on the effect of these, and increased public health awareness, on other circulating diseases. Here, Abd-Elshafy *et al.* (2022), discuss the impact of increased hygiene practices on common enteric virus infections at a location in Egypt. By monitoring wastewater during a period of high COVID-19 prevalence and comparing enteric virus data to a similar period prior to the pandemic, the authors suggest that the public health impact of this behaviour amongst the population indirectly lessens the incidence of enteric diseases.

The global public health threat of antimicrobial resistance (AMR) has meant there is an increased requirement for effective monitoring. This has elevated debate on the value of wastewater monitoring in response. Understanding the relationship between the temporal variation in Antimicrobial Resistance Gene (ARG) levels and the presence of resistance markers in wastewater is a key challenge within this value proposition. Steenbeek *et al.* (2022) report a significant positive correlation between several antibiotics and ARGs, and integrons (i.e., the measure of microbial adaptation and evolution), as measured in wastewater. However, the authors also highlight the need to design appropriate sampling schemes to reduce the risk of mischaracterising the burden of antibiotic resistance.

WBE has the potential to provide an increasing level of resolution on community or aggregated population health, as the quality, quantity, and diversity of data extracted from wastewater is realised. Soller *et al.* (2022) demonstrate the utility of wastewater-based mechanistic models for normalisation of wastewater data to ensure that the collated information is consistent under variable conditions. Hence, the integration of assured WBE programmes, including best practice approaches, within public health systems will enable authorities and decision-makers to respond with the appropriate level of knowledge and confidence. During the COVID-19 pandemic, small-scale, community-organised wastewater-based monitoring initiatives emerged in parallel with larger, well-funded government programmes. One such study (Brooks *et al.* 2023), conducted in Yarmouth, ME, USA (pop. 8,900) was established to assist with the public health response to disease outbreaks, providing end-to-end surveillance, data analyses and dissemination at community level and affect a rapid response at local scale.

Although, during the pandemic, WBE has been integrated into some public health surveillance programmes, several important research gaps need to be resolved to expand the utility and reliability of WBE in the future both for SARS-CoV-2 and other public health applications. Robins *et al.* (2022) review these gaps in the context of WBE for pathogenic viruses, antimicrobial resistance, and chemicals. Additional research is needed in the areas of markers in stool and urine, environmental factors, and methodologies for sample collection and quantification (Robins *et al.* 2022).

Compounding the wider need for both empirical research, standardisation of methods and protocols, and greater, open communication between disciplines and sectors, Hyllestad (Hyllestad *et al.* 2022) conducted a systematic review of the value proposition for environmental surveillance as an early-warning of public health threats. Although focused on the large and expanding data and information evidence generated within the COVID-19 pandemic, this review and the aforementioned research gaps analysis (Robins *et al.* 2022) suggest that WBE sits firmly at the frontier of global public health – one that requires appropriate resourcing, evidence development, and stronger alignment with the plethora of public health functions that underpin societal health resilience into the future.

Guest Editors
Matthew J. Wade
Analytics & Data Science Directorate, UK Health Security Agency, Nobel House, Smith Square, London SW1P 3JR, United Kingdom

Joshua T. Bunce
Department for Environment, Food and Rural Affairs, 2 Marsham Street, London SW1P 4DF, United Kingdom

Susan Petterson IWA
School of Medicine, Griffith University, North Sydney, 2060, Australia

Christobel Ferguson
The Water Research Foundation, West Quincy Ave, Denver, Colorado 80235, USA

Nohelia Castro del Campo
Centro de Investigación en Alimentación y Desarrollo, Carretera a Eldorado k.m. 5.5 Campo El Diez, C.P. 80110. Culiacán, Sinaloa, México

Erica Gaddis
SWCA Environmental Consultants, 257 East 200 South, Salt Lake City, Utah 84111, USA

Panagiotis Karanis
Unit of Anatomy and Morphology, University of Nicosia Medical School, 21 Ilia Papakyriakou, 2414 Engomi, CY-1700, Nicosia, Cyprus

REFERENCES

Abd-Elshafy, D. N., Nadeem, R. & Bahgat, M. M. 2022 Prevalence of enteric viruses in wastewater in Egypt after the COVID-19 pandemic. *J. Water Health* **20** (11), 1668–1672. doi:10.2166/wh.2022.223.

Anneser, E., Riseberg, E., Brooks, Y. M., Corlin, L. & Stringer, C. 2022 Modeling the relationship between SARS-CoV-2 RNA in wastewater or sludge and COVID-19 cases in three New England regions. *J. Water Health* **20** (5), 816–828. doi:10.2166/wh.2022.013.

Brooks, Y. M., Gryskwicz, B., Sidaway, E., Shelley, B., Coroi, L., Downing, M., Downing, T., McDonnell, S., Ostrye, D., Hoop, K. & Parrish, G. 2023 A case study of a community-organized wastewater surveillance in a small community: correlating weekly reported COVID-19 cases with SARS-CoV-2 RNA concentrations during fall 2020 to summer 2021 in Yarmouth, ME. *J Water Health* **21** (3), 329–342. doi: https://doi.org/10.2166/wh.2023.238

de Jonge, E. F., Peterse, C. M., Koelewijn, J. M., Anne-Merel R., van der Drift, Rudolf F. H. J. van der Beek, Nagelkerke, E. & Lodder, W. J. 2022 The detection of monkeypox virus DNA in wastewater samples in the Netherlands. *Sci. Total Environ.* **852**, 158265. doi:10.1016/j.scitotenv.2022.158265.

Fang, Z., Roberts, A. M. I., Mayer, C.-D., Frantsuzova, A., Potts, J. M., Cameron, G. J., Singleton, P. T. R. & Currie, I. 2022 Wastewater monitoring of COVID-19: a perspective from Scotland. *J Water Health* **20** (12), 1688–1700. doi: https://doi.org/10.2166/wh.2022.082

Faraway, J., Boxall-Clasby, J., Feil, E. J., Gibbon, M. J., Hatfield, O., Kasprzyk-Hordern, B. & Smith, T. 2022 Challenges in realising the potential of wastewater-based epidemiology to quantitatively monitor and predict the spread of disease. *J. Water Health* **20** (7), 1038–1050. doi:10.2166/wh.2022.020.

Hyllestad, S., Myrmel, M., Lomba, J. A. B., Jordhøy, F., Schipper, S. K. & Amato, E. 2022 Effectiveness of environmental surveillance of SARS-CoV-2 as an early warning system during the first year of the COVID-19 pandemic: a systematic review. *J. Water Health* **20** (8), 1223–1242. doi:10.2166/wh.2022.115.

Klapsa, D., Wilton, T., Zealand, A., Bujaki, E., Saxentoff, E., Troman, C., Shaw, A. G., Tedcastle, A., Majumdar, M., Mate, R., Akello, J. O., Huseynov, S., Zeb, A., Zambon, M., Bell, A., Hagan, J., Wade, M. J., Ramsay, M., Grassly, N. C., Saliba, V. & Martin, J. 2022 Sustained detection of type 2 poliovirus in London sewage between February and July, 2022, by enhanced environmental surveillance. *The Lancet* S0140673622018049. doi:10.1016/S0140-6736(22)01804-9.

Markt, R., Endler, L., Amman, F., Schedl, A., Penz, T., Büchel-Marxer, M., Grünbacher, D., Mayr, M., Peer, E., Pedrazzini, M., Rauch, W., Wagner, A. O., Allerberger, F., Bergthaler, A. & Insam, H. 2022 Detection and abundance of SARS-CoV-2 in wastewater in Liechtenstein, and the estimation of prevalence and impact of the B.1.1.7 variant. *J. Water Health* **20** (1), 114–125. doi:10.2166/wh.2021.180.

Naughton, C. C., Roman, F. A. Jr., Alvarado, A. G. F., Tariqi, A. Q., Deeming, M. A., Bibby, P. K., Bivins, A., Rose, J. B., Medema, G., Ahmed, W., Katsivelis, P., Allan, V., Sinclair, R., Zhang, Y. & Kinyua, M. N. 2021 *Show us the Data: Global COVID-19 Wastewater Monitoring Efforts, Equity, and Gaps*. Public and Global Health, preprint. doi:10.1101/2021.03.14.21253564.

Reyne, M. I., Allen, D. M., Levickas, A., Allingham, P., Lock, J., Fitzgerald, A., McSparron, C., Nejad, B. F., McKinley, J., Lee, A., Bell, S. H., Joshua Quick, J., Houldcroft, C. J., Bamford, C. G. G., Gilpin D. F. & McGrath, J. W. 2023 Detection of human adenovirus F41 in wastewater and its relationship to clinical cases of acute hepatitis of unknown aetiology. *Sci. Total Environ.* **857**, 159579. doi:10.1016/j.scitotenv.2022.159579.

Robins, K., Leonard, A. F. C., Farkas, K., Graham, D. W., Jones, D. L., Kasprzyk-Hordern, B., Bunce, J. T., Grimsley, J. M. S., Wade, M. J., Zealand, A. M. & McIntyre-Nolan, S. 2022 Research needs for optimising wastewater-based epidemiology monitoring for public health protection. *J. Water Health* **20** (9), 1284–1313. doi:10.2166/wh.2022.026.

Ryerson, A. B., Lang, D., Alazawi, M. A., Neyra, M., Hill, D. T., St. George, K., Fuschino, M., Lutterloh, E., Backenson, B., Rulli, S., Ruppert, P. S., Lawler, J., McGraw, N., Knecht, A., Gelman, I., Zucker, J. R., Omoregie, E., Kidd, S., Sugerman, D. E., Jorba, J., Gerloff, N., Ng, T. F. F., Lopez, A., Masters, N. B., Leung, J., Burns, C. C., Routh, J., Bialek, S. R., Oberste, M. S. & Rosenberg, E. S. 2022 Wastewater testing and detection of poliovirus type 2 genetically linked to virus isolated from a paralytic polio case – New York, March 9–October 11, 2022. *MMWR Morb. Mortal. Wkly. Rep.* **71** (44), 1418–1424. doi:10.15585/mmwr.mm7144e2.

Sakarovitch, C., Schlosser, O., Courtois, S., Proust-Lima, C., Couallier, J., Pétrau, A., Litrico, X. & Loret, J.-F. 2022 Monitoring of SARS-CoV-2 in wastewater: what normalisation for improved understanding of epidemic trends? *J. Water Health* **20** (4), 712–726. doi:10.2166/wh.2022.012.

Soller, J., Jennings, W., Schoen, M., Boehm, A., Wigginton, K., Gonzalez, R., Graham, K. E., McBride, G., Kirby, A. & Mattioli, M. 2022 Modeling infection from SARS-CoV-2 wastewater concentrations: promise, limitations, and future directions. *J. Water Health* **20** (8), 1197–1211. doi:10.2166/wh.2022.094.

Steenbeek, R., Timmers, P. H. A., van der Linde, D., Hup, K., Hornstra, L. & Been, F. 2022 Monitoring the exposure and emissions of antibiotic resistance: co-occurrence of antibiotics and resistance genes in wastewater treatment plants. *J. Water Health* **20** (8), 1157–1170. doi:10.2166/wh.2022.021.

The Economist 2022 'How covid-19 spurred governments to snoop on sewage', *The Economist*. Available from: https://www.economist.com/international/2022/09/07/how-covid-19-spurred-governments-to-snoop-on-sewage.

Wolfe, M. K., Duong, D., Bakker, K. M., Ammerman, M., Mortenson, L., Hughes, B., Arts, P., Lauring, A. S., Fitzsimmons, W. J., Bendall, E., Hwang, C. E., Martin, E. T., White, B. J., Boehm, A. B. & Wigginton, K. R. 2022 Wastewater-based detection of two influenza outbreaks. *Environ. Sci. Technol. Lett.* **9** (8), 687–692. doi:10.1021/acs.estlett.2c00350.

Zuckerman N. S., Itay, B.-O., Danit, S., Efrat, B., Hagar, M., Shulman L. M., Nofar, L., Leah, W., Irina, A., Zvi, C., Klil, K., Rinat, V., Michal, E., Fratty, I. S., Miranda, G., Marina, W., Oran, E., Ruth, Y., Lior, H.-S., Sharon, A.-P., Ella, M. & Merav, W. 2022 Emergence of genetically linked vaccine-originated poliovirus type 2 in the absence of oral polio vaccine, Jerusalem, April to July 2022. *Eurosurveillance* **27** (37). doi:10.2807/1560-7917.ES.2022.27.37.2200694.

doi: 10.2166/wh.2021.180

Detection and abundance of SARS-CoV-2 in wastewater in Liechtenstein, and the estimation of prevalence and impact of the B.1.1.7 variant

R. Markt [a,*], L. Endler [b], F. Amman [b], A. Schedl [b], T. Penz[b], M. Büchel-Marxer[c], D. Grünbacher[d], M. Mayr[a], E. Peer[a], M. Pedrazzini[c], W. Rauch [d], A. O. Wagner[a], F. Allerberger[e], A. Bergthaler[b] and H. Insam[a]

[a] Department of Microbiology, Universität Innsbruck, Technikerstraße 25d, Innsbruck 6020, Austria
[b] Research Center for Molecular Medicine of the Austrian Academy of Sciences, Lazarettgasse 14, Vienna 1090, Austria
[c] Ministry of Social Affairs and Culture, Peter-Kaiser-Platz 1, Vaduz 9490, Principality of Liechtenstein
[d] Unit of Environmental Engineering, Department of Infrastructure, Universität Innsbruck, Technikerstraße 13, Innsbruck 6020, Austria
[e] Austrian Agency for Health and Food Safety (AGES), Vienna 1220, Austria
*Corresponding author. E-mail: rudolf.markt@uibk.ac.at

RM, 0000-0003-0195-1249; LE, 0000-0002-9115-8756; FA, 0000-0002-8646-859X; AS, 0000-0002-1126-7726

ABSTRACT

The new coronavirus 2 (SARS-CoV-2) is known to be also shed through feces, which makes wastewater-based surveillance possible, independent of symptomatic cases and unbiased by any testing strategies and frequencies. We investigated the entire population of the Principality of Liechtenstein with samples from the wastewater treatment plant Bendern (serving all 39,000 inhabitants). Twenty-four-hour composite samples were taken once or twice a week over a period of 6 months from September 2020 to March 2021. Viral RNA was concentrated using the PEG centrifugation method followed by reverse transcription quantitative PCR. The aim of this research was to assess the suitability of SARS-CoV-2 fragments to relate the viral wastewater signal to the incidences and assess the impact of the emerging B.1.1.7. variant. The viral load in the wastewater peaked at almost 9×10^8 viral fragments per person equivalent (PE) and day on October 25, and showed a second peak on December 22 reaching a viral load of approximately 2×10^8 PE^{-1} d^{-1}. Individual testing showed a lag of 4 days and a distinct underestimation of cases at the first peak when testing frequency was low. The wastewater signal showed an immediate response to the implementation of non-pharmaceutical interventions. The new virus variant B.1.1.7. was first detected in wastewater on December 23, while it was first observed with individual testing on January 13, 2021. Further, our data indicate that the emergence of new virus variant may change the wastewater signal, probably due to different shedding patterns, which should be considered in future models.

Key words: COVID-19, inflow, SARS-CoV, sewer, wastewater surveillance

HIGHLIGHTS

- Wastewater-based epidemiology covering an entire country (Liechtenstein).
- Lucrative lead of 4 days of SARS-CoV-2 gene copy numbers versus individual testing.
- Variant of concern B.1.1.7. detected in wastewater 2 weeks prior to individual testing.
- Variant of concern B.1.1.7. increased the ratio of wastewater signal per incidence.

1. INTRODUCTION

The new coronavirus respiratory disease (called COVID-19) is caused by the severe acute respiratory syndrome coronavirus 2 (SARS-CoV-2) that is known to be also shed through feces (Jones *et al.* 2020). Since disease outbreak is now globally widespread and had caused a pandemic, the need for fast, sensitive and reliable surveillance systems has become increasingly important. Thus, the detection of SARS-CoV-2 cannot solely rely upon individual testing, which cannot be seen as a long-term solution both from an economical and environmental perspective. As soon as disease prevalence reaches low levels, alternative options for disease surveillance must be adopted. One such option is wastewater-based epidemiology (WBE; Medema *et al.* 2020) that has been for many years a powerful and effective tool in controlling and eradicating diseases such as polio (Berchenko *et al.* 2017). For many years though, WBE has been neglected as a tool in disease surveillance;

however, with the onset of the COVID-19 pandemic researchers started to reconsider this powerful tool for detecting SARS-CoV-2 in wastewater across the globe (e.g., La Rosa *et al.* 2020; Ahmed *et al.* 2021; Gerrity *et al.* 2021; Karthikeyan *et al.* 2021).

The primary mode of transmission of SARS-CoV-2 is via respiratory droplets that people produce when they exhale, cough or sneeze (Karia *et al.* 2020). It has been shown that when sputum or saliva is swallowed (approximately 1.0–1.5 L $person^{-1}$ d^{-1}), viral RNA enveloped in mucus may pass the stomach and intestines in a semi-protected state, thus avoiding degradation by gastric acid and pancreatic juices (Hirose *et al.* 2017). Apart from cells in the respiratory tract, SARS-CoV-2 may also infect other organs such as the gastrointestinal tract through the angiotensin-converting enzyme 2 receptor (ACE2), an important component of the angiotensin system (Pola *et al.* 2021). This results in diarrhea, vomiting and other gastrointestinal disorders (Xiao *et al.* 2020), which affect shedding viral particles through our wastewater system. It is estimated that up to 67% of infected persons shed SARS-CoV-2 viral particles with their stool (Parasa *et al.* 2020; Wu *et al.* 2020). As SARS-CoV-2 particles in sputum and saliva may also directly contribute to the viral load in wastewater (Wyllie *et al.* 2020), given the abundance of the virus genetic material in nasopharyngeal fluids (approximately 10^5–10^{11} gc mL^{-1}) is much higher than in both urine (approximately 10^2–10^5 gc mL^{-1}) and feces (approximately 10^2–10^7 gc mL^{-1}) (Jones *et al.* 2020), it is important to say that the virus load in wastewater may not only be a result of feces and urine shedding but also of sputum and saliva spreading of infected people.

For the above reason, monitoring of SARS-CoV-2 RNA in wastewater may serve the purpose of following and predicting important infection trends (Wu *et al.* 2020) during the course of the COVID-19 disease. It is important that the degree of fecal release of viral particles is not related to the severity of the disease (Zheng *et al.* 2020) as SARS-CoV-2-RNA was detected in fecal probes despite negative PCR-testing result (Chen *et al.* 2020). Thus, WBE may be viewed as a monitoring and surveillance method independent of symptomatic cases and unbiased by any selected testing strategies and frequencies. Testing of wastewater may also be beneficial for seasonal outcomes as well as predicting travel and commuting effects (Pearce *et al.* 2020).

However, like any other monitoring system, WBE also has drawbacks such as sampling procedure, concentration steps and RNA quantification method. Commonly, 24-h volumetric sampling is preferred. While earlier studies often have relied on frozen wastewater samples, it has been found that freezing and thawing of samples may drastically lower the yield of intact RNA (Markt *et al.* 2021). In general, low viral loads in wastewater require elaborated and well-considered strategies to concentrate the RNA, such as the polyethylene glycol (PEG) precipitation method, which has been chosen in many laboratories despite the disadvantage of losing the viral fragments bound to solid matter (around 50%) (Lu *et al.* 2020a, 2020b; Pérez-Cataluña *et al.* 2021). Extraction protocols and RT-qPCR bring additional uncertainties. However, it is clear and strongly recommended recently by the European Union, that WBE has huge potential to be used for COVID-19 surveillance (European Commission 2021). In particular, better estimates of the prevalence of COVID-19 in a population are required for optimizing and fine-tuning disease control strategies.

Smaller countries with only a few wastewater treatment plants (WWTPs) may particularly benefit from WBE during this pandemic. An example of such a country is the Principality of Liechtenstein which is here shown in our study. It is the fourth smallest state in Europe and geographically located in the alps between Switzerland and Austria. The country with approximately 39,000 inhabitants has only one WWTP (ARA – Abwasserreinigungsanlage Bendern) serving the whole country. Thus, intensive wide-scale individual testing combined with wastewater surveillance on SARS-CoV-2, both detection and sequencing, offers Liechtenstein a unique opportunity to shed light on an entire country's viral prevalence on COVID-19. The aim of this study is to show the dynamics of COVID-19 fragments in the inflow of the WWTP of ARA Bendern (technically configured for 105,000 population equivalents) and relate this to the COVID-19 disease incidence. It is hypothesized that WBE (i) shows rises in viral loads earlier than individual testing does, (ii) disease incidence is closely correlated with wastewater signal, (iii) new viral variants are detected earlier in wastewater than by individual testing and (iv) new variants change the viral wastewater signal.

2. MATERIALS AND METHODS

2.1. Sample collection and preparation

About 250 mL liquid influent samples were collected with a 24-h volume-equivalent sampler, cooled to 4 °C (Markt *et al.* 2021) and shipped to the laboratory in cooled packaging. In the laboratory, 70 g of the sample was transferred to an

80 mL centrifugation tube (Oak Ridge Type 3118, Nalgene™) and spiked with a transgenic virus-like particle based on the MS2-phage as internal control for sample preparation and RT-qPCR (real-time quantitative PCR) analysis. In order to remove particulate matter, the tube was then centrifuged for 30 min at 4,500 g (4508 R cooling centrifuge, Eppendorf, Hamburg, Germany) without brake (Medema *et al.* 2020). About 65 mL of the supernatant were then transferred to a centrifuge tube filled with 6.5 g of PEG (PEG 8000; ROTIPURAN®) and 1.46 g of sodium chloride. To dissolve the PEG, the tube was shaken with a tilt/roller mixer (RS-TR 05, Phönix Instrument, Garbsen, Germany) for approximately 10 min at room temperature. RNase-free water was used for balancing the centrifuge tubes. Samples were then further centrifuged at 12,000 g for 99 min without brake. The supernatant was carefully removed, leaving a small transparent pellet. About 800–1,000 μL of lysis buffer were added to the pellet, followed by vortexing for 15 s at 2,500 rpm and another centrifugation at 2,000 g for a few seconds (with stop) to collect all the lysis buffer droplets at the bottom of the falcon tube. The entire solution was then transferred to an Eppendorf micro reaction tube. RNA was then purified from the lysed pellet using the Monarch™ total RNA Miniprep Kit (New England Biolabs, Ipswich, USA).

2.2. RNA quantification

RNA copy numbers were determined using a duplex measurement assay for SARS-CoV-2 targeting the nucleocapsid (N1) gene primers (Lu *et al.* 2020a, 2020b) at the green channel and the process control based on MS2-phage at the yellow channel. A total of 20 μL RT-qPCRs contained 10 μL Luna Universal Probe One-Step Reaction Mix (2×), 1 μL Luna WarmStart® RT Enzyme Mix (20×) (both New England Biolabs), 0.8 μL primer (forward and reverse, final concentration 0.4 μM each target), 0.4 μL probe (final concentration 0.2 μM each target), 2 μL PCR grade water and 5 μL template. Analyses were conducted on a RotorGene cycler (Qiagen, Hilden, Germany). After an initial reverse transcription at 55 °C for 10 min followed by 95 °C for 1 min of denaturation, 45 cycles of 95 °C for 10 s and 60 °C for 40 s were performed running a two-step protocol. To calculate copy numbers, a plasmid standard containing the N-gene of SARS-CoV-2 (2019-nCoV_N_Positive Control, IDT, Leuven, Belgium) was used (Markt *et al.* 2021). Samples with dropouts or difference >2× the standard deviation per run of mean cycle threshold of the internal control were discharged.

2.3. Background data

The WWTP ARA Bendern (47.21202°N, 9.50106°E) serves and collects wastewater of all 11 communities of the Principality of Liechtenstein. The maximum length of stay of the wastewater within the catchment system is lower than 6 h, and the annual wastewater volume is 9.5×10^6 m^3. The daily load of chemical oxygen demand and NH_4-N (ammonium nitrogen) sums up to 11,000 and 386 kg, respectively. Wastewater SARS-CoV-2-data were obtained on a weekly basis from September 20 to October 25, 2020; later, the inflow was sampled twice a week.

Individual testing was exclusively based on PCR tests, resulting in incidences on a daily basis. Almost all of the positive samples were sequenced to check for variants of concern. Data are available at https://www.liechtenstein.li/land-und-leute/gesellschaft/

2.4. Virus signal – data processing

Wastewater-based epidemiology encompasses a range of complex processes and variability, i.e., from viral shedding of infected persons to virus transport and degradation in the sewer system, sampling and quantification in the laboratory. Therefore, we outline a three-step procedure for data processing.

2.4.1. Outlier detection

Since the sewer system is a combined system, rain events dilute the wastewater and result in an increased runoff. For extreme discharge conditions, part of the wastewater will be immediately directed into the receiving water via combined sewer overflows. For these cases, a significant part of the viral load will not appear at the sampling point due to coincidental loss. Measurements taken under such conditions are thus likely to yield a false negative or lower signal and are treated as an outlier. We used the inflow data from the WWTP ARA Bendern over a period of 10 years (2010–2020) where flow is recorded daily (Q in $m^3\ d^{-1}$) and NH_4-N every 5 days (c_{NH_4} in g NH_4-N m^{-3}). In the following, we apply NH_4 as a proxy for the virus as we assume that both are diluted and thus behave approximately equal in the sewer system. In Figure 1, we use the relation of the NH_4 load ($Q^*c_{NH_4}$) to the mean NH_4 load during dry weather conditions (the mean dry weather flow in 2020 was 24,300 $m^3\ d^{-1}$). Under ideal conditions, this should be a unit fraction, but the influence of combined sewer overflows is visible for

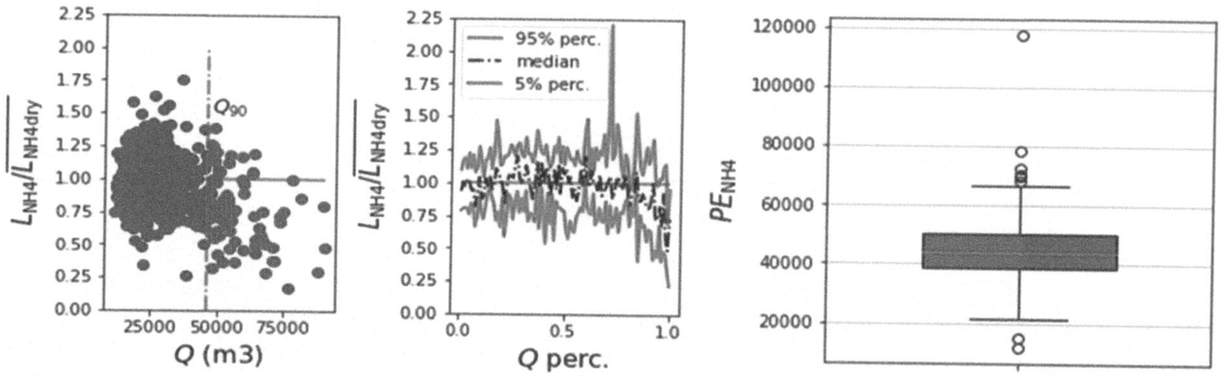

Figure 1 | Left – relative ammonium (NH_4) load versus inflow; center – relative NH_4 load plotted as the ordered list (percentiles) of inflow plus confidence intervals (5–95%); right – boxplot of PE computed with 8 g NH_4/PE/d from the inflow data 2010–2020.

higher inflows. From this, we conclude that samples taken while the inflow is higher than the 90th percentile of the recorded inflow data (here 46,250 $m^3 d^{-1}$) are potentially false and thus to be treated as outliers, which was the case for three samples.

2.4.2. Normalization

The use of population biomarkers in WBE is standard practice to account for the temporal variation in the population of the watershed (Been *et al.* 2014). While there are certainly more accurate biomarkers possible, in this case study we are restricted to the use of the parameter NH_4 in the inflow and apply a standard load of 8 g NH_4/PE (person equivalent)/day for an estimation such as:

$$\mathrm{PE}_{\mathrm{NH_4}} = \frac{c_{\mathrm{NH_4}} * Q}{8}$$

Figure 1 depicts the variation of the PE discharging into the WWTP from the inflow data over the past 10 years. The median (44,100 PE) matches closely with the actual population, including regular incoming commuters who are common for the country. We further estimate realistic boundaries of the variation with the boxplot features as 20,900 PE minimum and 67,800 PE maximum. Samples outside of that range are either potentially false numbers or caused by external NH_4 sources that do not relate to the population and are removed as outliers.

2.4.3. Smoothing

Arabzadeh *et al.* (2021) describe the background and necessity of data smoothing in relation to SARS-CoV-2 signals. Following their procedure, we transformed the raw signal (c_{virus} in copies mL^{-1}) to the specific viral load (L_{virus}) using NH_4-based normalization:

$$L_{\mathrm{virus}} = \frac{c_{\mathrm{virus}} * 8}{c_{\mathrm{NH_4}}}\left[\frac{\text{Mega gene copies}}{\mathrm{PE} * d}\right]$$

Mega gene copies (Mgc) refers to 10^6 gene copies. For data smoothing, we apply the SPLINE method (Arabzadeh *et al.* 2021).

2.5. Incidence modeling

Incidence is here defined as the epidemiological situation based on active cases documented by individual PCR tests. This measure is likewise common in pandemic surveillance as, e.g., 7-day incidence rate per 100,000. However, note that incidence deviates significantly from the actual infection status (usually denoted as prevalence) as there is always a significant number of undocumented cases which are not considered (Rippinger *et al.* 2021). Moreover, the incidence values are subject to test numbers and strategies and thus contain a significant uncertainty both in assessment and interpretation.

Despite the deficiencies, incidence modeling still serves as a valuable information and a prediction tool on epidemiological dynamics. In the following, we apply a simple regression model to estimate active cases (prevalence) from the viral load data (as given after data processing and smoothing). For calibration, we apply the period of the second pandemic wave from November 16, 2020 to December 31, 2020 and used the model both retrospectively (the first pandemic wave) and predictively (until the end of March 2021). The underlying assumption is that the relation of undocumented cases to recorded cases in the calibration period applies also to the whole dataset.

2.6. SARS-CoV-2 whole genome sequencing

Extracted RNA was transcribed into cDNA and amplified according to the ARTIC Network (https://artic.network/) protocol (https://github.com/artic-network/artic-ncov2019/tree/master/primer_schemes/nCoV-2019/V3). Sample pooling and sequencing was performed on an Illumina NovaSeq 6000 where quality filtering of raw reads, demultiplexing, mapping and low frequency calling for single-nucleotide polymorphisms (SNPs) was conducted according to a published protocol by Popa *et al.* (2020). SNPs-calling for SARS-CoV-2 variant assignment to the B.1.1.7 lineage is based on the mutations listed in https://cov-lineages.org/global_report_B.1.1.7.html. Thereby, the median frequency of all B.1.1.7-specific SNPs covered and assigned was used to estimate the relative abundance of the B.1.1.7 variant in wastewater. Wet-lab library preparation and bioinformatic analysis was conducted at the CeMM Research Center for Molecular Medicine of the Austrian Academy of Sciences, Vienna. Using this method, we reached a median of 514 not sufficiently covered nucleotides per Sars-CoV-2 genome. Consensus sequences were deposited on GISAID (Liechtenstein, Enviromental samples).

To investigate a variant-driven shedding behavior, we compared the ratio of virus load per 7-day incidence over two time periods using the Mann–Whitney *U*-test. The first time period (named wild-type) comprises data from November 6, 2020 to December 26, 2020, and the second time period (named B.1.1.7) comprises all data beginning with the occurrence of B.1.1.7 in wastewater starting on December 27, 2020.

2.7. Statistics

Single wastewater samples were taken weekly, respectively, twice a week. To be able to relate the wastewater data to daily incidences, the data were normalized on the basis of NH_4-N and data were smoothed as described above (Arabzadeh *et al.* 2021).

3. RESULTS

In Liechtenstein, over a period of 14 weeks from July 2020 to mid-September 2020, the number of tested individuals ranged around 200 PCR tests per week. During the following time, the PCR test frequency was sharply increased to an average of 800 PCR tests per week until the end of March 2021. The number of new positive-tested persons showed two peaks, one during the first week of November 2020 and a second during the last two weeks of December 2020 with 36 and 41 cases per day, respectively. Consistent with these numbers, the percentage of positive tests peaked during this same time, while later it decreased from the beginning of January onwards (Figure 2).

Based on the daily performed positive PCR tests, the 7-day incidence rate was calculated and showed two distinct 'wave' peaks (Figure 3). Similarly, the viral wastewater signals showed two distinct peaks. The first peak of the wastewater signal was about four times higher than the second one. While this first wastewater peak showed up earlier than the first peak of the 7-day incidence rate, which was not the case for the second peak. In contrast to the peaks, in both 'waves' the increase of the wastewater signal started earlier than the incidences did. During the two 'waves', the wastewater signal started to rise after September 28 and November 29, while the incidences started to rise around October 4 and December 17, respectively. Along with the incidences, the wastewater signal decreased in parallel as shown during the periods of November 2020 and January 2021. Effects of non-pharmaceutical interventions were reflected by both incidences and wastewater signals (Figure 3).

In Liechtenstein, most of the positive individual tests were analyzed for virus variant mutations through quality-tested, mutation-specific PCRs. This method allowed the detection of unique mutation constellations, such as the N501Y and E484 K mutations, within a few hours in contrast to genome sequencing, which requires several days (ÖGLMKC 2021). In the individual samples, the first variant of concern (VOC; variant B.1.1.7) was detected on January 13, 2021. The 7-day incidence rate of the VOC is shown (Figure 4). Conversely, sequencing of wastewater samples revealed a weak signal of the variant B.1.1.7, based on two of 16 marker mutations in late December 2022, followed by a higher than 20% signal in mid-January 2021 (12 of 16 marker mutations). A few weeks later the virus load signal was nearly up to 100%. The B.1.1.7 viral wastewater peaks on January 20 were followed by a peak based on individual testing 7-day incidence on January

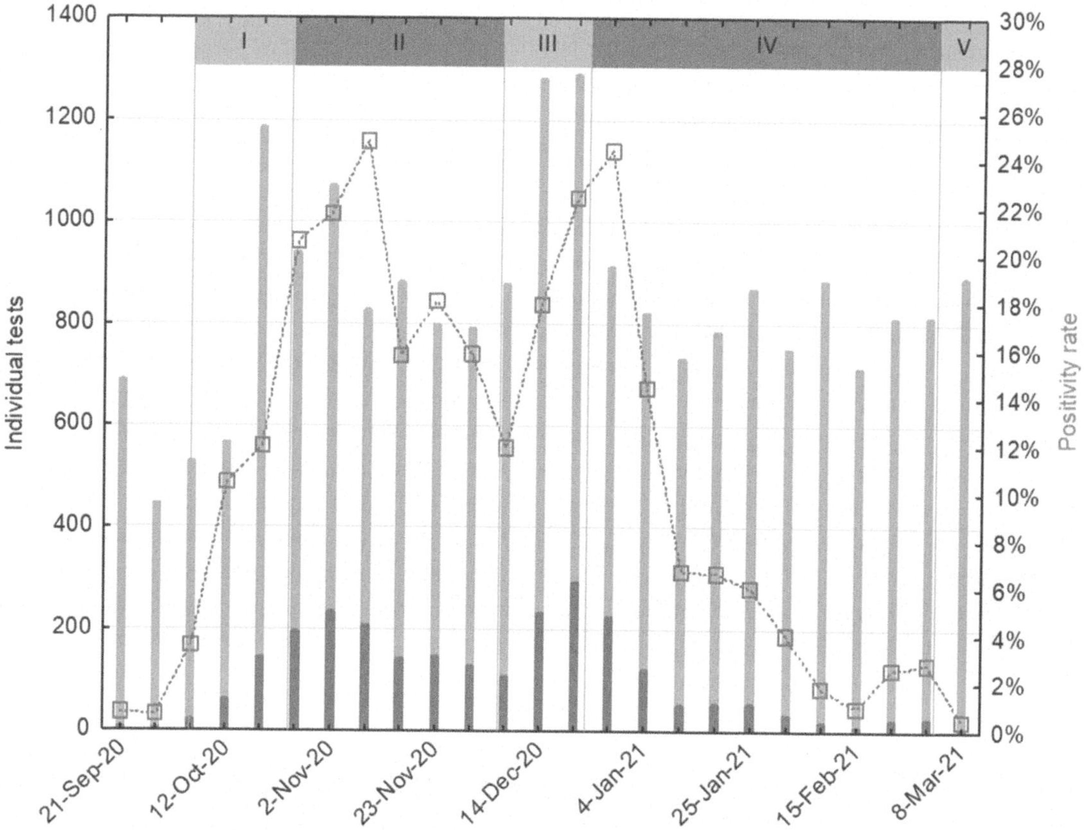

Figure 2 | Individual testing in Liechtenstein (weekly means). Number of positive (red bars), negative (green bars) and total number of tests (green+red bars) and positivity rate (red-dashed line). Non-pharmaceutical interventions in upper bar as: I. Oct 10 increased social distancing in public and private encounters <30 persons, II. Oct 24 gastronomy industry closed, III. Nov 30 gastronomy industry opened, IV. Dec 20 culture, sport, gastronomy industry closed, V. March 1 culture, sport reopened, encounters with 10 persons max.

24. While by February 18, the wastewater signal reached another peak, this was observed for the 7-day incidence only on March 8.

While the ratio of virus load per 7-day incidence was very high at the time when the test frequency was increased, this ratio decreased until the beginning of November 2020. Then, for nearly 2 months, the ratio of virus load per 7-day incidence remained constant at around 0.62 ± 0.14 Mgc PE^{-1} 7-day $incidence^{-1}$ (Figure 5). During this episode, 992 ± 187 tests per week were conducted. From December 27, 2020, when the variant B.1.1.7 was detected for the first time in wastewater, the ratio increased to an average of 1.24 ± 0.76 Mgc PE^{-1} 7-day $incidence^{-1}$ and showed two small peaks, one around mid-January and another one around the end of February 2021. Individual testing decreased during this period to 814 ± 64 tests per week. A comparison by the Mann–Whitney *U*-test indicated a statistically significantly higher ($n > 50$ days, $p < 0.005$) virus load per incidence for the B.1.1.7 period compared with the wild-type period (Davies *et al.* 2021).

Correlating active cases (proxied by the 14-day incidence) and the virus load in the wastewater, we derive

$$\text{Active cases}_{(t)} = L_{\text{virus}\,(t-\text{lag})} * 1.773$$

as the regression model ($R^2 = 0.97$, standard error = 37.3). The lag of the virus signal has here been determined as 4 days. The upper and lower 95% confidence interval values of the regression parameter are computed as 1.676 and 1.869, respectively. While the statistical comparison of the virus load per 7-day incidence rate of the wild type and the variant B.1.1.7 periods indicate a statistically significant difference, the difference in the shedding dynamics is not reflected in the model.

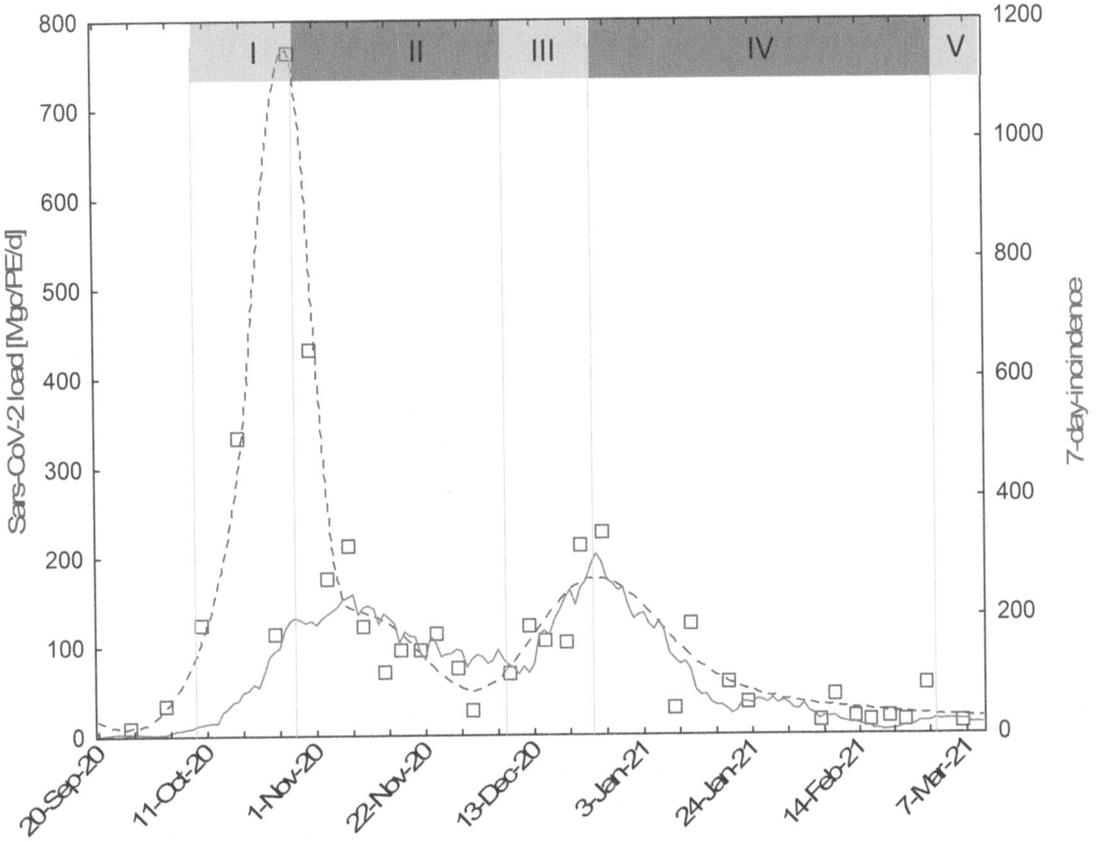

Figure 3 | Comparison of 7-day incidence from individual testing (red line) and the PE normalized virus load (blue squares) with its Spline-fit (dark blue-dashed line). Non-pharmaceutical interventions (shaded, header): see Figure 2.

4. DISCUSSION

The principality of Liechtenstein is a small country sharing a strong connected economy with the neighboring countries such as Austria and Switzerland; thus, high commuter traffic and intermixing of nationalities can be well observed. Throughout the pandemic Liechtenstein's borders to Switzerland remained open, while the border to Austria was closed through times of lockdown on either side, except for labor commuters (Regierung des Fürstentum Liechtenstein 2021). The policy of flattening the incidence curve in Liechtenstein was very successful from the start of the pandemic. Concerning the first wave, it is known that the population responded well to the governmental COVID-19 safety and preventive measurements, although they were considered less strict than many other countries (Thiel *et al.* 2020). In this paper, we hence present data on the second wave of cases in which two major peaks were observed and give insights into the viral abundance and shedding in wastewater across the country.

The major non-pharmaceutical measures are indicated in Figure 1, and both incidences and wastewater signals indicate their effectiveness. The response in the wastewater signal was faster than that observed in the 7-day incidence.

Confirming our first hypothesis (i), Figure 3 shows that the SARS-CoV-2 viral loads in Liechtenstein wastewater increased prior to rising numbers in individual tests, which are in accordance with previous findings that suggested WBE could be used as an early warning tool for SARS-CoV-2 detection and its spreading (e.g. Medema *et al.* 2020; Ahmed *et al.* 2021; Gerrity *et al.* 2021). The clarity of the finding was to some extent blurred by a change in testing strategy during the period from the end of September to mid-October 2020 when testing effort was doubled. Except for considerable testing intensification in mid-November to December 2020, the testing strategy and frequency remained rather constant. Increased testing frequency is known to lead to an increased detection of incidences (Wu *et al.* 2020), while disease incidences derived from wastewater are unbiased. Disease prevalence, however, is hard to deduct from wastewater testing since reliable data on shedding are yet only available from animal models (Mohandas *et al.* 2021).

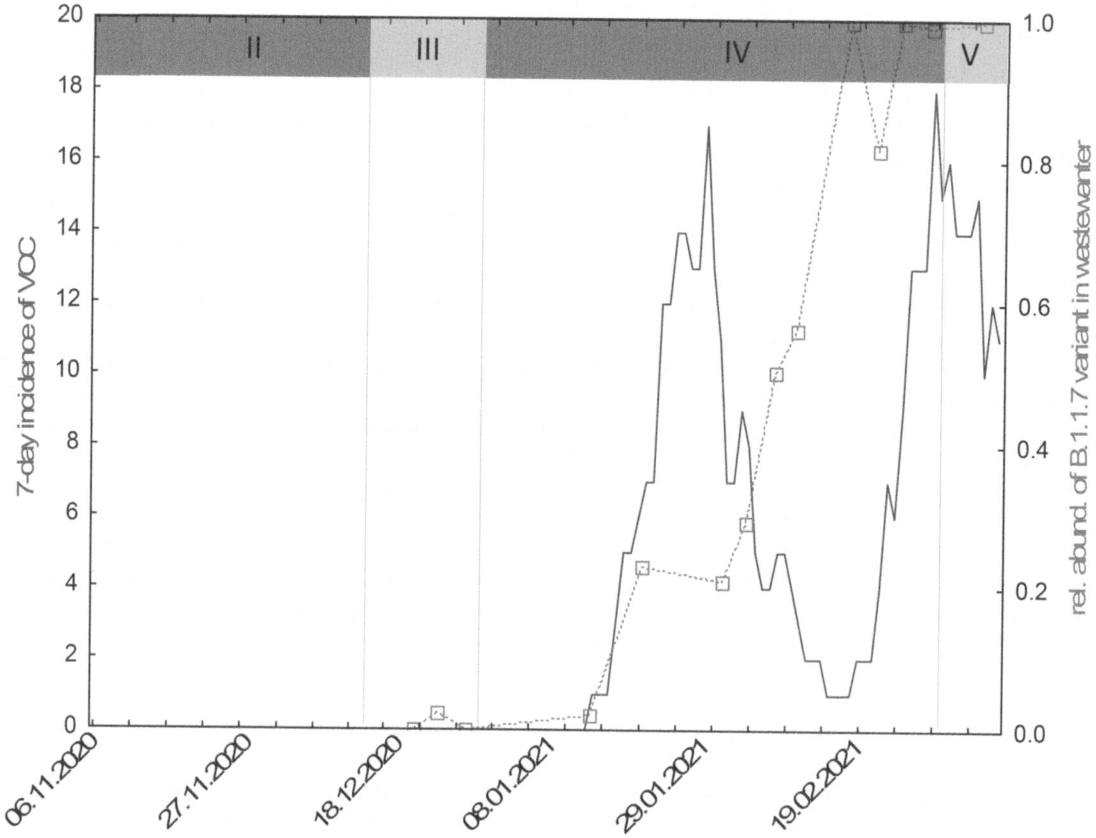

Figure 4 | Relative abundance of B.1.1.7 variant in wastewater based on median over found marker mutations (blue-dashed line) and 7-day incidence of VOC (red line). Non-pharmaceutical interventions (header, shaded): see Figure 2.

A linear regression model with an intercept of 0 was applied to correlate the incidences based on individual testing and wastewater signal. With a time-lag of 4 days, the highest correlations were achieved. Despite its simplicity, the incidence model exhibits a good prediction capability from the beginning of January 2021 onwards. While the spread of variants is likely to cause differences in the shedding dynamics, such effects are lost in the noise of the signal. Hence, the dynamics of both the wastewater signal and the epidemiologic situation are too low to identify such effects. The model clearly indicates a first pandemic wave approximately at the end of October 2020 that does not match the recorded active cases. However, since the predictions match the time period and also the severity (predicted 7-day incidence approximately 650 cases per 100,000) with the general pandemic situation in nearby countries (Austria/Switzerland), it is likely that the peak has been missed in the official recordings due to an insufficient number of tests. In summary, these data confirm both hypotheses (i) and (ii) by showing that the wastewater signal indicates changes in disease prevalence prior to individual testing and both signals are correlated as long as the testing strategy remains constant.

With the above presented results, the wastewater data suggest an underestimation of cases during the first peak of cases when the ratio of wastewater signal/incidence was extraordinarily high. Then, upon constant testing strategy, the ratio of wastewater/incidence remained constant. However, while testing frequency remained constant, the wastewater signal started to rise again in mid-January 2021. This was related to the beginning increase of the B.1.1.7 variant concentrations that were found both in individual tests and in the wastewater samples. It is estimated that the lineage B.1.1.7 has a 43–90% higher reproduction number than pre-existing variants, suggesting also higher shedding rates (Davies *et al.* 2021; Volz *et al.* 2021). Little is known about the behavior of B.1.1.7 in context with stool analysis, and there is a need for further research regarding this topic. Since the swallowing of sputum and saliva may considerably contribute to viral titers in stool, it seems evident that this also has an effect on viral wastewater load. Apart from the gut itself as a source of SARS-CoV-2 genes (Fitzgerald *et al.* 2021), viral genes contained in sputum and saliva may also ultimately contribute to the viral load in wastewater. The large amounts produced and swallowed per day (Rudney *et al.* 1995) and also directly excreted into

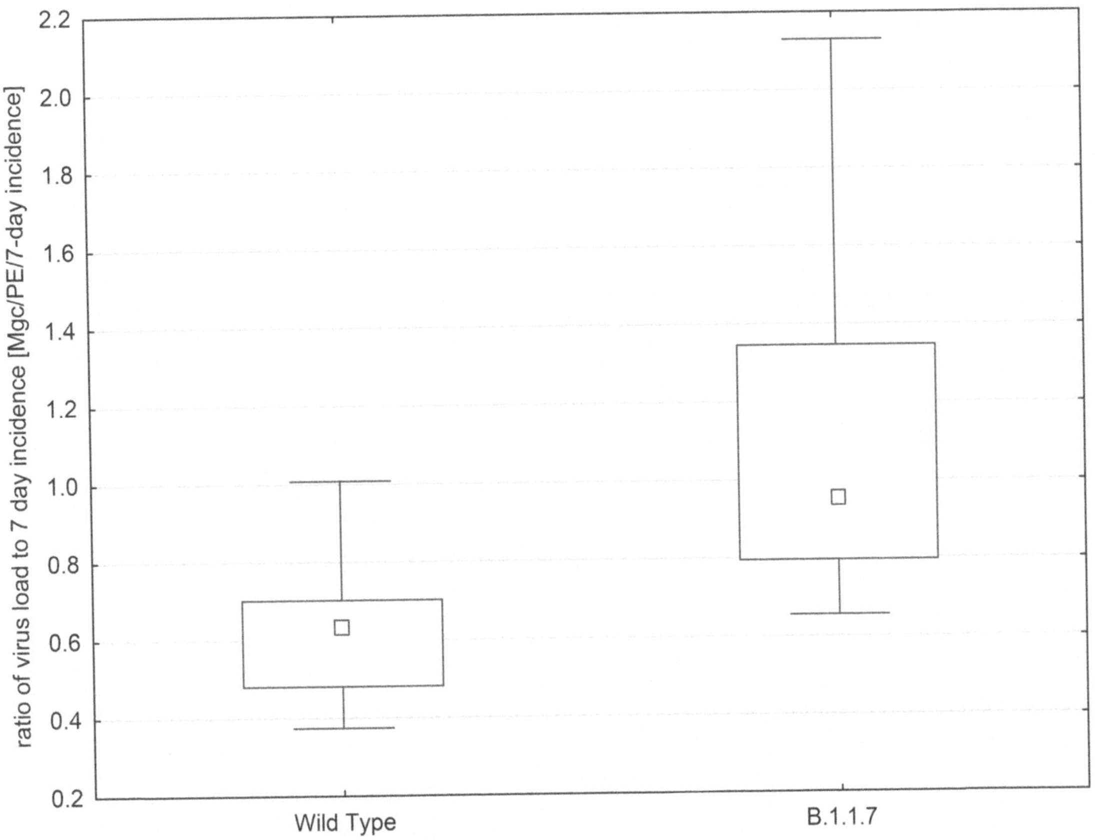

Figure 5 | Median of the ratio of the virus load per 7-day incidence during two periods of constant testing frequency: Nov. 6–Dec. 26 (wild-type period) and Dec. 27–Mar. 10, 2021 (B.1.1.7 period).

the sewer system, e.g. by toothbrushing, may considerably contribute to the gene copy numbers in wastewater. The observation of the early emergence of B.1.1.7 in wastewater as early as in December 2020 corroborates hypothesis (iii) in detecting new viral variants earlier than individual testing does. This is further supported by the emerging small peaks of the virus signal in wastewater per observed incidence around January 20 and February 19, 2021 (Figure 4).

The observations presented above could be challenged if there were changes in the wastewater temperature during the time of examination. The estimated half-life of SARS-CoV-2 ranges between 5 and 7 h at ambient (20 °C) conditions (Hart & Halden 2020). However, in the months of December 2020 to February 2021, the mean temperature of the inflow wastewater was as low as 10.9 ± 1.5 °C, and thus no temperature effect on virus fragment survival is expected to affect the results.

Finally, hypothesis (iv) is confirmed by the comparison of two periods characterized by constant testing frequency. After the first detection of the B.1.1.7 variant, the ration of WW signal per incidence increased and concurs with the view that variants like B.1.1.7 are characterized by enhanced fecal shedding dynamics (Davies *et al.* 2021; Volz *et al.* 2021). However, this could not be confirmed by our model that did not appear to be sensitive enough considering the low number of observations. This will have to be investigated in more detail with a broader dataset.

5. CONCLUSION

Presented data show a good correlation between the number of COVID-19 cases and SARS-CoV-2 gene concentration in the Liechtenstein wastewater system. Thus, the view that wastewater surveillance and monitoring offer an early indication for the spread of the SARS-CoV-2 is strongly supported. Particularly, it could be shown that wastewater data can better reflect changes in viral prevalence than individual testing. Similarly, wastewater viral loads may serve as an early indicator of mutations such as shown in the present case with the virus B.1.1.7 variant. For modeling prevalence and general interpretation of viral wastewater signals, it is suggested to accompany SARS-CoV-2 quantification in wastewater with genome sequencing. At low-case prevalence, wastewater monitoring and surveillance could serve as a crucial and efficient tool for

SARS-CoV-2 containment, requiring testing of asymptomatic individuals only when rises in the wastewater signal are evident. The lucrative lead of a couple of days would offer the opportunity to effectively set non-pharmaceutical interventions like social distancing or protective masks. To fully understand the changes of viral spreading and respond accordingly, it is important to identify where the virus variants are and thus, the inclusion of viral variants in predictive models will require a broader database and benefit the defeat of the virus across the globe.

ACKNOWLEDGEMENTS

The research was funded by Förderkreis 1669 of the University of Innsbruck and the Principality of Liechtenstein. Virus sequencing was supported by a collaboration with the Austrian Agency for Health and Food Safety (AGES). Publication fees were partly covered by the Publikationsfonds der Universität Innsbruck. We appreciate valuable laboratory assistance by Ljubica Begovic.

DATA AVAILABILITY STATEMENT

Data cannot be made publicly available; readers should contact the corresponding author for details.

REFERENCES

Ahmed, W., Tscharke, B., Bertsch, P. M., Bibby, K., Bivins, A., Choi, P., Clarke, L., Dwyer, J., Edson, J., Nguyen, T. M. H., O'Brien, J. W., Simpson, S. L., Sherman, P., Thomas, K. V., Verhagen, R., Zaugg, J. & Mueller, J. F. 2021 SARS-CoV-2 RNA monitoring in wastewater as a potential early warning system for COVID-19 transmission in the community: a temporal case study. *Sci. Total Environ.* **20** (761), 144216. doi:10.1016/j.scitotenv.2020.144216.

Arabzadeh, R., Grünbacher, D. M., Insam, H., Kreuzinger, N., Markt, R. & Rauch, W. 2021 Data filtering methods for SARS-CoV-2 wastewater surveillance. *Water Sci. Technol.* **84** (6), 1324–1339.

Been, F., Rossi, L., Ort, C., Rudaz, S., Delémont, O. & Esseiva, P. 2014 Population normalization with ammonium in wastewater-based epidemiology: application to illicit drug monitoring. *Environ. Sci. Technol.* **48**, 8162–8169.

Berchenko, Y., Manor, Y., Freedman, L., Kaliner, E., Grotto, I., Mendelson, E. & Huppert, A. 2017 Estimation of polio infection prevalence from environmental surveillance data. *Sci. Transl. Med.* **9** (383), eaaf6786. doi:10.1126/scitranslmed.aaf6786.

Chen, Y., Chen, L., Deng, Q., Zhang, G., Wu, K., Ni, L., Yang, Y., Liu, B., Wang, W., Wei, C., Yang, J., Ye, G. & Cheng, Z. 2020 The presence of SARS-CoV-2 RNA in the feces of COVID-19 patients. *J. Med. Virol.* **92** (7), 833–840. doi:10.1002/jmv.25825.

Davies, N. G., Abbott, S., Barnard, R. C., Jarvis, C. I., Kucharski, A. J., Munday, J. D., Pearson, C. A. B., Russell, T. W., Tully, D. C., Washburne, A. D., Wenseleers, T., Gimma, A., Waites, W., Wong, K. L. M., van Zandvoort, K., Silverman, J. D., Diaz-Ordaz, K., Keogh, R., Eggo, R. M., Funk, S., Jit, M., Atkins, K. E. & Edmunds, J. 2021 Estimated transmissibility and impact of SARS-CoV-2 lineage B.1.1.7 in England. *Science* **372** (6538), eabg3055. doi:10.1126/science.abg3055.

European Commission 2021 Commission recommendation on a common approach to establish a systematic surveillance of SARS-CoV-2 and its variants in wastewaters in the EU, Brussels, March 17, 2021. Available from: https://ec.europa.eu/environment/pdf/water/ecommendation_covid19_monitoring_wastewaters.pdf.

Fitzgerald, S. F., Rossi, G., Low, A. S., McAteer, S. P., O'Keefe, B., Findlay, D., Cameron, G. J., Pollard, P., Singleton, P. T. R., Ponton, G., Singer, A. C., Farkas, K., Jones, D., Graham, D. W., Quintela-Baluja, M., Tait-Burkard, C., Gally, D. L., Kao, R. & Corbishley, A. 2021 Site specific relationships between COVID-19 Cases and SARS-CoV-2 Viral Load in Wastewater Treatment Plant Influent. *Environ Sci Technol.* **55** (22), 15276–15286. doi:10.1021/acs.est.1c05029. Epub 2021 Nov 5. PMID: 34738785; PMCID: PMC8577191.

Gerrity, D., Papp, K., Stoker, M., Sims, A. & Frehner, W. 2021 Early-pandemic wastewater surveillance of SARS-CoV-2 in Southern Nevada: methodology, occurrence, and incidence/prevalence considerations. *Water Res.* **10**, 100086.

Hart, O. E. & Halden, R. U. 2020 Computational analysis of SARS-CoV-2/COVID-19 surveillance by wastewater-based epidemiology locally and globally: feasibility, economy, opportunities and challenges. *Sci. Total Environ.* **730**, 138875. doi:10.1016/j.scitotenv.2020.138875.

Hirose, R., Nakaya, T., Naito, Y., Daidoji, T., Watanabe, Y., Yasuda, H., Konishi, H. & Itoh, Y. 2017 Mechanism of human influenza virus RNA persistence and virion survival in feces: mucus protects virions from acid and digestive juices. *J. Infect. Dis.* **216**, 105–109. doi:10.1093/infdis/jix224.

Jones, D. L., Quintela Baluja, M., Graham, D. W., Corbishley, A., McDonald, J. E., Malham, S. K., Hillary, L. S., Connor, T. R., Gaze, W. H., Moura, I. B., Wilcox, M. H. & Farkas, K. 2020 Shedding of SARS-CoV-2 in feces and urine and its potential role in person-to-person transmission and the environment-based spread of COVID-19. *Sci. Total Environ.* **749**, 1879–1026. doi:10.1016/j.scitotenv.2020.141364.

Karia, R., Gupta, I., Khandait, H., Yadav, A. & Yadav, A. 2020 COVID-19 and its modes of transmission. *SN Compr. Clin. Med.* **2**, 1798–1801. doi:10.1007/s42399-020-00498-4.

Karthikeyan, S., Ronquillo, N., Belda-Ferre, P., Alvarado, D., Javidi, T., Longhurst, C. A., Knight, R. & Cristea, I. M. 2021 High-throughput wastewater SARS-CoV-2 detection enables forecasting of community infection dynamics in San Diego County. *mSystems* **6**, e00045–21. doi:10.1128/mSystems.00045-21.

La Rosa, G., Iaconelli, M., Mancini, P., Bonanno Ferraro, G., Veneri, C., Bonadonna, L., Lucentini, L. & Suffredini, E. 2020 First detection of SARS-CoV-2 in untreated wastewaters in Italy. *Sci. Total Environ.* **20** (736), 139652. doi:10.1016/j.scitotenv.2020.139652.

Lu, X., Wang, L., Sakthivel, S. K., Whitaker, B., Murray, J., Kamili, S., Lynch, B., Malapati, L., Burke, S. A., Harcourt, J., Tamin, A., Thornburg, N. J., Villanueva, J. M. & Lindstrom, S. 2020a US CDC real-time reverse transcription PCR panel for detection of severe acute respiratory syndrome coronavirus 2. *Emerg. Infect. Dis.* **26** (8), 1654–1665. https://doi.org/10.3201/eid2608.201246.

Lu, D., Huang, Z., Luo, J., Zhang, X. & Sha, S. 2020b Primary concentration – the critical step in implementing the wastewater-based epidemiology for the COVID-19 pandemic: a mini-review. ci. *Total Environ.* **10** (747), 141245. doi:10.1016/j.scitotenv.2020.141245.

Markt, R., Mayr, M., Peer, E., Wagner, A. O., Lackner, N. & Insam, H. 2021 Detection and stability of SARS-CoV-2 fragments in wastewater: impact of storage temperature. Preprint on MedRxiv. Available from: https://www.medrxiv.org/content/10.1101/2021.02.22.21250768v1.

Medema, G., Heijnen, L., Elsinga, G., Italiaander, R. & Brouwer, A. 2020 Presence of SARS-Coronavirus-2 RNA in sewage and correlation with reported COVID-19 prevalence in the early stage of the epidemic in The Netherlands. *Environ. Sci. Technol. Lett.* **7**, 511–516. doi:10.1021/acs.estlett.0c00357.

Mohandas, S., Yadav, P. D., Nyayanit, D., Deshpande, G., Shete-Aich, A., Sapkal, G., Kumar, S., Jain, R., Kadam, M., Kumar, A., Patil, D. Y., Sarkale, P., Gawande, P. & Abraham, P. 2021 Comparison of the pathogenicity and virus shedding of SARS CoV-2 VOC 202012/01 and D614G variant in hamster model. *BioRxiv*. doi:10.1101/2021.02.25.432136.

ÖGLMKC (Austrian Association for Quality Assurance and Standardization of Medical and Diagnostic Tests) 2021 Available from: https://oequasta.at/en/publications/cov-2 (accessed 30 May 2021).

Parasa, S., Desai, M., Thoguluva Chandrasekar, V., Patel, H. K., Kennedy, K. F., Roesch, T., Spadaccini, M., Colombo, M., Gabbiadini, R., Artifon, E., Repici, A. & Sharma, P. 2020 Prevalence of gastrointestinal symptoms and fecal viral shedding in patients with coronavirus disease 2019: a systematic review and meta-analysis. *JAMA Network Open* **3** (6), e2011335. doi:10.1001/jamanetworkopen.2020.11335.

Pearce, N., Vandenbroucke, J. P., VanderWeele, T. J. & Greenland, S. 2020 Accurate statistics on COVID-19 are essential for policy guidance and decisions. *Am. J. Public Health* **110**, 949–951.

Pérez-Cataluña, A., Cuevas-Ferrando, E., Randazzo, W., Falcó, I., Allende, A. & Sánchez, G. 2021 Comparing analytical methods to detect SARS-CoV-2 in wastewater. *Sci. Total Environ.* **758**, 143870. doi:10.1016/j.scitotenv.2020.143870.

Pola, A., Murthy, K. S. & Santhekadur, P. K. 2021 COVID-19 and gastrointestinal system: a brief review. *Biomed. J.* in press. doi:10.1016/j.bj.2021.01.001.

Popa, A., Genger, J.-W., Nicholson, M. D., Penz, T., Schmid, D., Aberle, S. W., Agerer, B., Lercher, A., Endler, L., Colaço, H., Smyth, M., Schuster, M., Grau, M. L., Martínez-Jiménez, F., Pich, O., Borena, W., Pawelka, E., Keszei, Z., Senekowitsch, M., Laine, J., Aberle, J. H., Redlberger-Fritz, M., Karolyi, M., Zoufaly, A., Maritschnik, S., Borkovec, M., Hufnagl, P., Nairz, M., Weiss, G., Wolfinger, M. T., von Laer, D., Superti-Furga, G., Lopez-Bigas, N., Puchhammer-Stöckl, E., Allerberger, F., Michor, F., Bock, C. & Bergthaler, A. 2020 Genomic epidemiology of superspreading events in Austria reveals mutational dynamics and transmission properties of SARS-CoV-2. *Sci. Transl. Med.* **12**, eabe2555.

Regierung des Fürstentum Liechtenstein. Medienmitteilung 17 March 2020: Generelles Veranstaltungsverbot und weitere Schliessungen. Available from: https://www.regierung.li/coronavirus/2020-3 (accessed 26 June 2020).

Rippinger, C., Bicher, M., Urach, C., Brunmeir, D., Weibrecht, N., Zauner, G., Sroczynski, G., Jahn, B., Mühlberger, N., Siebert, U. & Popper, N. 2021 Evaluation of undetected cases during the COVID-19 epidemic in Austria. *BMC Infect. Dis.* **21**, 70. doi:10.1186/s12879-020-05737-6.

Rudney, J. D., Ji, Z. & Larson, C. J. 1995 The prediction of saliva swallowing frequency in humans from estimates of salivary flow-rate and the volume of saliva swallowed. *Arch. Oral Biol.* **40**, 507–512. doi: 10.1016/0003-9969(95)00004-9.

Thiel, S., Weber, M., Risch, L., Wohlwend, N., Lung, D., Hillmann, D., Ritzler, M., Risch, M., Kohler, P., Vernazza, P., Kahlert, C. R., Fleisch, F., Cusin, A., Karajan, T. V., Copeland, S. & Paprotny, M. 2020 Flattening the curve in 52 days: characterization of the COVID-19 pandemic in the principality of Liechtenstein. *Swiss Med. Wkly.* **150**, w20361. doi:10.4414/smw.2020.20361.

Kenyon, A., Davis, T., Pybus, O. G., Du Plessis, L., Zarebski, A. E., Raghwani, J., Kraemer, M. U. G., Francois, S., Attwood, S. W., Vasylyeva, T. I., Zamudio, M. E., Gutierrez, B., Torok, M. E., Hamilton, W. L., Goodfellow, I. G., Hall, G., Jahun, A. S., Chaudhry, Y., Hosmillo, M., Pinckert, M. L., Georgana, I., Moses, S., Lowe, H., Bedford, L., Moore, J., Stonehouse, S., Fisher, C. L., Awan, A. R., BoYes, J., Breuer, J., Harris, K. A., Brown, J. R., Shah, D., Atkinson, L., Lee, J. C. D., Storey, N., Flaviani, F., Alcolea-Medina, A., Williams, R., Vernet, G., Chapman, M. R., Levett, L. J., Heaney, J., Chatterton, W., Pusok, M., Xu-McCrae, L., Smith, D. L., Bashton, M., Young, G. R., Holmes, A., Randell, P. A., Cox, A., Madona, P., Bolt, F., Price, J., Mookerjee, S., Nascimento, F. F., Siveroni, I., Johnson, R., Volz, E. M., Rowan, A., Taylor, G. P., Smollett, K. L., Quick, J., McMurray, C., Stockton, J., Nicholls, S., Rowe, W., Poplawski, R., McNally, A., Nunez, R. T. M., Mason, J., Robinson, T. I., O'Toole, E., Watts, J., Breen, C., Cowell, A., Sluga, G., Machin, N. W., Ahmad, S. S. Y., George, R. P., Halstead, F., Sivaprakasam, V., Hogsden, W., Illingworth, C. J., Jackson, C., Thomson, E. C., Shepherd, J. G., Asamaphan, P., Niebel, M. O., Li, K. K., Shah, R. N., Jesudason, N. G., Tong, L., Broos, A., Mair, D., Nichols, J., Carmichael, S. N., Nomikou, K., Aranday-Cortes, E., Johnson, N., Starinskij, I., Filipe, da Silva, Robertson, D. L., Orton, R. J., Hughes, J., Vattipally, S., Singer, J. B., Nickbakhsh, S., Hale, A. D., Macfarlane-Smith, L. R., Harper, K. L., Carden, H., Taha, Y., Payne, B. A. I., Burton-Fanning, S., Waugh, S., Collins, J., Eltringham, G., Rushton, S., O'Brien, S., Bradley, A., Maclean, A., Mollett, G., Blacow, R., Templeton, K. E., McHugh, M. P., Dewar, R., Wastenge, E., Dervisevic, S., Stanley, R., Meader, E. J., Coupland, L., Smith, L., Graham, C., Barton, E., Padgett, D., Scott, G., Swindells, E., Greenaway, J., Nelson, A., McCann, C. M., Yew, W. C., Andersson, M., Peto, T., Justice, A., Eyre, D., Crook, D., Sloan, T. J., Duckworth,

N., Walsh, S., Chauhan, A. J., Glaysher, S., Bicknell, K., Wyllie, S., Elliott, S., Lloyd, A., Impey, R., Levene, N., Monaghan, L., Bradley, D. T., Wyatt, T., Allara, E., Pearson, C., Osman, H., Bosworth, A., Robinson, E., Muir, P., Vipond, I. B., Hopes, R., Pymont, H. M., Hutchings, S., Curran, M. D., Parmar, S., Lackenby, A., Mbisa, T., Platt, S., Miah, S., Bibby, D., Manso, C. & The COVID-19 Genomics UK (COG-UK) 2021 Assessing transmissibility of SARS-CoV-2 lineage B.1.1.7 in England. *Nature* **593**, 266–269. doi:10.1038/s1586-021-0370-x.

Wu, F., Zhang, J., Xiao, A., Gu, X., Lee, W. L., Armas, F., Kauffman, K., Hanage, W., Matus, M., Ghaeli, N., Endo, N., Duvallet, C., Poyet, M., Moniz, K., Washburne, A. D., Erickson, T. B., Chai, P. R., Thompson, J. & Alm, E. J. 2020 SARS-CoV-2 titers in wastewater are higher than expected from clinically confirmed cases. *Preprint in medRxiv*. doi:10.1101/2020.0.05.2005150.

Wyllie, A. L., Fournier, J., Casanovas-Massana, A., Campbell, M., Tokuyama, M., Vijayakumar, P., Warren, J. L., Geng, B., Muenker, M. C., Moore, A. J., Vogels, C. B. F., Petrone, M. E., Ott, I. M., Lu, P., Venkataraman, A., Lu-Culligan, A., Klein, J., Earnest, R., Simonov, M., Datta, R., Handoko, R., Naushad, N., Sewanan, L. R., Valdez, J., White, E. B., Lapidus, S., Kalinich, C. C., Jiang, X., Kim, D. J., Kudo, E., Linehan, M., Mao, T., Moriyama, M., Oh, J. E., Park, A., Silva, J., Song, E., Takahashi, T., Taura, M., Weizman, O.-E., Wong, P., Yang, Y., Bermejo, S., Odio, C. D., Omer, S. B., Cruz, Dela, Farhadian, S., Martinello, R. A., Iwasaki, A., Grubaugh, N. D. & Ko, A. I. 2020 Saliva or nasopharyngeal swab specimens for detection of SARS-CoV-2. *N. Engl. J. Med.* **383**, 1283–1286. doi:10.1056/NEJMc2016359.

Xiao, F., Tang, M., Zheng, X., Liu, Y., Li, X. & Shan, H. 2020 Evidence for gastrointestinal infection of SARS-CoV-2. *Gastroenterology* **158**, 1831–1833.e3. doi:10.1053/j.gastro.2020.02.055.

Zheng, S., Fan, J., Yu, F., Feng, B., Lou, B., Zou, Q., Xie, G., Lin, S., Wang, R., Yang, X., Chen, W., Wang, Q., Zhang, D., Liu, Y., Gong, R., Ma, Z., Lu, S., Xiao, Y., Gu, Y., Zhang, J., Yao, H., Xu, K., Lu, X., Wei, G., Zhou, J., Fang, Q., Cai, H., Qiu, Y., Sheng, J., Chen, Y. & Liang, T. 2020 Viral load dynamics and disease severity in patients infected with SARS-CoV-2 in Zhejiang province, China. *BMJ* **369**, m13. doi:10.1136/bmj.m1443.

First received 20 July 2021; accepted in revised form 14 September 2021. Available online 30 November 2021

doi: 10.2166/wh.2022.082

Wastewater monitoring of COVID-19: a perspective from Scotland

Zhou Fang [a,*], Adrian M. I. Roberts[a], Claus-Dieter Mayer[a], Anastasia Frantsuzova[a], Jackie M. Potts[a], Graeme J. Cameron[b], Peter T. R. Singleton[b] and Iona Currie[c]

[a] Biomathematics & Statistics Scotland, James Clerk Maxwell Building, King's Buildings, Peter Guthrie Tait Road, Edinburgh EH9 3FD, UK
[b] Scottish Environment Protection Agency, Strathallan House, Castle Business Park, Stirling FK9 4TZ, UK
[c] Scottish Government, Rural and Environment Science and Analytical Services, 5 Atlantic Quay, Atlantic Quay, 150 Broomielaw, Glasgow G28LU, UK
*Corresponding author. E-mail: zhou.fang@bioss.ac.uk

ZF, 0000-0001-7692-5040

ABSTRACT

Scotland introduced wastewater monitoring for COVID-19 early in the pandemic. From May 2020, samples have been taken and analysed using quantitative polymerase chain reaction (qPCR). The programme was expanded to over 100 sites accounting for around 80% of the population. Data are presented publicly via a dashboard and regular reports are produced for both the public and health professionals. Wastewater-based epidemiology (WBE) offers opportunities and challenges. It offers an objective means of measuring COVID-19 prevalence and can be more practical or timely than other methods of mass testing. However, it also has substantial variability impacted by multiple environmental factors. Methods for data collection and analysis have developed significantly through the pandemic, reflecting the evolving situation and policy direction. We discuss the Scottish experience of wastewater monitoring for COVID-19, with a focus on the analysis of data. This includes our approach to flow normalisation, our experience of variability in measurements and anomalous values, and the visualisation and presentation of data to stakeholders. Summarising the Scottish methodology as of March 2022, we also discuss how wastewater data were used for informing policy and public health actions. We draw lessons from our experience and consider future directions for WBE in Scotland.

Key words: COVID-19, data analysis, wastewater-based epidemiology

HIGHLIGHTS

- The Scottish COVID-19 WBE programme covered around 80% of the population with regular COVID-19 wastewater testing for over a year.
- WBE reports complemented case data to assist policymakers and other stakeholders.
- Techniques and decisions (e.g. normalisation and visualisation) were driven by changing context and feedback.
- The practical pros and cons of WBE were made apparent, suggesting future directions.

GRAPHICAL ABSTRACT

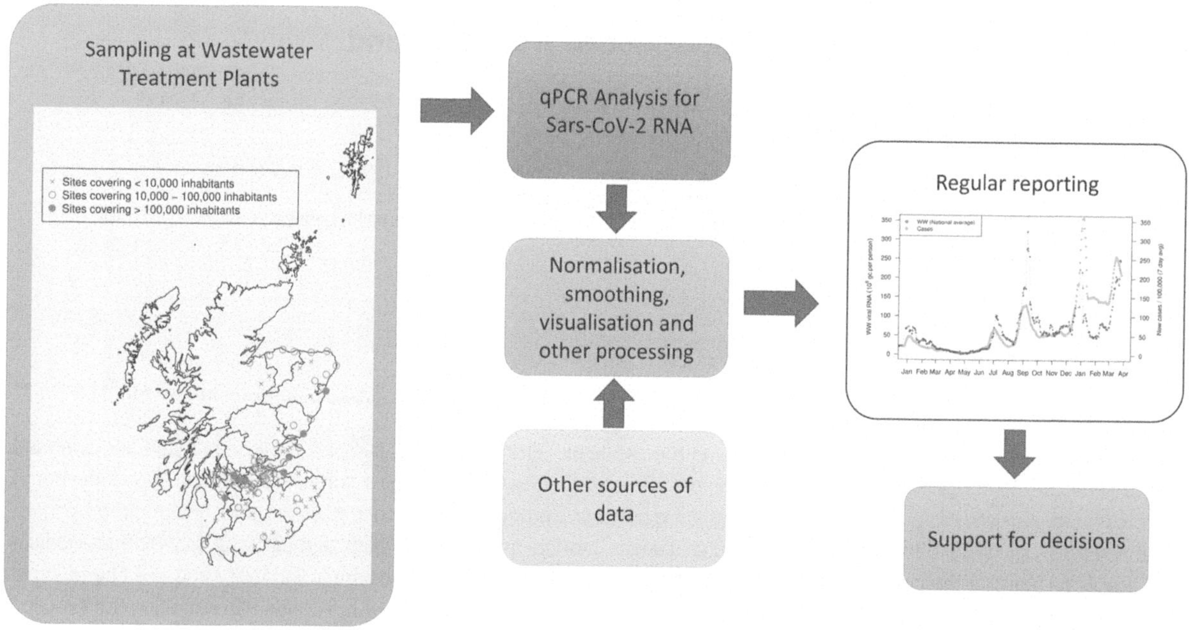

INTRODUCTION

In 2020, the COVID-19 pandemic rapidly spread across the world, with the first cases seen in Scotland on 1 March. By the end of April, cases exceeded 10,000. Monitoring of the pandemic and its effects was seen as essential for directing efforts to contain its spread.

While the changing rates of reported cases as well as randomised survey data offer some insight into the progression of the pandemic, one alternative approach was wastewater (WW) monitoring. Wastewater-based epidemiology (WBE) had previously been used to monitor diseases such as norovirus and rotavirus (Santiso-Bellón *et al.* 2020) and Aichi virus (Lodder *et al.* 2013) and holds potential advantages over traditional methods. During the pandemic, WBE has been used in many countries for monitoring COVID-19 levels (COVIDPoops19; Naughton *et al.* 2021), including more widely in the UK (Wade *et al.* 2022).

From May 2020, samples began to be collected and tested for the SARS-CoV2 virus in Scotland (Fitzgerald *et al.* 2021), with the programme expanding greatly over time. By 2022, the programme covered around 80% of the population with 120 sites being sampled. The programme is ongoing as of the time of writing, with weekly reports being produced.

Unlike typical research projects, the dynamic nature of the pandemic led to an equally dynamic set of project parameters, with requirements changing rapidly over time. It involves multiple organisations, including a water company, Scottish Water, to provide samples, the Scottish Environment Protection Agency (SEPA) to conduct qPCR and other chemical analyses, a group of statisticians, Biomathematics and Statistics Scotland (BioSS) to conduct data analysis and reporting, and the Scottish Government to coordinate, disseminate and provide policy direction. There was close collaboration with scientists in Scotland and the United Kingdom, and laboratories were identified to provide emergency capacity – with some duplicate samples sent to other laboratories to measure inter-lab variability. Results of the monitoring programme were distributed via dashboards and reports to stakeholders, including the Scottish Government, public health officials, researchers, and the public. Feedback from the Scottish Government drove many of the decisions and improvements in the work.

This paper discusses the Scottish COVID-19 WBE programme up to March 2022, the time of writing. This includes the evolving sampling regime but particularly focuses on the development of data analysis techniques in Scotland, and the production and use of data outputs. We focus on specific components which are particular to the Scottish programme. Despite being motivated by the unique circumstances of the COVID-19 pandemic, these methodologies have potential application in a range of environmental monitoring scenarios. A summary of the overall methodology is provided. We also discuss the pros and cons of WBE, the lessons learnt, and proposed future directions for WBE in Scotland.

Data produced from the programme are available online at BioRDM Team (2022).

WW SAMPLING IN SCOTLAND

Scotland adopted WW sampling for COVID-19 early in the pandemic, with an arrangement where sampling was performed by Scottish Water, and qPCR testing conducted by the SEPA. Initial trials attempting to detect SARS-CoV-2 virus fragments in WW began in April 2020 with the development of a national monitoring programme starting in late May 2020 (Fitzgerald *et al.* 2021).

The main sampling sites used in the Scottish programme consisted of sewage and WW treatment plants, where auto-samplers were installed. There are over 1,800 wastewater treatment works (WWTWs) in the Scottish Network, and it would be uneconomic to sample all of them. However, WWTWs are diverse in terms of the size of the population covered, ranging from fewer than a hundred to over half a million inhabitants. Selecting the larger sites allowed samples from an initial list of 28 sites to cover around half the sewered population. This was the setup for much of 2020 when this programme used resources released by the pandemic restrictions that interrupted other work at the SEPA and Scottish Water. Dedicated resources from the Scottish Government expanded the programme from the winter of 2020, rapidly exceeding 100 sampling sites, and growing to 120 sampling sites by 2022. These covered around 80% of the population. An overview of the WWTWs sampled and the resulting population coverage is shown in Figure 1.

Auto-samplers used across the network varied in model. A typical sampler would collect a fixed subsample of around 10 ml every 15 min, producing a time composite sample of influent WW over a 24-h period. Generally, samplers were refrigerated to

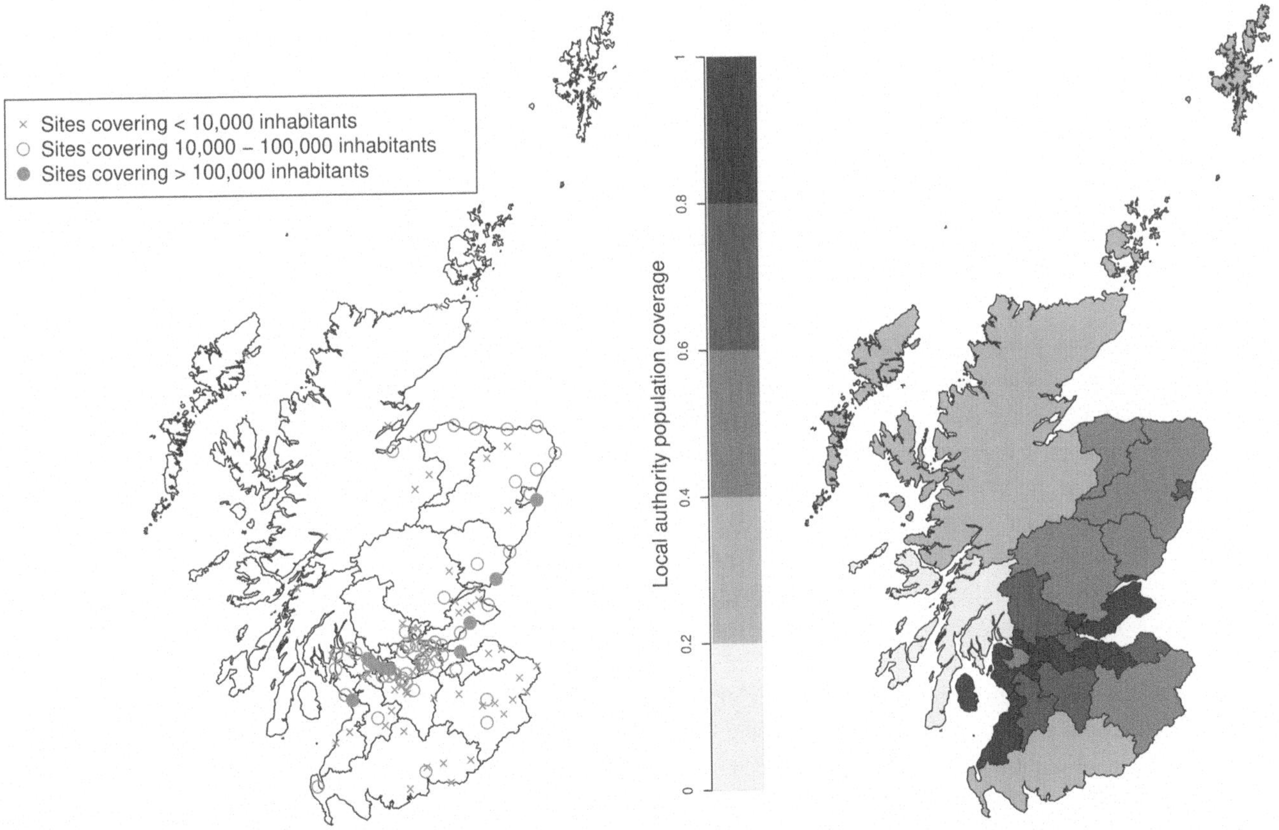

Figure 1 | Left: A map of wastewater treatment works sampled so far in the Scottish COVID-19 wastewater monitoring project. The type of point denotes the population covered by the site. Right: A map of local authority areas in Scotland with shading denoting the proportion of the population covered by WW COVID-19 testing on an average week.

best preserve the samples taken. A higher sampling frequency would reduce sampling variability by more evenly sampling from waste packets in the system but introduce practical issues due to the increased volume of samples produced and transported.

With the additional resources from the Scottish government, in addition to increased sampling at treatment plants, NHS health boards were able to request sampling from within the sewer network, allowing the monitoring to generate data for smaller areas, such as a university campus, street, or even a few individual buildings. These 'network samples' were so-called grab samples – samples taken directly from the WW stream at a specific position in the sewage network via manholes. The aim was to collect these samples at the same time (particular to each site) each day, though obtaining access in some cases complicated this. Around 44 such network sites have been sampled, with about 23 continuing to be sampled after the start of 2022. These samples were used as part of regular reporting, but for specific one-off work.

All samples were sent to SEPA's laboratory for analysis. To measure the level of COVID-19 RNA in samples, the virus in the sample was concentrated before extraction. The N1 gene for SARS-CoV-2 was then detected and measured using RT-qPCR (Fitzgerald *et al.* 2021). By March 2022, over 10,000 samples had been analysed. The sampling frequency varied over this period across WWTW sites, ranging between approximately weekly to three to four times a week at the most frequently sampled sites. Sampling was generally targeted to be most frequent at those sites with the most populated catchment areas in order to balance the need for reliable smoothed estimates with the need for high population coverage, both nationally and within local authority areas.

PROCESSING OF WW DATA

To extract useful information from the WW data, it was necessary to develop a statistical analysis strategy including a comparison to other data sources. In late 2020, BioSS was commissioned to process and produce WW data output. A data processing pipeline was developed, including functional elements that were in some cases particular to the Scottish programme.

This pipeline involved pre-processing the data, adjustments and computation of aggregate summaries, and the semi-automated production of reports, with all processing done using the R programming language (R Core Team 2021). Here, we describe normalisation – an important part of data pre-processing, adjustments required to produce data on different geographic bases, decisions made in data visualisation, and other methodologies we developed.

Normalisation for dilution

Measurements of viral levels by PCR were represented as a concentration estimate, specifically the estimated number of gene copies per litre. Early in the programme, the decision was made to normalise these values by the flow levels measured at each WWTW, and by an estimate of the population covered by the WWTW.

Rationale behind flow normalisation

The decision to normalise was driven by a mechanistic understanding of the processes leading to WW collection. In Scotland, WW at each WWTW usually combines both wastes collected from properties and rainwater. During periods of high precipitation, viral material in WW would be diluted, leading to unrepresentative low measurements. In some cases, data on each day's intake of WW at the WWTW, measured by meters operated by Scottish water, were available. In those cases, we multiplied the measured RNA concentrations by that day's flow to obtain a daily gene copy measure. In addition, because WWTWs serve a wide range of catchment sizes, we also made use of the population served by each WWTW, as estimated by Scottish Water internal methods. The flow-adjusted gene copy rate is thus divided by this population quantity, to facilitate easier cross-site comparisons. The result was a measurement of the gene copies per day per person.

A major component in flow changes was seasonal effects, with observed flow levels sometimes twice as high during winter as opposed to summer. There could also be notable effects during extreme weather, especially during periods of intensive rainfall, when dilution effects could be sizable for some WW systems. In periods of relatively stable weather, the effect of normalisation was small compared to the variability seen in RNA concentrations.

Flow normalisation is a relatively straightforward way of encompassing environmental covariates in the assessment of COVID-19 levels. Compared to more dynamic models (Wade *et al.* 2022), flow normalisation is simple to incorporate into dashboards, allowing easy in-time access to adjusted data. It is also easy to understand and gives values in gene copies per person per day that allow simple comparison between locations.

Estimation of flow from ammonia concentrations

In practice, flow measurements were frequently unavailable (especially later in the programme) or were published with such a delay after sampling that they could not be incorporated into a timely analysis. Instead, we used ammonia concentrations, which were subject to the same dilution processes as the viral material, as a covariate for predicting the flow rate. Measurements of ammonia, obtained through colorimetry at Scottish Water laboratories, were more readily available because they are both cheaper and quicker to obtain than for flow.

We fitted a cross-site linear mixed model to predict log flow from log population and log ammonia, allowing different WWTW sites to differ in intercept and slope coefficients. Though theoretically, a direct dilution relationship would imply that flow should be inversely proportional to ammonia concentration, estimation of the coefficients through modelling allowed us to compensate for uncertainty in the concentration measurements and the flow–ammonia relationship.

Even the availability of ammonia data was often delayed by a number of days compared to the RNA concentration. In the meantime, a temporary estimate was required for normalisation. Initially, we opted to estimate flow by simply taking an average of flow at that site. This approach did not adequately account for the seasonal patterns in precipitation, resulting in excessively low estimates of flow during winter months. Subsequently, we adopted an approach where a generalised additive model (GAM, Wood 2017) was fitted across sites. This included an overall non-linear time trend in ammonia levels with individual site effects added. This model allowed ammonia, and thus flow, to be interpolated from the available data.

The limitation of ammonia-based adjustments is that they assume that the total volume of ammonia in the WW remains fairly constant over time and is accurately measured. Thus, uncertainty may arise due to errors and variability in the laboratory process for measuring ammonia, or due to changes in ammonia unrelated to dilution (such as agricultural or industrial sources). Examination of individual WWTW sites showed that the strength of the relationship between flow and ammonia varied. In some locations, the relationship was poor, or not linear as the model assumed. However, in some cases, this may be the result of 'flow capping', where extremely high levels of precipitation lead to some WW flow being redirected instead of passing through treatment works. In these cases, it would be more accurate to use ammonia-based metrics to estimate dilution. However, we prioritised concerns about the uncertainty of the ammonia adjustment, and thus used measurements of flow where available.

In-network samples

The flow and ammonia adjustment mentioned was focused on autosampler outputs from WWTWs covering wide areas. In our work, we also looked at grab samples taken within the sewage network. In these cases, normalisation was complicated by two factors:

- Specific characteristics used for normalisation may be unavailable. In particular, we usually did not have a usable measure of flow, and in many cases, there could be no accurate population estimate because that population can change greatly from sample to sample.
- Individual small subnetworks may have dilution properties that do not respect the assumptions of the normalisation we otherwise use. For example, some subnetworks were closed systems where there is no dilution from rainwater, while others may only be in use at certain times.

For network samples, we normalised the RNA concentration measurements by dividing them by the ammonia concentration. This produced a ratio that should be independent of the degree of use of WW services or dilution, assuming that the ammonia concentration experiences the same environment as the RNA concentration. The nature of in-network samples varied widely, but measurements from them tended to be noisier than WWTW samples, probably due to sampling effects, or the presence of inhibitory substances.

Geographic matching and aggregation

For comparison with the WW viral level data, it was desirable to produce appropriately matching data sets – in particular, case data. These comparisons give a direct sense of the relationship between WW and case rates over time. For example, a comparison of the variability of WW viral levels to case levels may be used to determine if particular sites had an excessively high day-to-day variation in WW viral levels. Changes in WW trends may also be observed and compared to changes in case levels, acting as a confirmation of case trends or identifying locations where one or the other may have issues. Case data were desirable as a comparator because they were measured daily and available at a high degree of spatial resolution. Case data were sourced from an open data site, published by Public Health Scotland (PHS 2022).

However, making these comparisons was complicated by the fact that cases and WW viral levels were measured and reported on different geographic bases. For WW, the basic unit was the catchment area of a given WW treatment works (or portion of the network, in case of in-network samples), which covered between a few thousand and several hundred thousand inhabitants. Meanwhile, cases were associated with individual addresses, then for the sake of privacy, aggregated into larger groupings. In Scotland, these groupings formed a cascading hierarchy of levels, composed of data zones of around a thousand inhabitants, intermediate zones (more commonly called neighbourhoods) of around 5,000 inhabitants, local authority areas of 20,000 to 600,000 inhabitants, and health board areas of 20,000 to 1.2 million inhabitants. While there are strong restrictions on data use due to confidentiality for case data at the data zone level, censored and smoothed data were available at the intermediate zone level, and fuller data were present at larger aggregations. Due to this factor, and sample size issues, Office of National Statistics (ONS) survey data were only available on the level of large subregions, which each contain multiple health boards.

To match one to the other, we used a postcode and census-based methodology. By quantifying each WWTW catchment as a collection of postal districts with population data specified by the latest census, we were able to determine the degree (in terms of the number of inhabitants) by which each catchment intersected with each geographic grouping. This overlap population was used to weight averages of case rates across neighbourhoods to obtain corresponding case rates for each catchment area.

We used data zone-level data to evaluate the loss of precision arising from using catchment case rates based on aggregating intermediate zones. For most WWTW sites, correlations between measures based on data zone and intermediate zone aggregations were good (>0.95 in most cases). For a small number of very small or isolated sites, intermediate zone level aggregation gave a poor approximation compared to the use of data zones.

As well as showing results for individual sites, there was policy interest in summaries for local authorities, health boards, and nationally. To produce plots for these different political areas, data were aggregated over sites. In this case, the overlap population computed above was used instead to weight site-level WW viral data. A rolling weekly average was used to smooth over sampling day effects.

Production of visualisations

At the start of the pandemic, it was difficult to predict what the relationship between daily WW RNA gene copy levels and other monitoring metrics, such as positive test levels and daily hospitalisations/deaths, would be. Furthermore, changes in circumstances, such as the emergence of new variants, vaccination programmes, and changes in testing policy, were anticipated to affect the relationships between monitoring measurements. This meant that a more formal modelling framework may have been difficult to maintain effectively, as basic assumptions that hold at one time-point may not hold at another. While we produced more specific comparisons from time to time, visual representations of site-level data and aggregates for specific geographic areas formed the core component of the reporting of WW viral loads.

Comparative time series

The most common visualisation we used was to jointly plot normalised daily WW viral RNA levels and the daily new case levels against time on the same axes. An example is shown in Figure 2.

For these comparative time series plots, a number of considerations were made. First, we opted not to use logarithmic scaling for our graphs. This was because our outputs were to be directed potentially towards a less technically experienced audience who may not be skilled at interpreting such graphs (Romano *et al.* 2020). Furthermore, the presence of values below the limit of detection at some sites meant that an offset would be required for logarithmic transforms to make sense, and the choice of the offset might greatly impact interpretation. While a logarithmic axis may be justified by the greater variation seen with higher COVID-19 levels, this was offset by the needs of the stakeholders.

Second, the juxtaposition of case and WW viral levels required a choice of the relative scaling of WW and new case rates. Early on, we experimented with the use of the least squares fit to find the best scaling factor to relate cases to WW. However, soon it emerged that having different scaling factors in each graph created confusion, as a reader would have to refer to each graph's axis to determine whether WW viral levels were relatively high or low. Refitting these scaling factors with each new set of data would also create inconsistent graphing between reports. Coincidentally, across most of our sites, WW viral levels

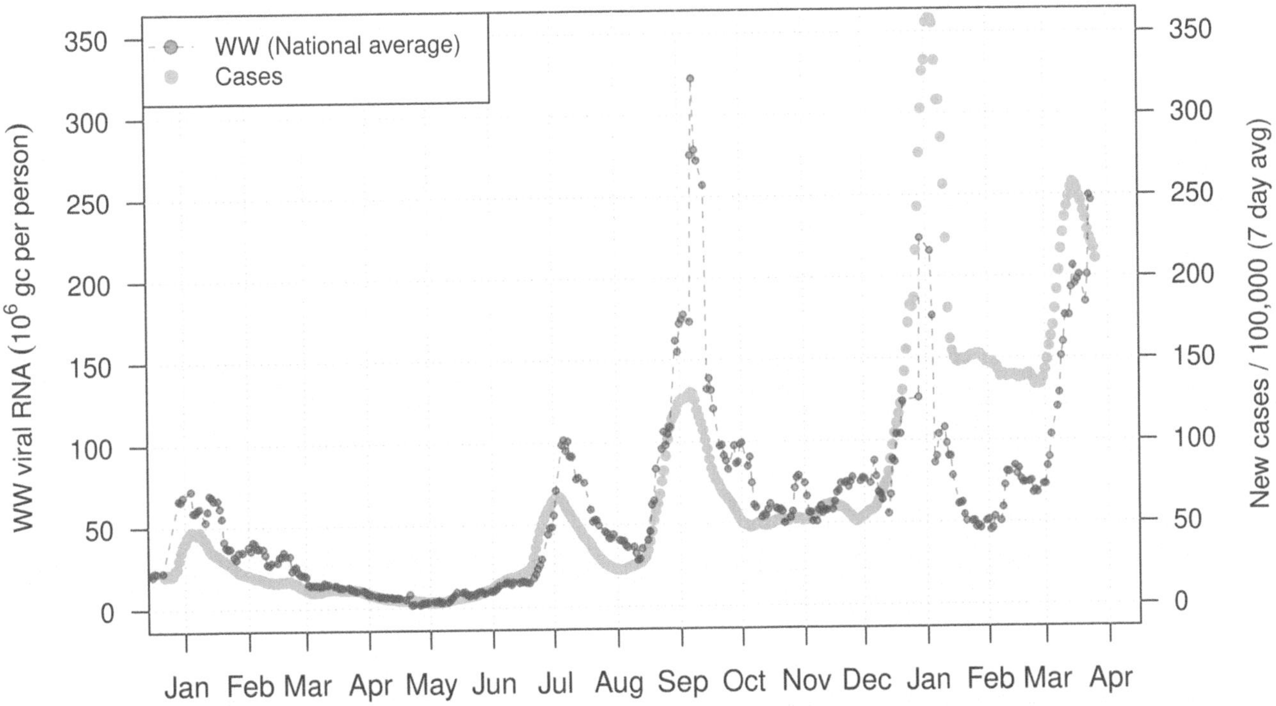

Figure 2 | An example plot produced during the Scottish WW COVID-19 project. This graph shows Scottish population-weighted daily averages for WW viral levels alongside daily case levels from January 2021 to mid-March 2022. Wastewater viral levels are fairly well aligned to case rates, up until December 2021, apart from peaks in case/WW viral levels where WW viral levels tend to overshoot.

as measured in millions of gene copies per person per day had an approximately one-to-one relationship with case rates, if the latter was given by the daily rate of new COVID-19 cases per 100,000 inhabitants. Therefore, we opted to adopt this constant scaling in our graphing. We found the relationship to be remarkably stable up until the appearance of the Omicron variant in December 2021.

Map outputs

We also used maps to display aspects of the data. Initially, we used bubble maps where the size of the points for each sampled location showed the level of COVID-19 found. These maps could be made using both WW and case levels. However, this visualisation became difficult to interpret when a larger number of sites are sampled, with too many points preventing specific locations from being identified. This was especially problematic in the densely populated central belt of Scotland. When COVID-19 levels were high, points also tend to overlap, and when levels were low, it can be difficult to identify differences.

Instead, we mapped population-weighted mean WW levels for each of Scotland's 32 local authorities using a colour scale. This was done initially only on the basis of 2-week aggregate values, though faster changes in COVID-19 levels led to us adopting weekly maps as well. Examples of 2-week maps are shown in Figure 3. Alongside maps of current WW COVID-19 levels, we also produced maps showing the changes in these levels relative to the previous 2-week period. We chose to represent absolute values for the changes, rather than relative values (such as percentage change) because relative values amplify the visual significance of changes at locations with very low levels. In practice these were the least reliable measurements, and unlikely to represent a consistent trend, and were not as important from a policy point of view as regions with higher COVID levels.

We used the Viridis and Turbo colour scales (Garnier *et al.* 2022) to reduce problems for those with colour blindness and ensure uniform perceptual distances. For both maps, we used a logarithmically scaled colour bar. Although the logarithmic scaling makes direct interpretation harder, this allowed the colour range on the map to cover a wide range of values without requiring constant re-calibration, thus ensuring consistency in plotting COVID-19 levels from report to report.

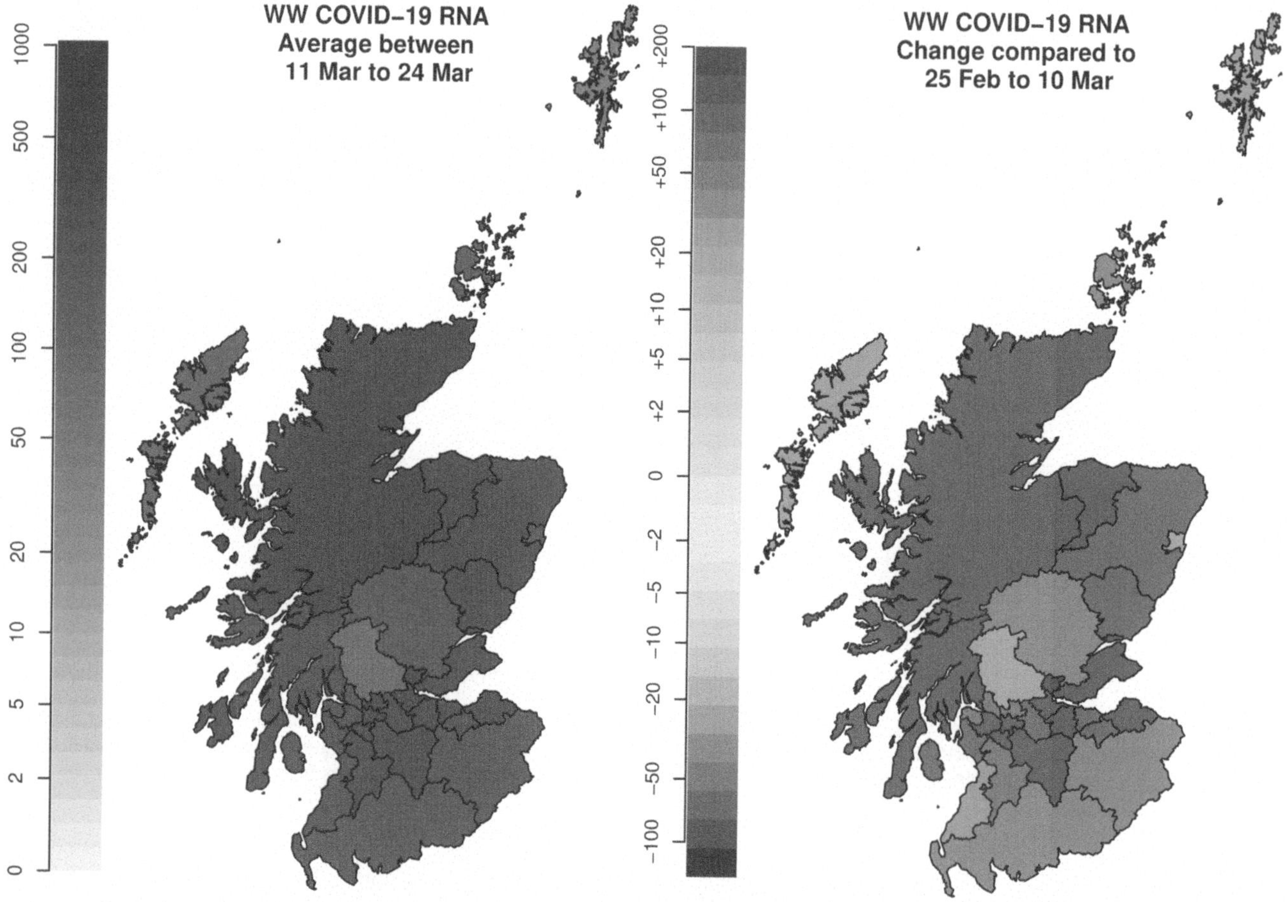

Figure 3 | An example map produced during the programme. The left map shows levels of wastewater COVID-19 at each local authority, aggregated within a 2-week period. The right compares the level to the prior 2-week period.

Interactive dashboard

To satisfy stakeholder requirements for live access to data, an interactive dashboard was developed by SEPA. This was implemented using TIBCO Spotfire® in the summer of 2020 and made publicly available in October 2020. This was gradually updated over time to enhance the data available to the public and for public health professionals, providing interactive graphs of WW levels, details on the monitoring network, and at-a-glance summaries of WW data across large regions of Scotland.

The data feeds for the dashboard came from both the SEPA laboratories which provided the raw concentration values and Scottish Water which provided the data required to normalise the raw concentrations. The data feeds were merged upon opening of the dashboard with the normalised results and various analyses being calculated each day.

The dashboard complemented WW COVID-19 reports, produced by BioSS on a weekly, then twice-weekly basis. While the dashboard output was more timely, the written reporting allowed individual commentary informed by specific analyses, to assist readers in identifying what aspects are most important or statistically reliable.

Additional methodological changes

A key aspect of the COVID-19 WW monitoring project is its dynamic nature. It was necessary to balance the need for consistent reporting with the need to readjust goals over time. Requirements changed in tandem with the state of the pandemic, through times of high and low COVID-19 levels, and as policy priorities changed. This resulted in a number of additional procedures being added, including:

- Smoothing of major sites: High levels of day-to-day variation in WW COVID-19 levels were seen at most sites. However, the WW sampling was generally not frequent enough to make site-level running-means useful. Thus, a smoothing model based on a GAM was used to smooth this variation and derive estimates of the underlying trend. A Tweedie distribution (Tweedie 1984) was used for the modelling because it is a flexible non-negative distribution that broadly captures the mean-variance characteristics we observed. The Tweedie GAM not only produces a smooth line to help interpretation but also gives a summary of the variability characteristics of each site. These in turn can be used to generate confidence intervals on the estimated trend and other probabilistic metrics to inform decision-making.
- Outlier detection: Occasionally, WW-based estimates of viral levels gave single individual values that were extremely high relative to levels immediately before and after, and which were not reflected in the case-level data. These spike events could potentially have an outsized effect on other methodologies and could dominate visualisations. We used a GAM model to classify points with a high propensity of being spike, taking into account case data and WW data up to each observation. A threshold was used to remove some of these points according to a desired level of false positives and negatives on manually labelled data.

By March 2022, pandemic monitoring was influenced by additional concerns. Firstly, the appearance of the new Omicron variant has led to reductions in detected virus per infected individual across the world (Rasenberg 2022), probably due to reductions in the COVID-19 shedding rate. At the same time, case-based metrics have been impacted by the discontinuation of PCR testing for certain types of cases and the incorporation of Lateral Flow Device (LFD) testing. In response to these challenges, we have worked on incorporating other sources of data into the ongoing analyses, including the UK Office for National Statistics (ONS) Coronavirus Infection Survey (CIS; ONS 2022), as well as sequencing of WW samples for detection and quantification of COVID-19 variants. Sample analysis methodology will also pursue greater automation to better handle the volume of analyses required.

METHODOLOGICAL SUMMARY

The methodology for COVID-19 WBE in Scotland may then be summarised in the following steps.

A network of auto-samplers operated at WWTW sites across Scotland, collecting composite WW samples. At regular times according to the sampling plan, these samples were collected and transported to the SEPA for laboratory qPCR analysis. Separately, samples were taken for ammonia analysis.

Once at the SEPA laboratories, samples were analysed for SARS-CoV-2 RNA according to 'Method 6' of Fitzgerald *et al.* (2021). This comprised spiking each sample with porcine reproductive and respiratory syndrome virus (PRRSv) for a check for chemical inhibition. Next, samples were clarified and then concentrated by centrifuge. Viral RNA was extracted using the QiAmp RNA extraction kit. After elution, one-step qPCR reactions were conducted using Luna® Universal One-Step RT-qPCR Kit, focusing on the N1 gene. From this, concentrations of viral RNA were computed.

Twice a week, the data from SEPA laboratories together with available sample flow or ammonia data from Scottish Water were taken by BioSS for processing and reporting. The first step involved adjusting sample RNA concentrations for dilution using flow or ammonia data, using coefficients previously derived. Data from other sources, such as case data, were also obtained at this stage, and geographically aggregated to produce comparable datasets to the WW data. Next, a preliminary report was produced, which allowed the data to be checked for errors, both via automatic means and by visual inspection. This report also included smoothing of WW outputs at major sites and mapping, which allowed the identification of significant patterns and changes in the data with respect to the estimation of COVID-19 prevalence. A report was then drafted, reviewed, and submitted together with a variety of data outputs as requested by stakeholders.

USE OF WW-DERIVED OUTPUTS

WW data have been integral in providing insight to better inform decisions made by the Scottish Government during the pandemic. Estimates based on WW data help the Scottish Government, local authorities and Scottish Health boards plan and put into place what is needed to keep the population of Scotland safe. Results from WW go to Ministers in a weekly report, providing additional information to help inform decision-making around the use of restrictions. Results also go to the general public as part of the weekly Research Findings output on COVID-19 (Scottish Government 2020–2022), and to scientists aiming to model COVID-19.

WW testing is used in Scotland to complement existing case testing, to confirm case trends, or provide an alternative means of surveillance as it is a COVID-19 indicator that is independent of healthcare-seeking behaviours and access to clinical testing. In some parts of Scotland, WW data have been used to inform decisions to deploy mobile testing. This has been most useful when prevalence levels have been low, such as in early summer 2021.

The Scottish Government's use of data from WW sampling allows for assessment of the progress of the pandemic in Scotland, including estimates of the R value and growth rate (Scottish Government 2020–2022). Scotland has been submitting estimates to be considered, discussed, and combined at the Epidemiology Modelling Review Group (EMRG) within The UK Health Security Agency (UKHSA) using WW data since June 2021.

EVALUATION OF WBE

At the time of writing, the COVID-19 pandemic persists in Scotland and around the world. WBE epidemiology still plays a critical role. From our experience of using WBE for COVID-19, we see several advantages for WW monitoring:

- WW data are efficient to collect for large areas or large populations, depending on the WW network. We were able to monitor the majority of Scotland's population on a regular basis, sufficient to capture some intra-week effects despite RNA analysis being done for only 100–300 samples per week. Information at this scale and resolution is very difficult to match through population surveys. For example, the ONS CIS survey requires testing several thousand participants for Scotland each week and can only give weekly estimates at the resolution of subregions (which are typically made up of several health boards).
- In contrast to other indicators, WW appears relatively unaffected by behavioural changes made in response to the state of the epidemic and policy changes. We saw this clearly in early 2022 when policy changes in individual PCR testing protocols greatly affected case-level measurements. When LFD-based statistics were introduced, an artificial increase appeared in certain areas, with large differences between genders or different socioeconomic statuses. We do not expect there to be similar effects with WW data.
- WW can produce accurate and timely data. While there are significant levels of noise in the observations, trends in WW measurements of COVID-19 levels closely follow case-level data up until December 2021, especially once smoothed, with only occasional exceptions during rapid rises in viral activity. While trends in WW viral levels do not consistently precede case data, this depends on the practicalities of how rapidly each data set can be collated. Independent of reporting lag, WW trends do appear to precede those seen in survey and hospitalisation data.

These aspects of WBE are generally independent of the particulars of the COVID-19 pandemic. If in any future epidemic situation, infected individuals shed pathogenic DNA/RNA into WW, WBE-based monitoring should be similarly effective.

Our work has also highlighted a number of issues with WBE:

- While the methodology described here tackles some of the uncertainty, WW-based measurements were noisier than other measures of disease levels. There was a substantial amount of day-to-day variability, and thus determining whether the virus is increasing or decreasing cannot be based on an individual measurement. Rather, it requires a week or more of data, preferably across several sites in a region. This sets a limit on the degree of immediacy that can be offered by WW data. In addition, we observed individual measurements that gave very high readings that do not appear reflective of real processes. The reasons for such anomalies are still not entirely understood. The variability of WW-based measurements can be ameliorated by more frequent sampling, but this has cost implications.
- Some regions were much easier to sample from than others. While larger WWTW catchments can cover several hundred thousand inhabitants, there were many smaller WWTW in Scotland that, in some cases, only cover a few hundred inhabitants. While a small number of sample sites can cover the majority of the Scottish population, monitoring for smaller more remote locations is unlikely to be cost effective. In turn, large-catchment sample sites which are more cost effective do not offer a great deal of spatial resolution. In-network sampling can supplement these measurements and add resolution to the data but suffer from further increased uncertainties.
- The relationship between WW viral levels and other metrics of disease epidemiology was affected by some external factors. While for the majority of the period so far, the case-level to WW-level ratio has remained constant, the hospitalisation to WW (and case) ratio fell rapidly in early 2021, probably due to increased vaccination. The case-to-WW ratio itself changed during the rise in prevalence of the Omicron strain. Leads and lags are also not necessarily consistent. At least part of these

changes represents real changes in the nature of the pandemic. Cases, hospitalisation, and survey-based measures themselves present different trends over time, and hence no single measure represents a gold standard that others can be judged. While comparisons between different measures are useful for validation purposes, deviations can also provide useful information.

More formal evaluations and reviews are in progress, and we expect more sophisticated research to be conducted once the immediate pressures of the pandemic are over.

CONSIDERATIONS ON THE FUTURE OF WBE

From the work done in the programme, we can derive a number of broad lessons about the future use of WBE.

- The general methodology used here, involving normalisation, aggregation, and visualisation is an effective means of displaying and interpreting WBE data.
- That said, further methodological developments may help, in particular focusing on the derivation and communication of uncertainty. This is a challenge due to the multiple sources of uncertainty present in WBE data.
- In an epidemic context, WBE projects require flexibility to adapt to changing requirements. Effective analysis methods and sampling strategies in low virus-level contexts may differ from high virus-level contexts.
- A framework of validation based on other data sources is beneficial to authenticate the quality of WW data and to generate confidence in its usage.
- The emergence of variant strains in a pandemic could impact WW-based prevalence estimates. Analyses should be integrated with data on variant prevalence to allow adjustment for differing levels of shedding. Using WW to quantify the levels of variants in time and space is still experimental.
- Sampling effort should consider the objectives required. Large numbers of samples at major population sites allow for more accurate assessments of national-level disease prevalence, while sampling at smaller sites may provide more useful data on local outbreaks. Some medium-sized sites are important to improve population coverage in certain locations, while spare sampling capacity may allow experimental development.
- To inform a better, and more rapid, response for future epidemics, retrospective research based on the data gathered in Scotland and elsewhere is critical, without the pressure of immediate policy requirements. Research focussing on developing an improved understanding of WBE and improving methodology in this area is an important aspect of this. To this end, the data and protocols have been curated and shared (BioRDM Team 2022; Roberts *et al.* 2022; Scorza *et al.* submitted). We aim to make our code for modelling and visualisations available in the future.

In the medium term, as the COVID-19 pandemic continues, WW-based monitoring is likely to continue to see use in many countries, potentially taking a greater role as other data sources face budget-based limitations. In addition, WW-based sequencing work to detect emerging variants of COVID-19 is expected to come to the forefront.

Retrospective analyses of WW data are also likely to be done. These would, as mentioned earlier, facilitate better evaluation of WBE outputs. They may also apply more sophisticated analysis methodologies coupled with other measures and understanding of behaviour to reveal, for example, the changing dynamics of the COVID-19 virus at different points of the pandemic.

Further into the future, post-pandemic, the sampling infrastructure established during the COVID-19 pandemic could be extended to monitoring indicators of other disease in the population, including both viruses and bacteria. WBE opens the potential to begin or to expand the simultaneous monitoring of the spread of antimicrobial resistance, drug use including antidepressants, or other chemicals (OHBP 2021; Sims *et al.* 2021). Such non-epidemic monitoring of population health will require careful consideration of ethical and political concerns.

ACKNOWLEDGEMENTS

The Scottish Government (Rural & Environmental Science and Analytical Services division) funds the sampling, testing and data analysis of wastewater in Scotland. We are particularly grateful for advice and direction from Professor Andrew Millar, University of Edinburgh.

DATA AVAILABILITY STATEMENT

All relevant data are available from an online repository or repositories (https://covid-ww-scotland.github.io/PrevalenceData).

CONFLICT OF INTEREST

The authors declare there is no conflict.

REFERENCES

BioRDM Team. 2022 *COVID Wastewater Scotland – Open Outputs of Monitoring COVID in Wastewater in Scotland.* Available from: https://biordm.github.io/COVID-Wastewater-Scotland/

COVIDPoops19: *Summary of Global SARS_CoV-2 Wastewater Monitoring Efforts.* Available from: https://www.arcgis.com/apps/dashboards/c778145ea5bb4daeb58d31afee389082

Fitzgerald, S. F., Rossi, G., Low, A. S., McAteer, S. P., O'Keefe, B., Findlay, D., Cameron, G. J., Pollard, P., Singleton, P. T. R., Ponton, G., Singer, A. C., Farkas, K., Jones, D., Graham, D. W., Quintela-Baluja, M., Tait-Burkard, C., Gally, D. L., Kao, R. & Corbishley, A. 2021 Site specific relationships between COVID-19 cases and SARS-CoV-2 viral load in wastewater treatment plant influent. *Environmental Science & Technology* **55** (22), 15276–15286. https://doi.org/10.1021/acs.est.1c05029.

Garnier, S., Ross, N., Rudis, B., Sciaini, M., Camargo, A. P. & Scherer, C. 2022 *Viridis – Colorblind-Friendly Color Maps for R.* https://doi.org/10.5281/zenodo.4679424.

Lodder, W. J., Rutjes, S. A., Takumi, K. & de Roda Husman, A. M. 2013 Aichi virus in sewage and surface water, the Netherlands. *Emerging Infectious Diseases* **19** (8), 1222–1230. https://doi.org/10.3201/eid1908.130312.

Naughton, C. C., Roman, F. A., Alvarado, A. G. F., Tariqi, A. Q., Deeming, M. A., Bibby, K., Bivins, A., Rose, J. B., Medema, G., Ahmed, W., Katsivelis, P., Allan, V., Sinclair, R., Zhang, Y. & Kinyua, M. N. 2021 Show us the Data: Global COVID-19 Wastewater Monitoring Efforts, Equity, and Gaps. medRxiv 2021.03.14.21253564. https://doi.org/10.1101/2021.03.14.21253564.

Office for National Statistics 2022 *Coronavirus (COVID-19) Infection Survey: Methods and Further Information.* Available from: https://www.ons.gov.uk/peoplepopulationandcommunity/healthandsocialcare/conditionsanddiseases/methodologies/covid19infectionsurveypilotmethodsandfurtherinformation

OHBP 2021 *Chemical Investigation Project 3 Scotland (CIP3 Scotland).* Available from: https://ohbp.org/2021/07/01/chemical-investigation-project-3-scotland-cip3-scotland-started-in-april-2021/

PHS 2022 *Scottish Health and Social Care Open Data.* Available from: https://www.opendata.nhs.scot/

Rasenberg, E. 2022 *Water News Europe: Omicron Hard to Detect in Wastewater.* Available from: https://www.waternewseurope.com/omicron-harder-to-detect-in-wastewater/

R Core Team. 2021 *R: A Language and Environment for Statistical Computing.* R Foundation for Statistical Computing, Vienna, Austria. Available from: https://www.R-project.org/

Roberts, A., Fang, Z., Mayer, C.-D., Frantsuzova, A., Cameron, G. J. & Scorza, L. C. T. 2022 *Data Normalisation of RT-qPCR Data for Detection of SARS-CoV-2 in Wastewater.* Available from: https://www.protocols.io/view/data-normalisation-of-rt-qpcr-data-462for-detection-o-b4eqqtdw

Romano, A., Sotis, C., Dominioni, G. & Guidi, S. 2020 The scale of COVID-19 graphs affects understanding, attitudes, and policy preferences. *Forthcoming, Health Economics* https://dx.doi.org/10.2139/ssrn.3588511.

Santiso-Bellón, C., Randazzo, W., Pérez-Cataluña, A., Vila-Vicent, S., Gozalbo-Rovira, R., Muñoz, C., Buesa, J., Sanchez, G. & Rodríguez, J. R. 2020 Epidemiological surveillance of norovirus and rotavirus in sewage (2016–2017) in Valencia (Spain). *Microorganisms* **8** (3). https://doi.org/10.3390/microorganisms8030458.

Scorza, L., Cameron, G., Murray-Williams, R., Findlay, D., Bolland, J., Cerghizan, B., Campbell, K., Thomson, D., Corbishley, A., Gally, D., Fitzgerald, S., Low, A., McAteer, S., Roberts, A., Fang, F., Mayer, C., Frantsuzova, A., Baby, A., Zieliński, T. & Millar, A. J. 2022 SARS CoV-2 RNA levels in Scotland's wastewater. *Scientific Data* **9**, 713. https://doi.org/10.1038/s41597-022-01788-3.

Scottish Government 2020–2022 *Coronavirus (COVID-19): Modelling the Epidemic.* Available from: https://www.gov.scot/collections/coronavirus-covid-19-modelling-the-epidemic/

Sims, N., Avery, L. & Kasprzyk-Hordern, B. 2021 *Review of Wastewater Monitoring Applications for Public Health and Novel Aspects of Environmental Quality (CD2020_07).* Scotland's Centre of Expertise for Waters (CREW). ISBN: 978-0-902701-85-4. Available from: crew.ac.uk/publications

Tweedie, M. C. K., 1984 An index which distinguishes between some important exponential families. In: *Statistics: Applications and New Directions. Proceedings of the Indian Statistical Institute Golden Jubilee International Conference* (Ghosh, J. K. & Roy, J. eds.). Indian Statistical Institute, Calcutta, pp. 579–604.

Wade, M. J., Lo Jacomo, A., Armenise, E., Brown, M. R., Bunce, J. T., Cameron, G. J., Fang, Z., Farkas, K., Gilpin, D. F., Graham, D. W., Grimsley, J. M. S., Hart, A., Hoffmann, T., Jackson, K. J., Jones, D. L., Lilley, C. J., McGrath, J. W., McKinley, J. M., McSparron, C., Nejad, B. F., Morvan, M., Quintela-Baluja, M., Roberts, A. M. I., Singer, A. C., Souque, C., Speight, V. L., Sweetapple, C., Walker, D., Watts, G.,

Weightman, A. & Kasprzyk-Hordern, B. 2022 Understanding and managing uncertainty and variability for wastewater monitoring beyond the pandemic: lessons learned from the United Kingdom national COVID-19 surveillance programmes. *Journal of Hazardous Materials* **424**, 127456. https://doi.org/10.1016/j.jhazmat.2021.127456.
Wood, S. N. 2017 *Generalized Additive Models: an Introduction with R*, 2nd edn. Chapman and Hall, Boca Raton.

First received 31 March 2022; accepted in revised form 28 November 2022. Available online 14 December 2022

doi: 10.2166/wh.2022.012

Monitoring of SARS-CoV-2 in wastewater: what normalisation for improved understanding of epidemic trends?

Charlotte Sakarovitch[a,*], Olivier Schlosser[b], Sophie Courtois[b], Cécile Proust-Lima[c], Joanne Couallier[a], Agnès Pétrau[d], Xavier Litrico[e] and Jean-François Loret[b]

[a] SUEZ, LYRE, 15 av Léonard de Vinci, 33600 Pessac, France
[b] SUEZ, CIRSEE, 38 rue du Président Wilson, 78230 Le Pecq, France
[c] Université de Bordeaux, INSERM, Bordeaux Population Health Center, UMR1219, F-33000 Bordeaux, France
[d] SUEZ Rivages Pro Tech, Technopôle Izarbel, 2 Allée Théodore Monod, 64210 Bidart, France
[e] SUEZ, CB21, 16 Place de l'Iris, 92040 Paris La Défense, France
*Corresponding author. E-mail: charlotte.sakarovitch@suez.com

ABSTRACT

SARS-CoV-2 RNA quantification in wastewater has emerged as a relevant additional means to monitor the COVID-19 pandemic. However, the concentration can be affected by black water dilution factors or movements of the sewer shed population, leading to misinterpretation of measurement results. The aim of this study was to evaluate the performance of different indicators to accurately interpret SARS-CoV-2 in wastewater. Weekly/bi-weekly measurements from three cities in France were analysed from February to September 2021. The concentrations of SARS-CoV-2 gene copies were normalised to the faecal-contributing population using simple sewage component indicators. To reduce the measurement error, a composite index was created to combine simultaneously the information carried by the simple indicators. The results showed that the regularity (mean absolute difference between observation and the smoothed curve) of the simple indicators substantially varied across sampling points. The composite index consistently showed better regularity compared to the other indicators and was associated to the lowest variation in correlation coefficient across sampling points. These findings suggest the recommendation for the use of a composite index in wastewater-based epidemiology to compensate for variability in measurement results.

Key words: COVID-19, normalisation, SARS-CoV-2, wastewater-based epidemiology, wastewater monitoring

HIGHLIGHTS

- SARS-CoV-2 RNA quantification in wastewater shows high variability and need to be normalized.
- The performance of various normalized indicators, including a composite index was assessed via two criteria: regularity and correlation with the incidence rate.
- The composite index showed the best performance.
- Except for the F-specific RNA bacteriophage normalisation index, correlation of all indicators with the incidence rate was substantial.

GRAPHICAL ABSTRACT

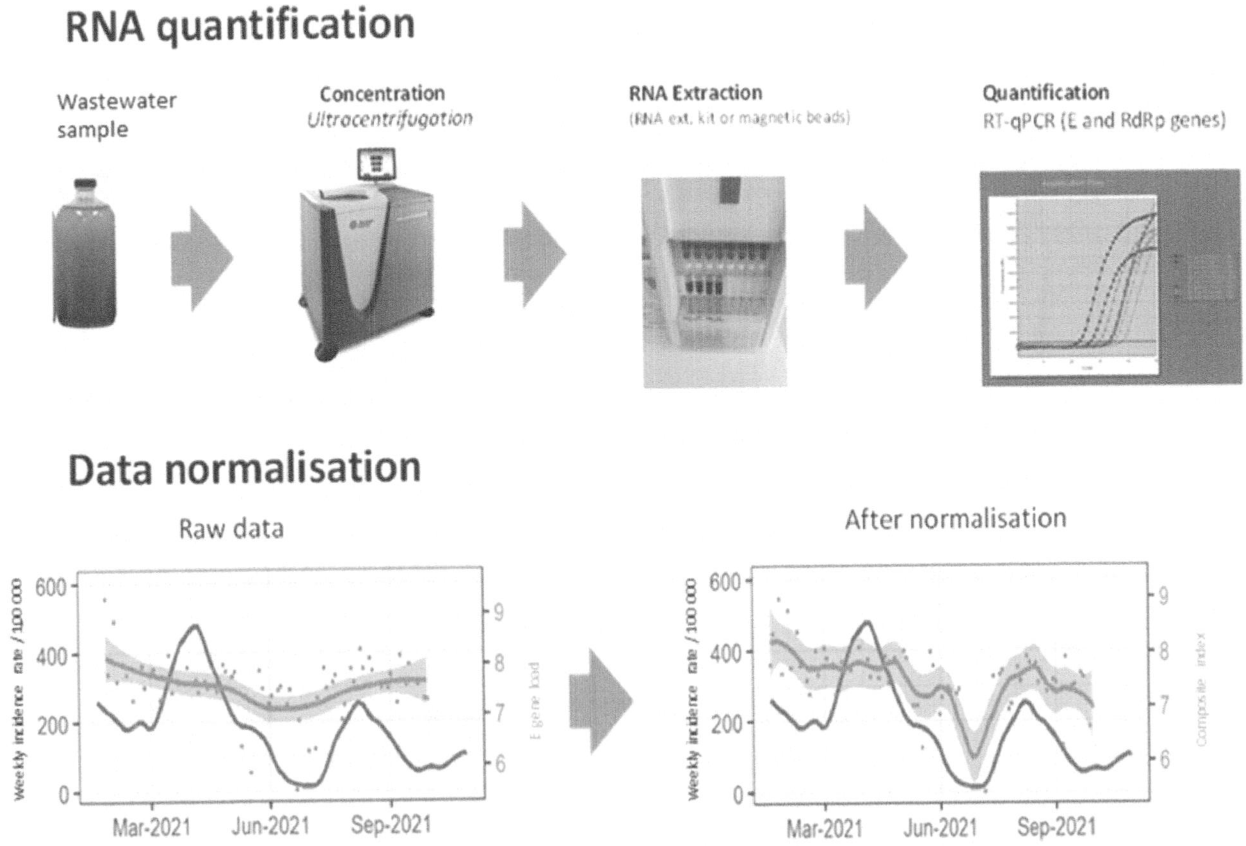

INTRODUCTION

The COVID-19 pandemic affecting the world since the end of 2019 has generated a scientific activity unprecedented in the history of mankind, and has reactivated the interest for wastewater epidemiology (Medema *et al.* 2020a; Polo *et al.* 2020). The genome of SARS-CoV-2, the coronavirus causing the disease, is indeed excreted in large quantities and during several weeks in the faecal material of infected individuals (Cheung *et al.* 2020; Gupta *et al.* 2020; Lescure *et al.* 2020; van Doorn *et al.* 2020; Wölfel *et al.* 2020; Zheng *et al.* 2020). It has been demonstrated that this virus is rapidly inactivated in wastewater (Varbanov *et al.* 2021; Wurtzer *et al.* 2021), and several studies report an absence of detection of viable SARS-CoV-2 in this medium (Döhla *et al.* 2020; Rimoldi *et al.* 2020; Albert *et al.* 2021; Westhaus *et al.* 2021; Robinson *et al.* 2022). In contrast, the viral genome is sufficiently stable in this medium (Varbanov *et al.* 2021; Wurtzer *et al.* 2021) to make it a good indicator of the spread of the disease in the population. This approach has been successfully used in the past to monitor the evolution of different infectious diseases, notably to survey the circulation of the poliovirus (WHO 2003).

With the objective of a large-scale surveillance of the circulation of infectious agents, considering viral detection in wastewater presents clear advantages, in comparison with a surveillance based on clinical testing, since it enables to cover large populations with a single analysis, takes into account pre-symptomatic and asymptomatic individuals, and offers a rapid answer enabling an early detection of the epidemic trends, potentially a few days ahead of more conventional health indicators (Medema *et al.* 2020b; Nemudryi *et al.* 2020; Bibby *et al.* 2021; D'Aoust *et al.* 2021a). For these reasons, the European Commission published in March 2021 a recommendation encouraging the European Member States to establish a wastewater surveillance system for SARS-CoV-2, focused primarily on sanitation systems of cities with over 150,000 inhabitants (CE 2021).

However, SARS-CoV-2 RNA concentration in wastewater can be affected by black water dilution factors in the sewer and movements of the population connected to the sewer shed due to commuting or vacations, thus leading to misinterpretation of measurement results. To take these factors into account, it is possible to correct the quantification results by using indicators reflecting the amount of population connected to the sanitation system. This approach is commonly referred to as 'normalisation'. In its document, the European Commission primarily recommends to normalise the quantification results by the population and the wastewater flow, as already used successfully by some authors (Rusiñol *et al.* 2021) and applied, e.g. in The Netherlands (https://coronadashboard.government.nl/landelijk/rioolwater). Additional normalisation means by using cross-assembly phage or pepper mild mottle virus is also recommended in this document, since both viruses have been described as good indicators of human faecal contamination and have been used for such purpose (Rosario *et al.* 2009; Park *et al.* 2020).

With the aim of rapidly detecting epidemic trends, it is however necessary to use normalisation means based on parameters that (i) reflect the amount of population connected to the sanitation system, (ii) are simultaneously measured on the water sample used to determine the concentration of viral RNA and (iii) are fast and easy to measure, and ideally cheap in order to enable a high frequency of analysis. With this aim, we compared in this study the following approaches:

- A normalisation by the population and the wastewater flow, as recommended by the European Commission.
- A normalisation based on the quantification of F-specific RNA bacteriophages. These phages are routinely quantified in our laboratory, together with SARS-CoV-2 RNA, to verify the recovery of viral genetic material. They are excreted by humans and usually present in large concentrations in urban wastewaters (McMinn *et al.* 2017).
- A normalisation based on the quantification of ammoniacal nitrogen. This parameter is routinely analysed as part of the regulatory monitoring, and ammonium has previously been shown to be a good indicator of population changes, and as such applicable for normalisation purposes in wastewater-based epidemiology (Been *et al.* 2014). More recently, it has been applied to normalise SARS-CoV-2 concentration in a monitoring programme in England (Sweetapple *et al.* 2021).

We also considered a composite index combining the information carried by all the indicators generated by the previous approaches. Indeed, each SARS-CoV-2 RNA indicator is subject to substantial random measurement error and combining the indicators into a composite score could thus reduce the measurement uncertainty (von Oertzen *et al.* 2010).

MATERIALS AND METHODS

Study sites and sample collection

This study used the data produced in the framework of the COVID-19 wastewater surveillance programme driven by the SUEZ for communities in France (https://www.suez.fr/fr-fr/actualites/covid-city-watch). Three French cities (named A, B and C) presenting a minimum of 6-month follow-up were selected for this study. The sewer system is combined at Sites A and C, leading to a significant increase in collected flows during rain events, whereas the sewer system at site B is separate and presents a much more constant flow (flow distributions are presented in Supplementary Material, Figure S2 and Table S2).

The sampling period covered the third epidemic wave (from March to June 2021) and fourth epidemic wave (from July to September 2021) that affected France (sites A and B), or only the third wave (site C). The sampling frequency ranged from twice a week to once every 2 weeks depending on the site and period (Table 1). The samples consisted of 24 h flow-proportional composite samples collected with automated samplers, preserved and transported at 5 °C, and delivered to the laboratory for analysis within 24 h. The sampling points were located either at the inlet of the wastewater treatment plant (A.1, A.2, B.1, B.2 and C.2) or on the collection network (C.1) (Table 1).

Health surveillance data

The weekly incidence rate (the total number of positive tests performed in a 7-day rolling period divided by the number of inhabitants) was extracted from the French institute for public health surveillance database (https://geodes.santepublique-france.fr/) at the metropolis level for sites A and B and at the department level for site C. In this national database, the new cases are reported by sampling date and residential address. During all the study period, the tests were free of charge for French residents. The date of entry in our database was the end-term of the 7-day period.

Table 1 | Sites and sampling points analysed

Site	Sampling point	First sample	Last sample	Sampling frequency	Total samples analysed	Population connected
A	1	Jan 18	Sept 29	2/week	72	152,000
	2	Jan 18	Sept 29	2/week	70	45,000
B	1	Feb 3	Aug 30	1–2/week	41	426,000
	2	Feb 3	Aug 30	1–2/week	41	182,000
C	1	Jan 18	June 28	1/week	24	10,000
	2	Jan 18	June 28	1/week	24	36,000
	2	July 5	Sept 27	1/2 weeks	7	36,000

Quantification of SARS-CoV-2 and normalisation parameters

For each wastewater sample, the concentration of RdRp-IP4 gene, E gene and F-specific RNA bacteriophage GGII genome were quantified by the same laboratory (details are provided in Supplementary Material A1). Genomes were quantified using one-step qRT-PCR assays.

F-specific RNA bacteriophages were used as internal standards for our analytical method to estimate the recovery rate of viruses. The reason for using these phages rather than crAssphage or PMMoV as recommended by the European Commission is that our method was developed in 2020, at the beginning of the pandemic, 1 year before the publication of the EU recommendation of March 2021, so we could not take into account this recommendation at that time.

For each sampling point, the size of the population connected to the sewer was estimated using data from the National Institute of Statistics and Economics Studies (INSEE, https://www.insee.fr). The wastewater daily flow rate was measured at each sampling point on each sampling day.

The instruments used for the measurement of ammonium concentrations were those operated routinely and differed from site to site. For site A, the ammonium concentration was measured by the Hach Lange LCK303 cuvette test using the spectrophotometer Hach Lange DR 2800. The measuring range extends from 2 to 47 mg/L of N-NH_4. For site C, the ammonium concentration was measured by the Nanocolor cuvette test using the spectrophotometer Nanocolor UV/VIS II. The measuring range extends from 4 to 80 mg/L of N-NH_4. These two alternative methods were validated according to the ISO 8466-1 and DIN 38402 A51. For site B, the post-steam distillation titrimetric method was used, according to the NF T90-015-1 standard. This method is suited for concentrations higher than 4 mg/L.

Normalised indicator candidates

Three quantifications of SARS-CoV-2 in wastewater were considered: RdRp-IP4 gene RNA concentration, E gene RNA concentration and the arithmetic mean between the two gene concentrations (referred to as RdRp, E and mean hereafter). For each of them, three different adjustments for the amount of faecal matter or shedding population were investigated in addition to the raw concentration:

Viral load per day per inhabitants (gc/day/inhabitants) $= [\text{SARS}-\text{CoV}-2\ \text{RNA}](\text{gc/L}) \times \text{dailyflow}(\text{m}^3/\text{d}) \times 1000(\text{L/m}^3)/\text{Population(inh)}$; referred to as RdRp load, E load and Mean load.

Phage normalisation index (without unit) $= [\text{SARS}-\text{CoV}-2\ \text{RNA}](\text{gc/L})/[\text{Phages}](\text{gc/L})$, referred to as RdRp_phages; E_phages and Mean_phages.

Ammoniacal nitrogen normalisation index (gc/N-NH_4 mg) $= [\text{SARS}-\text{CoV}-2\ \text{RNA}](\text{gc/L})/[\text{Ammoniacal nitrogen}](\text{mg/L})$; referred to as RdRp_$NH_4$, E_$NH_4$ and Mean_NH_4.

This resulted in 12 indicators, referred to as simple indicators. All quantities were log-transformed to approach a normal distribution.

In addition, we developed a composite index to take into account the information from all the previous indicators simultaneously. This composite indicator had to be a simple formula (simple enough to be implemented in a spreadsheet) and to be the same for every site (allowing to be used on any new site starting the surveillance without the need to learn a new model on specific historical data or to maintain different models for each site). After a first analysis of the 12 simple indicators, the three normalisation indexes using phages were excluded from our composite index model due to high variability (see 'Results' section). Furthermore, since the mean of RdRp-IP4 and E genes is a combination of the other indicators, we did not consider the three related simple indicators.

We considered a latent process mixed model (Proust-Lima *et al.* 2013) to analyse the repeated measures of the selected six simple indicators (the log-transformed of RdRp-IP4 gene concentration or E gene concentration and the log-transformation of their normalised values according to flow per inhabitant and NH_4 concentration) assuming that they all measured the same underlying quantity: the virus quantity in wastewater (see Supplementary Material A2 for details on the statistical model).

The composite score $\hat{\Lambda}(t)$ was derived as a weighted sum of the six simple indicators noted $Y_k(t)$ for indicator k ($k = 1$–6) and time t according to the following formula:

$$\hat{\Lambda}(t) = \frac{1}{K}\sum_{k=1}^{6}\frac{Y_k(t) - \hat{\eta}_{1k}}{\hat{\eta}_{2k}}$$

where $\hat{\eta}_{1k}$ and $\hat{\eta}_{2k}$ were optimised using the latent process mixed model.

An Excel spreadsheet is provided in the appendix to calculate this composite index (see Supplementary Material A0). The formula parameters have been calculated so as to be used generically.

To avoid overfit and over-optimistic results, we assessed the performance of the composite score in a leave-one-site-out cross-validation (Figure 1).

Measurement data analysis

The concentrations of SARS-CoV-2 RNA, bacteriophages, N-NH_4 and daily flow rates were described for all sites and related sampling points. In a first step, the association between the weekly incidence rate (at the end-term of the 7-day period) and each of the 12 simple indicators (log-transformed values) and the composite index was estimated using the Pearson correlation coefficient. Only the days with all 12 indicators available were used to calculate the correlation. To inform on the uncertainty of estimates and enable comparisons, the correlation coefficient r was associated with its 95% confidence interval. Concentration values below the limit of detection (LOD) or limit of quantification (LOQ), were omitted from calculation of correlation coefficients due to their uncertainty. For visualisation in figures and for the composite index statistical model calibration, these concentrations were substituted by values equal to half the corresponding limit.

Assessment of indicator performance

We compared the performance of the indicators according to the following two criteria: (i) their regularity (since the virus concentration in wastewater is expected to change smoothly over time) and (ii) their association with the weekly incidence rates.

For each sampling point and each indicator, the LOESS (locally estimated scatterplot smoothing) method was applied to summarise the observations into a smooth curve over time. The selection of the best fitting LOESS smoothed curve was based

Figure 1 | Schematic representation of cross-validation for the composite index calculation.

on the Akaike information criterion (AIC). The regularity was assessed through the mean absolute difference (MAD) between the observed values and the AIC-selected LOESS smoothed curve. A smaller value of MAD indicates a smaller measurement variability. Since all the indicators did not vary within the same scale, we rescaled the MAD (rMAD) to obtain values that can be compared, by dividing each MAD by the range of the indicator smooth curve. To assess the potential gain of using the composite index over the simple indicators, we computed the percentage of rMAD reduction (rMADred) defined as the difference between the rMAD of the simple indicator and of the composite indicator, divided by the rMAD of the simple indicator. We considered that $-10\% < \text{rMADred} \leq 10\%$ means regularity performances similar to those of the composite index, $10\% < \text{rMADred} \leq 30\%$ is considered irregular and $\text{rMADred} > 30\%$ very irregular.

The association with clinical cases was assessed through the strength of the correlation (the Pearson correlation coefficient) between the AIC-selected LOESS smoothed curve and the weekly incidence rate curve.

RESULTS

Descriptive analysis

Health surveillance data

Figure 2 presents the evolution of the weekly incidence rate at the three sites, during the study period.

The magnitude of each wave was different from one site to the other. Site A experienced a severe second wave in October/November 2020 with a weekly incidence rate greater than 1,000 per 100,000, that was twice that of the third wave and four times that of the fourth wave. Sites B and C had a similar third wave but differed in the magnitude of waves 2 and 4.

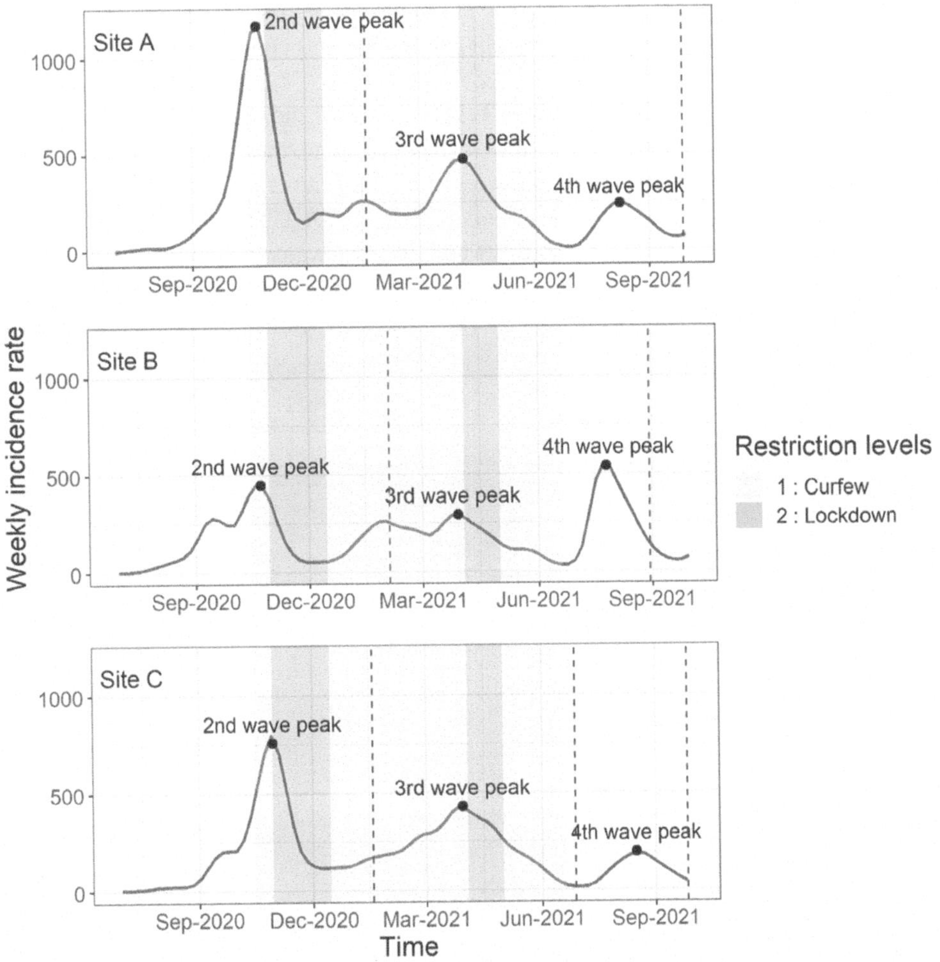

Figure 2 | Evolution of the weekly incidence rate per 100,000 inhabitants associated with sites A and B (metropolis level) and site C (department level). The periods delimited by the dashed lines are the periods of sampling in wastewater for each site.

Quantification of SARS-CoV-2 and normalisation parameters

Over the sampling period, RdRp and E genes were quantified in 94–100% and 79–100% of the samples, respectively. RdRp gene concentration reached the level of 10^6 gc/L on sites A and C and the level of 10^7 gc/L on site B. The highest level of E gene concentration was 10^6 gc/L on sites A and B, and 10^5 gc/L on site C (Figure 3 and Table S1). Bacteriophages RNA was quantified in all samples, with concentrations between 10^7 and 10^9 gc/L orders of magnitude (Figure S1 and Table S1 in Supplementary Material).

The order of magnitude of the flow rate at the inlet of the sewage treatment plants was 10^4 and 10^2 m^3/day at the collection network sampling point (C.1) (Table S2). At sites A and C, the daily flow measurement showed large variations between dry periods and high rainfall events (measured flows at the 99th percentile represented, respectively, about seven and nine times the median flow, during rain events). Conversely, at site B, which is equipped with a separate sewer system, the factor of variation did not exceed 2 (Figure S2).

The median concentration of $N\text{-}NH_4$ in wastewater ranged between 28 and 62 mg/L, depending on the sampling point (Table S3). Measurements at sites A and C showed minimum values much lower than at site B, mainly due to the dilution effect in combined sewer pipes during high rainfall events (Figure S3).

Correlations between the weekly incidence rate and measurement data-based indicators

The analysis of the correlation between the weekly incidence rate and indicators based on measurement data shows some common trends between sites:

At all sampling points except C.1, the weekly incidence rate and the raw concentration were significantly correlated (all $P \leq 0.003$), both for RdRp gene and E gene (Figure 4). The correlation was also statistically significant for the viral load, the $N\text{-}NH_4$ normalisation index and the composite index. The correlation was consistently lower with the phage normalisation index.

The strength of the correlation did not significantly differ between the four indicators, i.e. the raw concentration, the viral load, the $N\text{-}NH_4$ normalised index and the composite index. On the other hand, the strength of the correlation was most often lower for the phage normalisation index as compared to the four other indicators. For both sampling points on each site, no substantial difference in the strength of correlation was observed between targeted genes, whatever the indicator.

On site C, the correlation between the weekly incidence rate and the wastewater indicators clearly differed between the two sampling points. At C.2, all indicators except E_phages significantly correlated with the weekly incidence rate. On the other hand, at the sampling point C.1, where only 19 samples with all indicators available were analysed, the correlation coefficients were low with large confidence intervals.

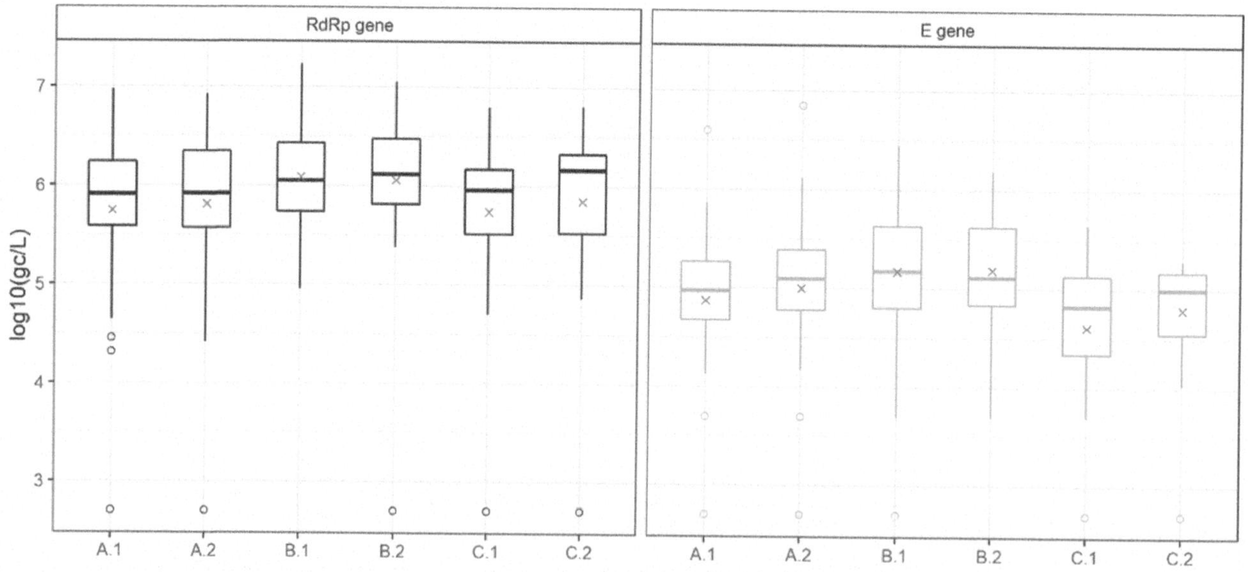

Figure 3 | Distribution of the RdRp gene and E gene concentrations in wastewater at different sampling points.

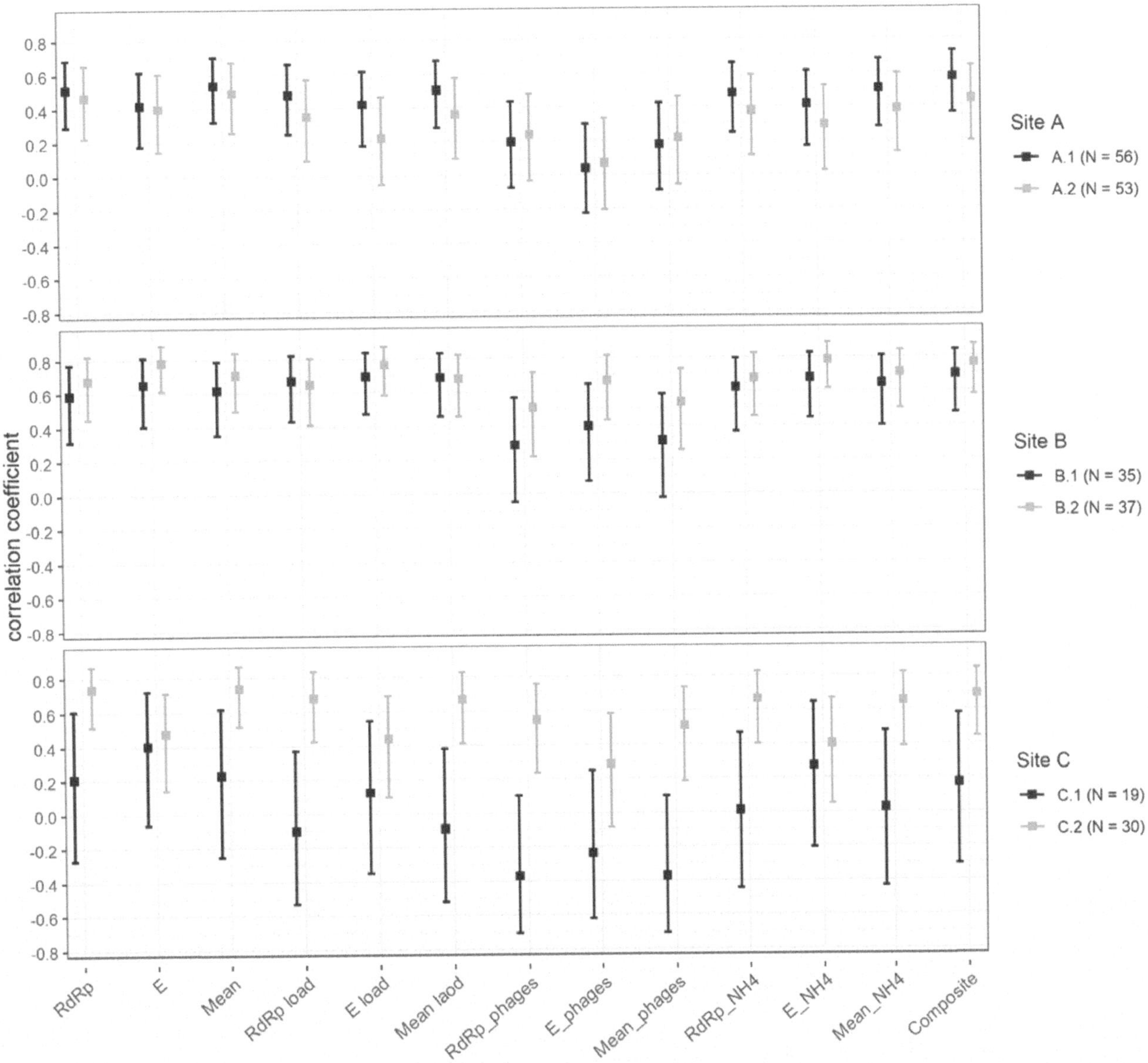

Figure 4 | Pearson correlation coefficient (*r*) and 95% confidence interval between the weekly incidence rate and the wastewater indicators.

Comparison of wastewater indicator performance

Indicator regularity

As shown by Figure 5 where the sampling points A.2 and B.1 were taken as examples, the profile of the smoothed curve and the dispersion of the observation points around the smoothed curve is different depending on the selected indicator. The graphs for all the other sites are available in Supplementary Material (Figures S4–S6).

As indicated by the rMAD displayed in Figure 6, the regularity of the indicators differed substantially across indicators and across sampling points. The composite index showed the overall best regular performance (mean rMAD = 0.12) with low between-site variability (standard deviation = 0.017).

The comparison of the regularity of the simple indicators with that of the composite index (Figure 7) further underlined the variability of the performance of indicators across sampling points: each indicator is very irregular (rMADred>30%) for at least one sampling point and 'irregular' for most of them; E load and E_NH_4 are 'regular' for B.1, B.2 and C.1 (rMADred≤10%) but 'very irregular' for A.1 and A.2 (rMADred>30%); Mean_NH_4 is 'regular' for B.2 but 'very irregular' for A.2.

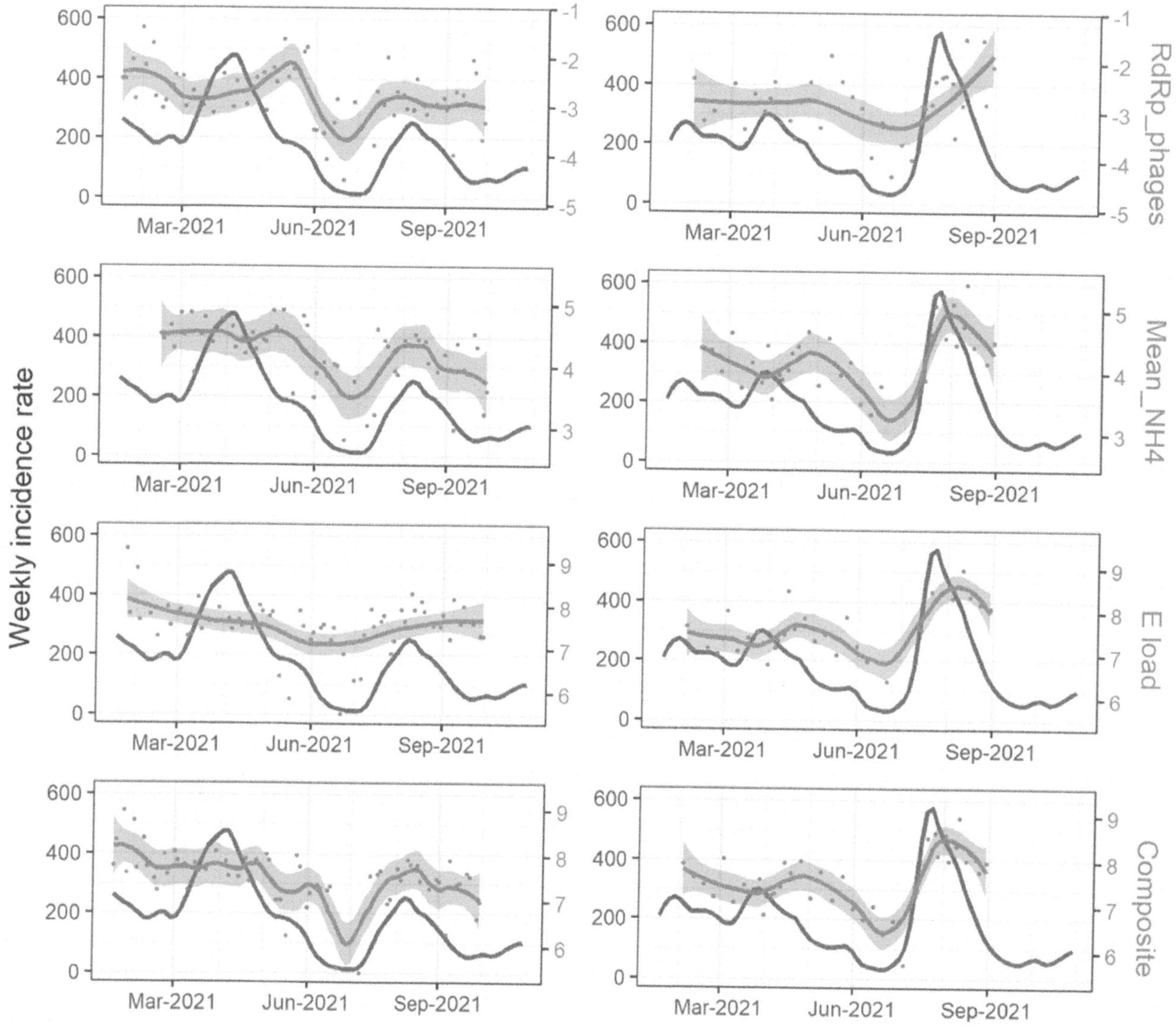

Figure 5 | Indicator observations (blue points) with the associated AIC-selected LOESS smoothed curve (blue line). The black curve is the corresponding weekly incidence rate per 100,000 inhabitants. The left column is for the sampling point A.2 and the right column is for the sampling point B.1. Please refer to the online version of this paper to see this figure in colour: http://dx.doi.org/10.2166/wh.2022.012.

At C.2, the simple indicators Mean and Mean load outperformed the regularity of the composite index by, respectively, 30 and 20%. Since the sampling frequency was very low at C.2 during the last 3 months of the observation period (only one sample every 2 weeks), the smoothed curve constructed from these points might be misleading and results from that site should be interpreted with caution.

Association of the weekly incidence rate with the LOESS curve-based indicators

Pearson correlation strength used to assess the association between the weekly incidence rate curve and the AIC-selected LOESS smoothed curve is indicated in Figure 8.

Except for the normalisations using phages, correlation coefficients were higher than 0.5 (with $P < 0.0001$) for all indicators and all sites and higher than 0.6 for all indicators except E and E load on site A. For the composite index, the correlation coefficients were among the highest.

Applying the LOESS method increased the strength of the correlation for almost all indicators. The variability between sampling points measured by the standard deviation ranged between 0.04 and 0.18 (Figure 8). The standard deviation was

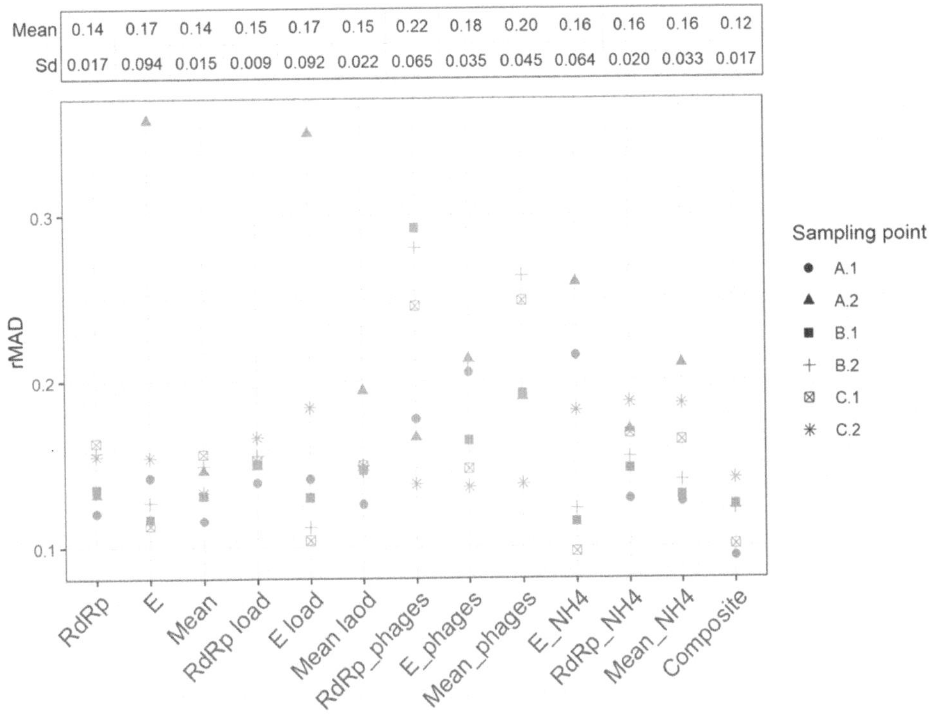

Figure 6 | Rescaled mean absolute difference between the observed values and the AIC-selected LOESS smoothed curve (rMAD) of each indicator for each sampling point. Mean, standard deviation and relative standard deviation for each indicator across sampling points.

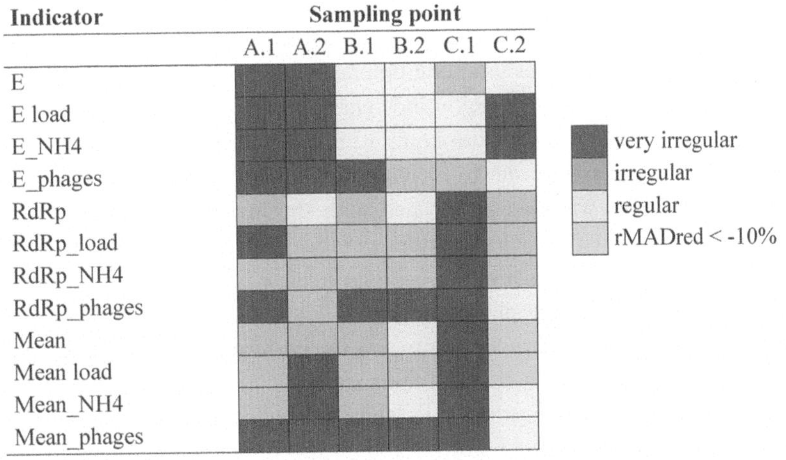

Figure 7 | Normalised indicator regularity performance classification using rMAD reduction (rMADred) obtained by the use of the composite index – light blue: rMADred≤10%, regularity performances alike to the composite index; regular blue: 10%<rMADred≤30% irregular indicator and dark blue: rMADred>30% very irregular indicator. Please refer to the online version of this paper to see this figure in colour: http://dx.doi.org/10.2166/wh.2022.012.

the lowest for the composite index. Nevertheless, for some of the other indicators such as RdRp, RdRp load, RdRp_NH_4, Mean, Mean load and Mean_NH_4, the standard deviation was too close to represent a significant difference.

DISCUSSION

SARS-CoV-2 RNA quantification in wastewater presents a good potential for monitoring the COVID-19 epidemics, but it shows high variability due to several factors responsible for noise in measurement, such as analysis uncertainty and variation

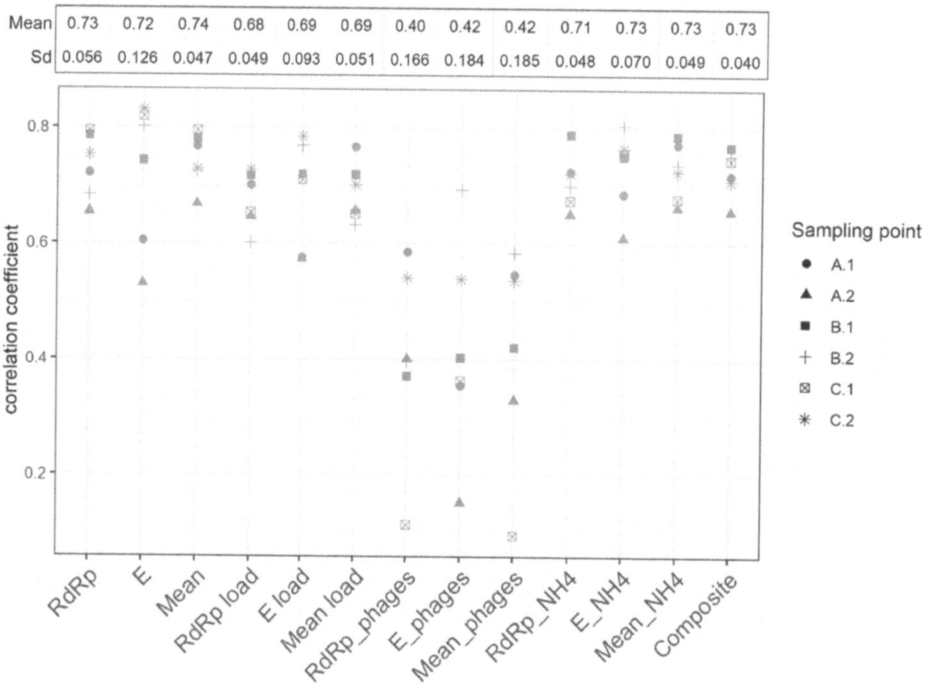

Figure 8 | Pearson correlation coefficient between each LOESS-based indicator and the corresponding weekly incidence rate for each sampling point. Mean, standard deviation and relative standard deviation for each indicator across sampling points.

in the total amount of human faecal matter in wastewater (Medema *et al.* 2020a; Crank *et al.* 2021; Zhu *et al.* 2021). In this work, we compared the performance of 13 indicators derived from RdRp and E genes quantification as candidates for the monitoring of the COVID-19 epidemics. At most sampling points, gene concentration and derived simple indicators were significantly, and moderately, correlated with the weekly incidence rate. However, between sampling points, variations in the strength of the correlation were high, even for the composite index. Applying the LOESS smoothing method to the measurement results and derived indicators in order to focus on the underlying signals substantially increased the strength of the correlation. The composite index was then associated with the lowest standard deviation estimates between the sampling points; however, the difference with other indicators was too small to be significant. Importantly, the performance comparison indicated that the composite index showed the best regularity (i.e. smallest errors) as compared to simple indicators that showed unstable regularity across sampling points.

We chose to rely on two complementary metrics for comparing the performance of different indicators and normalisation methods: regularity and correlation strength with weekly incidence rate.

By quantifying the distance between the observations and the smoothed curve of virus concentration with time, the regularity metric assesses the amount of measurement error in the indicator observations. Indeed, since the quantity of virus in wastewater is the result of the spread of the infection within the population, it is expected to change smoothly with time. Consequently, the further the observations from the smoothed curve, the poorer the indicator performs at estimating the 'true' variations with time we are looking for. Except for C.2 which should be interpreted with caution due to the low sampling frequency, the composite index was the only indicator that consistently showed better regularity, whatever the sampling point. Each simple indicator was 'very irregular' according to rMADred for at least one sampling point and 'irregular' for several. These variabilities highlight differences in simple indicator regularity performance between sites and the interest in having an indicator that consistently demonstrated the best regularity performance, whatever the site. This evidence indicates the value of using the composite index as a wastewater indicator.

The second metric we used to assess the performance of the wastewater indicator was the strength of the correlation of the AIC-selected LOESS smoothed curve with the weekly incidence rate. SARS-CoV-2 RNA is shed in faeces by infected individuals but the proportion of faecal shedders among them and the duration and magnitude of faecal shedding after infection are not very well known (Jones *et al.* 2020; van Doorn *et al.* 2020; Li *et al.* 2021). There is still uncertainty on the averaged excretion rate distribution, especially since it might depend on several factors such as the severity of the infection, the type of variant and the vaccination status (Kitajima *et al.* 2020; Wölfel *et al.* 2020; Bibby *et al.* 2021; Li *et al.* 2021; Miura *et al.* 2021; Zhu *et al.* 2021). Nevertheless, some studies provided evidence for a maximum excretion quantity within 1 week after infection (Cavany *et al.* 2021; Hoffmann & Alsing 2021; Miura *et al.* 2021). Considering that the cases detected through recorded laboratory tests in France are within a week after the infection, we assumed that the weekly incidence rate and the concentration of SARS-CoV-2 RNA in wastewater were synchronously correlated. Except for normalisation with phages, the correlation strength was good for all indicators at each site (r>0.6). In addition, as mentioned above, the composite index was the most stable indicator across sampling points with the lowest variation in correlation coefficient.

These findings demonstrated a lowest performance for the phage normalisation index. Regularity was low or very low, correlation was either not statistically significant or substantially lower than for the other indicators. Some hypotheses may be raised. First, the phage normalisation index is affected by biomolecular analysis uncertainty for both SARS-COV-2 and phages, resulting in higher total uncertainty compared to the other indicators. Second, even if F-specific RNA bacteriophages have been found in large quantities in raw sewage, in the order of 10^3–10^5 plaque forming units/mL (Havelaar *et al.* 1990; Cole *et al.* 2003; Blanch *et al.* 2006), studies demonstrated that their prevalence in human faeces was low, in the order of 10–20% (Grabow *et al.* 1995; Schaper *et al.* 2002). In addition, measurement of F-specific RNA bacteriophages in the wastewater samples in our study indicated large variations in phage load over time, up to 2 log (results not shown). These findings do not support the use of F-specific RNA bacteriophages as a reliable quantitative marker of the amount of faecal matter in wastewater samples. As mentioned in the 'Materials and Methods' section, F-specific RNA bacteriophages were selected before the publication of the EU recommendations. CrAssphage and PMMoV are other phages and viruses that have been proposed as potential useful markers for normalisation of SARS-CoV-2 quantification in wastewater (Wu *et al.* 2020; Heijnen *et al.* 2021; Wilder *et al.* 2021). Although F-specific RNA bacteriophages may present a higher variability in wastewater compared to crAssphage and PMMoV, the analytical uncertainty related with their quantification by PCR is probably in the same order. To date, there is no or limited evidence of improvement of results with this approach (Ai *et al.* 2021; D'Aoust *et al.* 2021a; Feng *et al.* 2021).

Although the strength of correlation with incidence rate is an important first-intention performance criterion to assess wastewater indicators (D'Aoust *et al.* 2021b; Sweetapple *et al.* 2021), it suffers several unavoidable flaws. First, the incidence rate estimation depends on the behaviour toward screening and this behaviour evolves with time as illustrated by the positivity rate of tested people (this rate varied between 1% to almost 10% in France during the study period). Therefore, for instance, if asymptomatic infected people are encouraged to be widely tested or not, if people are vaccinated and more often asymptomatic, if it is necessary to get a negative test to access some entertainment, then for the same level of the epidemic the incidence rate estimated by health authorities might vary. Second, the use of metropolis or departmental incidence rate is one of the limitations in this study. Within each site A and B, the two sampling points corresponded to two different populations, which may be associated to two different incidence rates. These specific incidence rates were unknown, and correlations involved sampling point-specific wastewater indicators and the metropolis incidence rate. This feature may have led to inaccuracy and uncertainty in correlation coefficient estimates. For site C, correlations involved the departmental incidence rate, which may have not been representative of the local incidence of infection.

Another limitation of this study is the relative low size of samples at each sampling point. Although the use of one sample per week is frequent in this field, this limitation led to large uncertainty in the correlation coefficient estimates and made it difficult to find differences in correlation strength between wastewater indicators. More samples are needed to improve this analysis.

CONCLUSION

Except for normalisation with bacteriophages, we observed a substantial association between simple wastewater indicators and the weekly incidence rate of SARS-CoV-2 infection. However, the strength of this correlation greatly varied between the sampling points, making it difficult to identify the best performing indicator. We proposed a composite index that was consistently more regular than the simple indicators across the six sampling points, and was highly correlated to the weekly incidence rate and showed one of the lowest between-sampling points variability in correlation coefficient. The composite index can be easily computed in any new sampling site based on observations of RdRp and E gene concentrations, N-NH_4 and daily flow per inhabitant (spreadsheet provided in Supplementary Material).

These findings highlight the interest of using the composite index in wastewater-based epidemiology to mitigate variability in measurement results and facilitate identification of trends in the evolution of SARS-CoV-2 RNA concentration in wastewater. Further studies should be performed to confirm these findings. Furthermore, the strong correlations observed between the indicators after smoothing and the weekly incidence rate indicates that the wastewater surveillance of COVID-19 is very valuable. Normalisation using this composite index and curve smoothing may improve the understanding of epidemic trends. Further efforts to model the complex relation between the number of new cases and the wastewater indicators should be conducted to enhance indicator relevance and interpretation of short-term changes in SARS-CoV-2 RNA quantification in wastewater.

DATA AVAILABILITY STATEMENT

Data cannot be made publicly available; readers should contact the corresponding author for details.

REFERENCES

Ai, Y., Davis, A., Jones, D., Lemeshow, S., Tu, H., He, F., Ru, P., Pan, X., Bohrerova, Z. & Lee, J. 2021 Wastewater SARS-CoV-2 monitoring as a community-level COVID-19 trend tracker and variants in Ohio, United States. *The Science of the Total Environment* **801**, 149757. https://doi.org/10.1016/j.scitotenv.2021.149757.

Albert, S., Ruíz, A., Pemán, J., Salavert, M. & Domingo-Calap, P. 2021 Lack of evidence for infectious SARS-CoV-2 in feces and sewage. *European Journal of Clinical Microbiology & Infectious Diseases*. https://doi.org/10.1007/s10096-021-04304-4.

Been, F., Rossi, L., Ort, C., Rudaz, S., Delémont, O. & Esseiva, P. 2014 Population normalization with ammonium in wastewater-based epidemiology: application to illicit drug monitoring. *Environmental Science & Technology* **48** (14), 8162–8169. https://doi.org/10.1021/es5008388.

Bibby, K., Bivins, A., Wu, Z. & North, D. 2021 Making waves: plausible lead time for wastewater based epidemiology as an early warning system for COVID-19. *Water Research* **202**, 117438. https://doi.org/10.1016/j.watres.2021.117438.

Blanch, A. R., Belanche-Muñoz, L., Bonjoch, X., Ebdon, J., Gantzer, C., Lucena, F., Ottoson, J., Kourtis, C., Iversen, A., Kühn, I., Mocé, L., Muniesa, M., Schwartzbrod, J., Skraber, S., Papageorgiou, G. T., Taylor, H., Wallis, J. & Jofre, J. 2006 Integrated analysis of established and novel microbial and chemical methods for microbial source tracking. *Applied and Environmental Microbiology* **72** (9), 5915–5926. https://doi.org/10.1128/AEM.02453-05.

Cavany, S., Bivins, A., Wu, Z., North, D., Bibby, K. & Perkins, T. A. 2021 *Inferring SARS-CoV-2 RNA Shedding Into Wastewater Relative to Time of Infection (P. 2021.06.03.21258238)*. https://doi.org/10.1101/2021.06.03.21258238.

CE 2021 *Commission Recommendation of 17.3.2021 on A Common Approach to Establish A Systematic Surveillance of SARS-CoV-2 and Its Variants in Wastewaters in the EU. European Commission.*

Cheung, K. S., Hung, I. F. N., Chan, P. P. Y., Lung, K. C., Tso, E., Liu, R., Ng, Y. Y., Chu, M. Y., Chung, T. W. H., Tam, A. R., Yip, C. C. Y., Leung, K.-H., Fung, A. Y.-F., Zhang, R. R., Lin, Y., Cheng, H. M., Zhang, A. J. X., To, K. K. W., Chan, K.-H., Yuen, K.-Y. & Leung, W. K. 2020 Gastrointestinal manifestations of SARS-CoV-2 infection and virus load in fecal samples from a Hong Kong cohort: systematic review and meta-analysis. *Gastroenterology* **159** (1), 81–95. https://doi.org/10.1053/j.gastro.2020.03.065.

Cole, D., Long, S. C. & Sobsey, M. D. 2003 Evaluation of F+ RNA and DNA coliphages as source-specific indicators of fecal contamination in surface waters. *Applied and Environmental Microbiology* **69** (11), 6507–6514. https://doi.org/10.1128/AEM.69.11.6507-6514.2003.

Crank, K., Chen, W., Bivins, A., Lowry, S. & Bibby, K. 2021 Contribution of SARS-CoV-2 RNA shedding routes to RNA loads in wastewater. *Science of The Total Environment* **806**, 150376. https://doi.org/10.1016/j.scitotenv.2021.150376.

D'Aoust, P. M., Graber, T. E., Mercier, E., Montpetit, D., Alexandrov, I., Neault, N., Baig, A. T., Mayne, J., Zhang, X., Alain, T., Servos, M. R., Srikanthan, N., MacKenzie, M., Figeys, D., Manuel, D., Jüni, P., MacKenzie, A. E. & Delatolla, R. 2021a Catching a resurgence: increase in SARS-CoV-2 viral RNA identified in wastewater 48 h before COVID-19 clinical tests and 96 h before hospitalizations. *The Science of the Total Environment* **770**, 145319. https://doi.org/10.1016/j.scitotenv.2021.145319.

D'Aoust, P. M., Mercier, E., Montpetit, D., Jia, J. J., Alexandrov, I., Neault, N., Baig, A. T., Mayne, J., Zhang, X., Alain, T., Langlois, M. A., Servos, M. R., MacKenzie, M., Figeys, D., MacKenzie, A. E., Graber, T. E. & Delatolla, R. 2021b Quantitative analysis of SARS-CoV-2 RNA from wastewater solids in communities with low COVID-19 incidence and prevalence. *Water Research* **2021** (188), 116560. doi: 10.1016/j.watres.2020.116560.

Döhla, M., Wilbring, G., Schulte, B., Kümmerer, B. M., Diegmann, C., Sib, E., Richter, E., Haag, A., Engelhart, S., Eis-Hübinger, A. M., Exner, M., Streeck, H. & Schmithausen, R. M. 2020 SARS CoV-2 in environmental samples of quarantined households. *medRxiv*.

Feng, S., Roguet, A., McClary-Gutierrez, J. S., Newton, R. J., Kloczko, N., Meiman, J. G. & McLellan, S. L. 2021 Evaluation of sampling, analysis, and normalization methods for SARS-CoV-2 concentrations in wastewater to assess COVID-19 burdens in Wisconsin Communities. *ACS ES&T Water* **1** (8), 1955–1965. https://doi.org/10.1021/acsestwater.1c00160.

Grabow, W. O. K., Neubrech, T. E., Holtzhausen, C. S. & Jofre, J. 1995 Bacteroides fragilis and Escherichia coli bacteriophages: excretion by humans and animals. *Water Science and Technology* **31** (5), 223–230. https://doi.org/10.1016/0273-1223(95)00270-W.

Gupta, S., Parker, J., Smits, S., Underwood, J. & Dolwani, S. 2020 Persistent viral shedding of SARS-CoV-2 in faeces – a rapid review. *Colorectal Disease: The Official Journal of the Association of Coloproctology of Great Britain and Ireland* **22** (6), 611–620. https://doi.org/10.1111/codi.15138.

Havelaar, A. H., Pot-Hogeboom, W. M., Furuse, K., Pot, R. & Hormann, M. P. 1990 F-specific RNA bacteriophages and sensitive host strains in faeces and wastewater of human and animal origin. *The Journal of Applied Bacteriology* **69** (1), 30–37. https://doi.org/10.1111/j.1365-2672.1990.tb02908.x.

Heijnen, L., Elsinga, G., de Graaf, M., Molenkamp, R., Koopmans, M. P. G. & Medema, G. 2021 Droplet digital RT-PCR to detect SARS-CoV-2 signature mutations of variants of concern in wastewater. *Science of The Total Environment* **799**, 149456. https://doi.org/10.1016/j.scitotenv.2021.149456.

Hoffmann, T. & Alsing, J. 2021 *Faecal Shedding Models for SARS-CoV-2 RNA Amongst Hospitalised Patients and Implications for Wastewater-Based Epidemiology (p. 2021.03.16.21253603)*. https://doi.org/10.1101/2021.03.16.21253603.

Jones, D., Baluja, M. Q., Graham, D., Corbishley, A., McDonald, J., Malham, S., Hillary, L., Connor, T., Gaze, W., Moura, I., Wilcox, M. & Farkas, K. 2020 *Fecal Shedding of SARS-CoV-2 and Its Potential Role in Person-To-Person Transmission and the Environment-Based Spread of COVID-19*. https://doi.org/10.20944/preprints202007.0471.v1.

Kitajima, M., Ahmed, W., Bibby, K., Carducci, A., Gerba, C. P., Hamilton, K. A., Haramoto, E. & Rose, J. B. 2020 SARS-CoV-2 in wastewater: state of the knowledge and research needs. *Science of The Total Environment* **739**, 139076. https://doi.org/10.1016/j.scitotenv.2020.139076.

Lescure, F.-X., Bouadma, L., Nguyen, D., Parisey, M., Wicky, P.-H., Behillil, S., Gaymard, A., Bouscambert-Duchamp, M., Donati, F., Le Hingrat, Q., Enouf, V., Houhou-Fidouh, N., Valette, M., Mailles, A., Lucet, J.-C., Mentre, F., Duval, X., Descamps, D., Malvy, D., Timsit, JF., Lina, B., Van-der-Werf, S. & Yazdanpanah, Y. 2020 Clinical and virological data of the first cases of COVID-19 in Europe: a case series. *The Lancet Infectious Diseases* **6**, 697–706. https://doi.org/10.1016/S1473-3099(20)30200-0.

Li, X., Zhang, S., Shi, J., Luby, S. P. & Jiang, G. 2021 Uncertainties in estimating SARS-CoV-2 prevalence by wastewater-based epidemiology. *Chemical Engineering Journal* **415**, 129039. https://doi.org/10.1016/j.cej.2021.129039.

McMinn, B. R., Ashbolt, N. J. & Korajkic, A. 2017 Bacteriophages as indicators of fecal pollution and enteric virus removal. *Letters in Applied Microbiology* **65** (1), 11–26. https://doi.org/10.1111/lam.12736.

Medema, G., Been, F., Heijnen, L. & Petterson, S. 2020a Implementation of environmental surveillance for SARS-CoV-2 virus to support public health decisions: opportunities and challenges. *Current Opinion in Environmental Science & Health* **17**, 49–71. https://doi.org/10.1016/j.coesh.2020.09.006.

Medema, G., Heijnen, L., Elsinga, G., Italiaander, R. & Brouwer, A. 2020b Presence of SARS-Coronavirus-2 RNA in sewage and correlation with reported COVID-19 prevalence in the early stage of the epidemic in The Netherlands. *Environmental Science & Technology Letters* **7** (7), 511–516. https://doi.org/10.1021/acs.estlett.0c00357.

Miura, F., Kitajima, M. & Omori, R. 2021 Duration of SARS-CoV-2 viral shedding in faeces as a parameter for wastewater-based epidemiology: re-analysis of patient data using a shedding dynamics model. *Science of The Total Environment* **769**, 144549. https://doi.org/10.1016/j.scitotenv.2020.144549.

Nemudryi, A., Nemudraia, A., Wiegand, T., Surya, K., Buyukyoruk, M., Cicha, C., Vanderwood, K. K., Wilkinson, R. & Wiedenheft, B. 2020 Temporal detection and phylogenetic assessment of SARS-CoV-2 in municipal wastewater. *Cell Reports Medicine* **1** (6), 100098. https://doi.org/10.1016/j.xcrm.2020.100098.

Park, G. W., Ng, T. F. F., Freeland, A. L., Marconi, V. C., Boom, J. A., Staat, M. A., Montmayeur, A. M., Browne, H., Narayanan, J., Payne, D. C., Cardemil, C. V., Treffiletti, A. & Vinjé, J. 2020 Crassphage as a novel tool to detect human fecal contamination on environmental surfaces and hands. *Emerging Infectious Diseases* **26** (8), 1731–1739. https://doi.org/10.3201/eid2608.200346.

Polo, D., Quintela-Baluja, M., Corbishley, A., Jones, D. L., Singer, A. C., Graham, D. W. & Romalde, J. L. 2020 Making waves: wastewater-based epidemiology for COVID-19–approaches and challenges for surveillance and prediction. *Water Research* **186**, 116404. https://doi.org/10.1016/j.watres.2020.116404.

Proust-Lima, C., Amieva, H. & Jacqmin-Gadda, H. 2013 Analysis of multivariate mixed longitudinal data: a flexible latent process approach. *British Journal of Mathematical and Statistical Psychology* **66** (3), 470–487. doi: 10.1111/bmsp.12000.

Rimoldi, S. G., Stefani, F., Gigantiello, A., Polesello, S., Comandatore, F., Mileto, D., Maresca, M., Longobardi, C., Mancon, A., Romeri, F., Pagani, C., Cappelli, F., Roscioli, C., Moja, L., Gismondo, M. R. & Salerno, F. 2020 Presence and infectivity of SARS-CoV-2 virus in wastewaters and rivers. *Science of The Total Environment* **744**, 140911. doi:10.1016/j.scitotenv.2020.140911.

Robinson, C. A., Hsieh, H.-Y., Hsu, S.-Y., Wang, Y., Salcedo, B. T., Belenchia, A., Klutts, J., Zemmer, S., Reynolds, M., Semkiw, E., Foley, T., Wan, X., Wieberg, C. G., Wenzel, J., Lin, C.-H. & Johnson, M. C. 2022 Defining biological and biophysical properties of SARS-CoV-2 genetic material in wastewater. *Science of The Total Environment* **807**, 150786. https://doi.org/10.1016/j.scitotenv.2021.150786.

Rosario, K., Symonds, E. M., Sinigalliano, C., Stewart, J. & Breitbart, M. 2009 Pepper mild mottle virus as an indicator of fecal pollution. *Applied and Environmental Microbiology*. https://doi.org/10.1128/AEM.00410-09.

Rusiñol, M., Zammit, I., Itarte, M., Forés, E., Martinez Puchol, S., Gironés, R., Borrego, C., Corominas, L. & Bofill-Mas, S. 2021 Monitoring waves of the COVID-19 pandemic: inferences from WWTPs of different sizes. *Science of The Total Environment* **787**, 147463. https://doi.org/10.1016/j.scitotenv.2021.147463.

Schaper, M., Jofre, J., Uys, M. & Grabow, W. O. K. 2002 Distribution of genotypes of F-specific RNA bacteriophages in human and non-human sources of faecal pollution in South Africa and Spain. *Journal of Applied Microbiology* **92** (4), 657–667. https://doi.org/10.1046/j.1365-2672.2002.01600.x.

Sweetapple, C., Wade, M. J., Grimsley, J. M. S., Bunce, J. T., Melville-Shreeve, P. & Chen, A. S. 2021 Population normalisation in wastewater-based epidemiology for improved understanding of SARS-CoV-2 prevalence: A multi-site study. *MedRxiv*, 2021.08.03.21261365. https://doi.org/10.1101/2021.08.03.21261365.

van Doorn, A. S., Meijer, B., Frampton, C. M. A., Barclay, M. L. & de Boer, N. K. H. 2020 Systematic review with meta-analysis: SARS-CoV-2 stool testing and the potential for faecal-oral transmission. *Alimentary Pharmacology & Therapeutics* **52** (8), 1276–1288. https://doi.org/10.1111/apt.16036.

Varbanov, M., Bertrand, I., Philippot, S., Retourney, C., Gardette, M., Hartard, C., Jeulin, H., Duval, R. E., Loret, J.-F., Schvoerer, E. & Gantzer, C. 2021 Somatic coliphages are conservative indicators of SARS-CoV-2 inactivation during heat and alkaline pH treatments. *Science of The Total Environment* **797**, 149112. https://doi.org/10.1016/j.scitotenv.2021.149112.

von Oertzen, T., Hertzog, C., Lindenberger, U. & Ghisletta, P. 2010 The effect of multiple indicators on the power to detect inter-individual differences in change. *British Journal of Mathematical and Statistical Psychology* **63** (Pt 3), 627–646. doi: 10.1348/000711010X486633.

Westhaus, S., Weber, F. A., Schiwy, S., Linnemann, V., Brinkmann, M., Widera, M., Greve, C., Janke, A., Hollert, H., Wintgens, T. & Ciesek, S. 2021 Detection of SARS-CoV-2 in raw and treated wastewater in Germany – suitability for COVID-19 surveillance and potential transmission risks. *Science of The Total Environment* **751**, 141750. doi: 10.1016/j.scitotenv.2020.141750.

WHO 2003 *Guidelines for Environmental Surveillance of Poliovirus Circulation (WHO/V & B/03.03)*. World Health Organization. Available from: https://apps.who.int/iris/handle/10665/67854.

Wilder, M. L., Middleton, F., Larsen, D. A., Du, Q., Fenty, A., Zeng, T., Insaf, T., Kilaru, P., Collins, M., Kmush, B. & Green, H. C. 2021 Co-quantification of crAssphage increases confidence in wastewater-based epidemiology for SARS-CoV-2 in low prevalence areas. *Water Research X* **11**, 100100. https://doi.org/10.1016/j.wroa.2021.100100.

Wölfel, R., Corman, V. M., Guggemos, W., Seilmaier, M., Zange, S., Müller, M. A., Niemeyer, D., Jones, T. C., Vollmar, P., Rothe, C., Hoelscher, M., Bleicker, T., Brünink, S., Schneider, J., Ehmann, R., Zwirglmaier, K., Drosten, C. & Wendtner, C. 2020 Virological assessment of hospitalized patients with COVID-2019. *Nature* **581**, 465–469.

Wu, F., Zhang, J., Xiao, A., Gu, X., Lee, W. L., Armas, F., Kauffman, K., Hanage, W., Matus, M., Ghaeli, N., Endo, N., Duvallet, C., Poyet, M., Moniz, K., Washburne, A. D., Erickson, T. B., Chai, P. R., Thompson, J. & Alm, E. J. 2020 SARS-CoV-2 titers in wastewater are higher than expected from clinically confirmed cases. *mSystems* **5** (4), e00614–20. https://doi.org/10.1128/mSystems.00614-20.

Wurtzer, S., Waldman, P., Ferrier-Rembert, A., Frenois-Veyrat, G., Mouchel, J. M., Boni, M., Maday, Y., Marechal, V. & Moulin, L. 2021 Several forms of SARS-CoV-2 RNA can be detected in wastewaters: implication for wastewater-based epidemiology and risk assessment. *Water Research* **198**, 117183. https://doi.org/10.1016/j.watres.2021.117183.

Zheng, S., Fan, J., Yu, F., Feng, B., Lou, B., Zou, Q., Xie, G., Sha, Lin, Wang, R., Yang, X., Chen, W., Wang, Q., Zhang, D., Liu, Y., Gong, R., Ma, Z., Lu, S., Xiao, Y., Gu, Y., Zhang, J., Yao, H., Xu, K., Lu, X., Wei, G., Zhou, J., Fang, Q., Cai, H., Qiu, Y., Sheng, J., Chen, Y. & Liang, T. 2020 Viral load dynamics and disease severity in patients infected with SARS-CoV-2 in Zhejiang province, China, January-March 2020: retrospective cohort study. *BMJ* **369**, m1443. https://doi.org/10.1136/bmj.m1443.

Zhu, Y., Oishi, W., Saito, M., Kitajima, M. & Sano, D. 2021 Early warning of COVID-19 in Tokyo via wastewater-based epidemiology: how feasible it really is? *Journal of Water and Environment Technology* **19** (3), 170–183. https://doi.org/10.2965/jwet.21-024.

First received 14 January 2022; accepted in revised form 25 February 2022. Available online 14 March 2022

doi: 10.2166/wh.2022.013

Modeling the relationship between SARS-CoV-2 RNA in wastewater or sludge and COVID-19 cases in three New England regions

Elyssa Anneser[a,†], Emily Riseberg [a,*,†], Yolanda M. Brooks[b], Laura Corlin[a,c] and Christina Stringer[d]

[a] Department of Public Health and Community Medicine, Tufts University School of Medicine, Boston, MA, USA
[b] Department of Sciences, St Joseph's College of Maine, Standish, ME, USA
[c] Department of Civil and Environmental Engineering, Tufts University School of Engineering, Medford, MA, USA
[d] New England Interstate Water Pollution Control Commission, Lowell, MA, USA
*Corresponding author. E-mail: emilyriseberg@fas.harvard.edu
†These authors contributed equally to the work.

ER, 0000-0003-4024-2533

ABSTRACT

Background: We aimed to compare statistical techniques estimating the association between SARS-CoV-2 RNA in untreated wastewater and sludge and reported coronavirus disease 2019 (COVID-19) cases. Methods: SARS-CoV-2 RNA concentrations (copies/mL) were measured from 24-h composite samples of wastewater in Massachusetts (MA) (daily; 8/19/2020–1/19/2021) and Maine (ME) (weekly; 9/1/2020–3/2/2021) and sludge samples in Connecticut (CT) (daily; 3/1/2020–6/1/2020). We fit linear, generalized additive with a cubic regression spline (GAM), Poisson, and negative binomial models to estimate the association between SARS-CoV-2 RNA concentration and reported COVID-19 cases. Results: The models that fit the data best were linear [adjusted R^2=0.85 (MA), 0.16 (CT), 0.63 (ME); root-mean-square error (RMSE)=0.41 (MA), 1.14 (CT), 0.99 (ME)), GAM (adjusted R^2=0.86 (MA), 0.16 (CT) 0.65 (ME); RMSE=0.39 (MA), 1.14 (CT), 0.97 (ME)], and Poisson [pseudo R^2=0.84 (MA), 0.21 (CT), 0.52 (ME); RMSE=0.39 (MA), 0.67 (CT), 0.79 (ME)]. Conclusions: Linear, GAM, and Poisson models outperformed negative binomial models when relating SARS-CoV-2 RNA in wastewater or sludge to reported COVID-19 cases.

Key words: COVID-19, SARS-CoV-2, sludge, statistical model comparisons, wastewater, wastewater-based epidemiology

HIGHLIGHTS

- Comparisons of statistical models can help inform public health departments on how to implement wastewater-based epidemiology.
- We compared the performance of linear, generalized additive, Poisson, and negative binomial models relating SARS-CoV-2 RNA in wastewater or sludge to reported COVID-19 cases.
- Linear, generalized additive, and Poisson models performed best and negative binomial models performed worst.

GRAPHICAL ABSTRACT

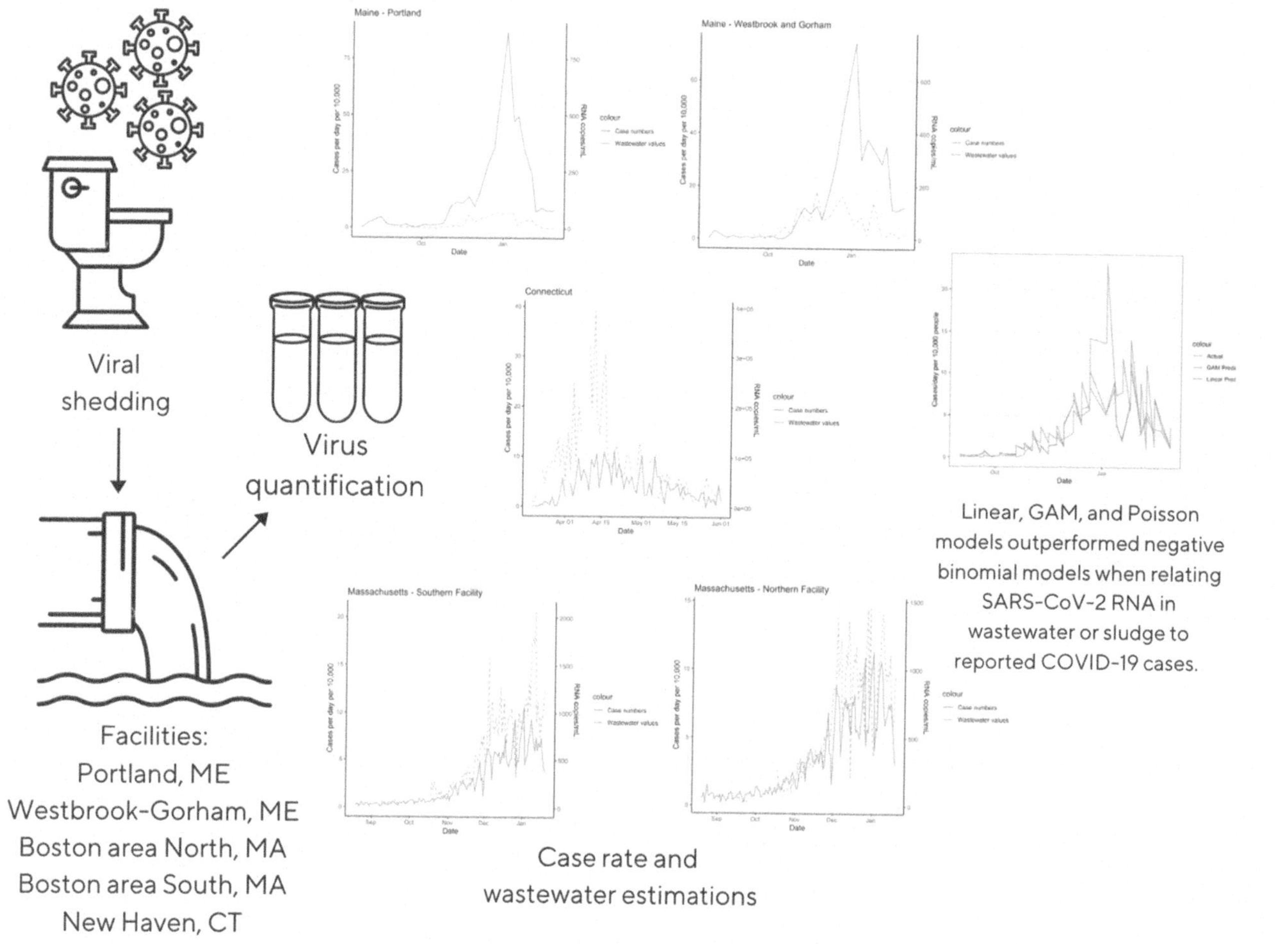

INTRODUCTION

Coronavirus Disease 2019 (COVID-19), caused by the SARS-CoV-2 virus, emerged in late 2019 and quickly escalated into a global health emergency and ongoing pandemic. The manifestation of symptoms and severity of COVID-19 vary among individuals but often include fever, shortness of breath, and loss of sense of taste or smell (CDC 2021). Asymptomatic carriers are estimated to contribute to approximately a quarter of positive cases and spread (He *et al.* 2021), and unless these individuals opt to receive a test or are part of a routine surveillance testing program, their cases will be undetected. Therefore, individual-level testing alone will generally underestimate the community prevalence of COVID-19 (Omori *et al.* 2020).

To complement individual-level testing, public health officials can take advantage of routine community-level wastewater surveillance since SARS-CoV-2 is shed fecally by symptomatic and asymptomatic carriers (La Rosa *et al.* 2020; Polo *et al.* 2020). Wastewater screening is a tool of wastewater-based epidemiology (WBE) and has been used to detect the community prevalence of other fecally shed viruses and illicit drug use (Sinclair *et al.* 2008; Zarei *et al.* 2020). Since the beginning of the COVID-19 pandemic, WBE has been increasingly used as a surveillance technique to monitor the state of the pandemic by localities throughout the world (Goulding *et al.* 2020; Feng *et al.* 2021; Shah *et al.* 2022). For example, previous studies found associations between the reported occurrence of COVID-19 and the concentration of SARS-CoV-2 ribonucleic acid (RNA) in wastewater in the Boston, Massachusetts (MA) metro area and in sludge in New Haven, Connecticut (CT) (Peccia *et al.* 2020; Wu *et al.* 2020). Studies in the Boston area suggest that the actual COVID-19 prevalence was approximately 4–200 times higher than the confirmed clinical case counts indicated (Wu *et al.* 2020). Additionally, WBE has been useful in predicting future trends in COVID-19 incidence since RNA concentrations of SARS-CoV-2 in sludge were found to lead clinical case trends by 2–10 days (Peccia *et al.* 2020).

Several statistical modeling techniques have been proposed to estimate the association between viral genetic markers in wastewater and sludge and COVID-19 cases in a community (McMahan *et al.* 2020; Medema *et al.* 2020; Peccia *et al.* 2020; Vallejo *et al.* 2020; Weidhaas *et al.* 2021). Linear regression and simple correlations have been used in multiple settings (Medema *et al.* 2020; Vallejo *et al.* 2020; Weidhaas *et al.* 2021); however, a standard framework to account for flow rate, environmental effects, and other confounding factors has not been established. Similarly, these methods fail to account for the potential for a nonlinear relationship between SARS-CoV-2 in wastewater and COVID-19 case counts. A second approach uses generalized additive models. One study analyzing data from Spain suggested that generalized additive models performed similarly to models that did not have distributional assumptions (Vallejo *et al.* 2020). A third common modeling strategy fits Poisson models to estimate the count of reported new cases. For example, these have been used in New Haven, CT to estimate trends of reported case numbers and hospitalization numbers (Peccia *et al.* 2020). One limitation of Poisson regression, however, is the assumption that the mean and variance of the outcome (i.e., case counts) are equal. Negative binomial models relax this assumption. Negative binomial models have been used to estimate COVID-19 cases in China based on proxies for water-borne SARS-CoV-2 (Wang *et al.* 2021). To our knowledge, negative binomial regression models have not yet been utilized to assess the association between SARS-CoV-2 RNA in wastewater or sludge and COVID-19 case counts. Given the building body of literature on potential models to estimate community COVID-19 prevalence from wastewater surveillance, these modeling approaches should be quantitatively compared and contextualized across multiple communities for implementation by public health officials.

In this study, we compared four statistical modeling approaches (linear, generalized additive, Poisson, negative binomial) to find an operational model for the associations between SARS-CoV-2 RNA concentrations in wastewater influent or sludge and COVID-19 for four communities in New England, USA (Boston, MA; Portland, Maine (ME); Gorham and Westbrook, ME; and New Haven, CT) serviced by distinct wastewater treatment facilities.

METHODS

SARS-CoV-2 sample collection and characteristics of sewersheds

Wastewater and sludge samples were collected from the Massachusetts Water Resource Authority [MWRA; Boston (and surrounding area), MA], East Shore Water Pollution Abatement Facility (ESWPAF; New Haven, CT), and the Portland Water District (PWD; Portland, Westbrook, and Gorham, ME) (Figure 1). Wastewater samples from MA and ME were collected via 24-h (1 L) composite samples of influent, with at least eight samples collected over the course of 24-h to represent the flow over the course of the day. In MA, samples were collected three to seven times per week both in the Northern and Southern facilities of the watershed system (MWRA – Wastewater COVID-19 Tracking n.d.). We analyzed MA data from 8/19/2020 to 1/19/2021. In ME, samples were collected weekly, and we analyzed data from 9/1/2020 to 3/2/2021. Primary sludge samples (40 mL) from CT were collected daily. We analyzed data from 3/1/2020 to 6/1/2020.

Quantification of SARS-CoV-2

The quantification methods differed slightly among the geographic regions. For MA, the laboratory methods for SARS-CoV-2 concentration estimation were reported previously (Wu *et al.* 2020). Briefly, MA used the TRIzol-chloroform approach to extract RNA and reverse transcription quantitative real-time polymerase chain reaction (RT-qPCR) to quantify SARS-CoV-2 RNA via the N1, N2, and N3 markers (Biobot Analytics, Inc. Cambridge, MA) (Wu *et al.* 2020). The TRIzol reagent was mixed with 300 μL of chloroform for 1 min and incubated at room temperature for 5 min. Then 600 μL of aqueous phase were transferred to new 1.50 mL tubes and mixed with an equal volume of isopropanol. After centrifuging for 10 min, the pellet was washed twice with 75% ethanol and then with 30 μL of diethyl pyrocarbonate (DEPC) water to recover the RNA (Wu *et al.* 2020). Values were normalized to the population using the RNA virus pepper mild mottle virus, a highly prevalent fecal marker (Wu *et al.* 2020; Feng *et al.* 2021). The limit of detection for the MA samples was 4.80 copies/mL of sewage after normalization, and samples below this limit were reported as nondetects.

For CT, the laboratory methods were also previously reported (Peccia *et al.* 2020). Total RNA was isolated using a commercial kit (RNeasey PowerSoil Total RNA Kit, Qiagen), and the RNA pellets were dissolved in 50 μL of ribonuclease-free water. Total RNA concentrations were measured using spectrophotometry (Peccia *et al.* 2020). CT evaluated the presence of the human *RP* gene marker in the sludge to establish the presence of human material (Peccia *et al.* 2020). The Bio-Rad iTaq Universal Probes One-Step kit was used, with 20 μL reactions run at 50 °C for 10 min, 95 °C for 1 min, and 40 cycles at 95 and 60 °C for 10 and 30 s, respectively, as recommended by the manufacturer. SARS-CoV-2 concentrations were calculated using

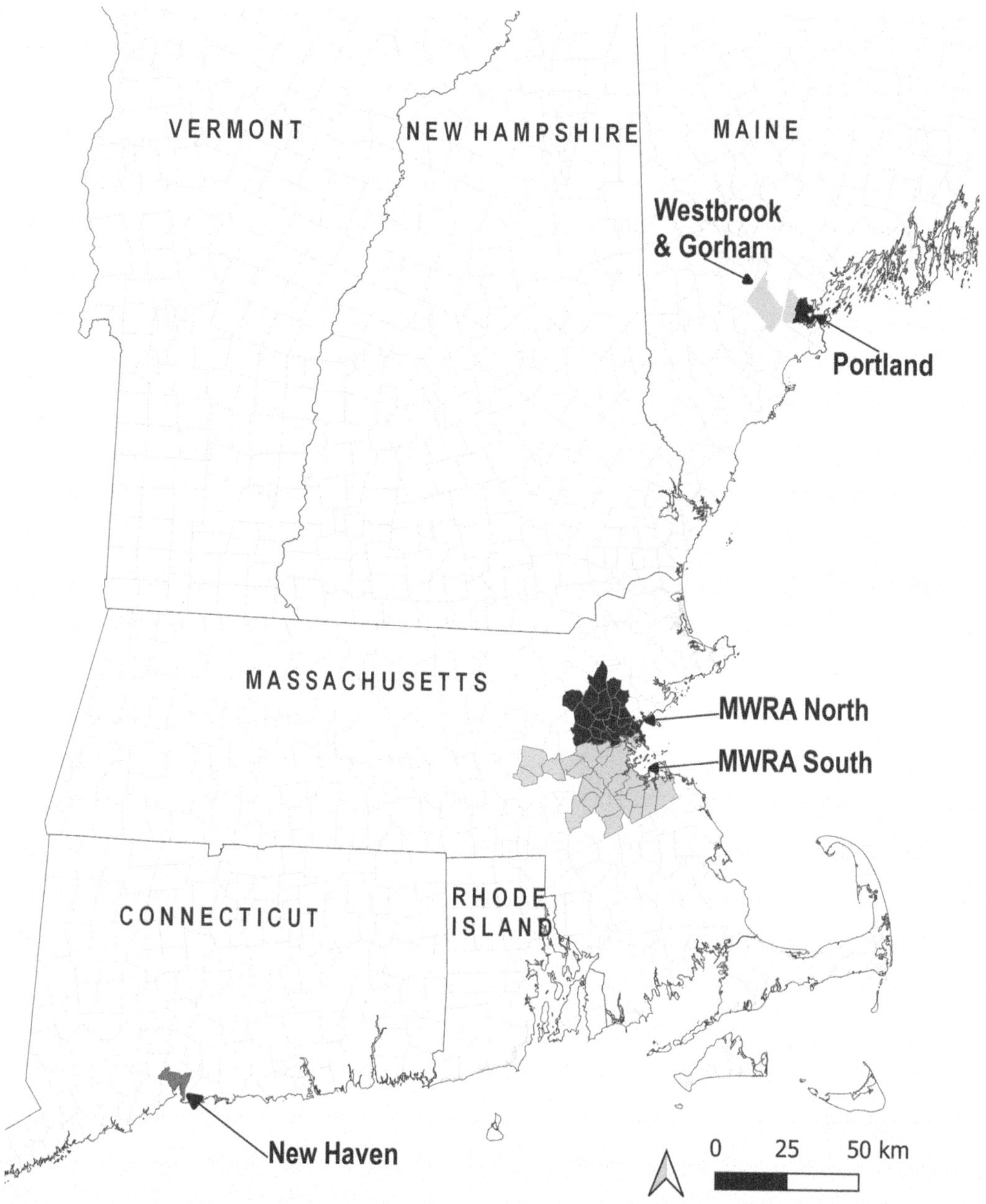

Figure 1 | Map of wastewater and sludge sampling areas in New England. Massachusetts wastewater data was collected from the Massachusetts Water Resources Association (MWRA). CT sludge data was collected from the East Shore Water Pollution Abatement Facility. Maine wastewater data was collected from the Portland Water District.

a standard curve, in which DNA from the WA1-USA strain of SARS-CoV-2 RNA was used as a template to determine transcripts of the SARS-CoV-2 *N* gene (Vogels *et al.* 2020). SARS-CoV-2 RNA copies were normalized to the total RNA in the sample (Peccia *et al.* 2020). Values below 39 copies/mL were not used.

ME utilized the SARS-CoV-2 RT-PCR test (IDEXX Laboratories, Inc. Westbrook, ME), a test of RT-qPCR. The reaction volume was 25 μL, and the template volume was 5 μL. SARS-CoV-2 RNA was quantified in all samples via the N1 and N2 markers (CDC 2020). Absolute concentrations of N1 and N2 genetic markers were calibrated with an external standard curve that was run every four runs. Each step of the standard curve on the external standard curve was run in triplicate. In each run, there was a no-template control, negative extraction control, and one step from the standard curve (2.0×10^3 or 2.0×10^4 copies per reaction). The recovery of SARS-CoV-2 was measured via spikes of bovine respiratory syncytial virus (BRSV) in the wastewater sample. Each sample was run in duplicate. More details on SARS-CoV-2 RNA quantification have been reported previously (Brooks *et al.* 2021). ME samples were adjusted using the 2019 American Community Survey population size estimates. The limit of detection and quantification for ME samples was 0.76 copies/mL before adjustment. Values below this limit were estimated to be half the concentration of the limit.

COVID-19 case count data

In MA, daily county-level case counts were available for each of the counties served by the MWRA (Middlesex, Suffolk, Norfolk, and Plymouth) (COVID-19 Response Reporting n.d.). Since not all cities and towns in these counties are served by the MWRA, we estimated the percentage of each county's population served by the MWRA using population size estimates (*U.S. Census Bureau QuickFacts n.d.*). As Middlesex County is served by both the Northern and Southern MWRA systems, we estimated the population served by each MWRA system in this county. Supplemental Table 1 shows the percentage of each county serviced by the Northern or Southern MWRA systems. The percentage of the county population served by the MWRA was multiplied by new daily laboratory-confirmed cases in each county to obtain the estimated daily case counts in the areas served by the Northern and Southern systems (Archive of COVID-19 Cases in Massachusetts 2021). We assumed that the total number of new cases per population had constant spatial distribution across each city and town within a county. For CT, we used the number of laboratory-confirmed new cases per day in New Haven as reported by the Connecticut Department of Public Health (COVID-19 Daily Report 2020). For ME, we obtained COVID-19 case counts (number of reported cases per week in ZIP codes within Portland and Westbrook/Gorham) from the Maine Center for Disease Control and Prevention (Coronavirus Disease 2019 (COVID-19) 2021). For consistency with the MA and CT data, we estimated daily cases in ME as the weekly reported case count divided by seven. Population sizes in all cities and towns were estimated using the 2019 American Community Survey (*U.S. Census Bureau QuickFacts n.d.*).

Covariate data

Models for each geographic region were fit separately. The covariates in each model were ambient temperature (all regions), flow rate (MA and ME), and facility (Northern or Southern for MA; Portland or Westbrook/Gorham for ME; not a covariate for CT since only one facility was included). We included temperature because it can affect the ability to detect SARS-CoV-2 RNA in wastewater (Mandal *et al.* 2020). We included flow rate because this has been suggested as a potential confounding factor in previous wastewater-based models of COVID-19 (Weidhaas *et al.* 2021). Average daily temperature data were obtained from the National Oceanic and Atmospheric Administration database (National Centers for Environmental Information 2021). Daily temperature in Boston was used to estimate both Northern and Southern facility temperatures for MA. Daily temperature in Portland was used to estimate the temperature in Portland, Westbrook, and Gorham, ME. Flow rate of influent wastewater was recorded at the time of sample collection at the MWRA (MA) and PWD (ME). Flow rate did not apply to the ESWPAF because the SARS-CoV-2 RNA concentration was calculated using normalized values from sludge instead of wastewater.

Statistical analysis

We compared case counts, SARS-CoV-2 RNA concentrations, and flow rates for facilities within states using Wilcoxon matched pair sign rank tests. We evaluated the correlation between daily average temperature and copies/mL of SARS-CoV-2 RNA at each facility using Spearman rank correlation. We assessed the association between SARS-CoV-2 RNA copies/mL in wastewater (or sludge) and daily COVID-19 measures using four modeling approaches (linear, generalized additive with a cubic regression spline, Poisson, and negative binomial models). SARS-CoV-2 RNA copies/mL data were log-transformed. Weeks with zero reported cases or zero copies detected/mL were removed from the analysis (n=3 in ME; 7/27/2020, 10/12/2020, and 2/21/2021). For the linear and generalized additive models, the outcome measure was natural log-transformed new COVID-19 cases per day. For the Poisson and negative binomial models, the outcome measure was the count of reported new cases per day. Each model was estimated twice: once unadjusted for potential confounders and once adjusted for average daily temperature (degrees Celsius), flow rate (millions of gallons per day; MA and ME only), and facility (MA and ME only). Adjusted models were assessed for multicollinearity, and variance inflation factors were less than or equal to three. The adjusted models (including flow rate and facility) were computed as follows:

$$\text{Linear model: } \ln(\text{new cases/population}) = \beta_0 + \beta_1 \ln(\text{SARS} - \text{CoV} - 2) + \beta_2 \text{temperature} + \beta_3 \text{ flow rate} + \beta_4 \text{ facility}$$

$$\text{Generalized additive model: } \ln(\text{new cases/population}) = \gamma_0 + f(\ln(\text{SARS} - \text{CoV} - 2)) + \gamma_1 \text{temperature} + \gamma_2 \text{flow rate} + \gamma_3 \text{ facility}$$

$$\text{Poisson and negative binomial model: new cases} = \alpha_0 + \alpha_1 \ln(\text{SARS} - \text{CoV} - 2) + \alpha_2 \text{temperature} + \alpha_3 \text{flow rate} + \alpha_4 \text{ facility}$$

We conducted a sensitivity analysis using weekly COVID-19 outcome measures for ME only. Additionally, we incorporated time lags with the MA data to assess whether associations were stronger when case counts were measured four, five, and six days prior. For all models, we compared model fit using adjusted R^2 (linear and generalized additive models), McFadden's pseudo-R^2 (Poisson and negative binomial models), and root-mean-square error (RMSE; all models). All analyses were conducted in R 4.10 using the 'mgcv', 'DescTools', 'MASS', 'ggplot2, and 'foreign' packages. All data were obtained from publicly available sources, and the study was considered Not Human Subjects Research by the Tufts Institutional Review Board.

RESULTS

The COVID-19 case counts, SARS-CoV-2 RNA concentrations, ambient temperature, and flow rate in the communities served by the watersheds under study are presented in Table 1. The estimated populations served were 1,605,054 people in the Northern MWRA district of MA; 781,829 in the Southern MWRA district of MA; 130,250 in the New Haven district of CT; 37,052 residents in Westbrook and Gorham, ME; and 66,215 in Portland, ME. The trend in the number of reported new cases per day per 10,000 population (Supplemental Figure 1) generally reflected the trends in the SARS-CoV-2 concentration in wastewater and sludge (Figure 2). Copies/mL of SARS-CoV-2 were significantly different in Portland versus Westbrook and Gorham, and the documented number of cases was significantly different for the MWRA Northern and Southern facilities (Supplemental Table 2). The variation in copies/mL among sampling locations depends on sampling methods (wastewater in MA and ME versus sludge in CT; sampling frequency) and number of cases (more cases in MA than ME). In both MA and ME, flow rate varied significantly by facility (Supplemental Table 2). The correlation coefficients between average daily temperature and wastewater values of SARS-CoV-2 copies/mL were −0.79 (MA, Northern facility), −0.81 (MA, Southern facility), −0.68 (ME, Portland), −0.29 (ME, Westbrook/Gorham), and −0.36 (CT).

Log-transformed SARS-CoV-2 concentrations were significantly associated with case count measures in every model (Table 2). For MA and ME (but not CT), adjusting for covariates generally attenuated effect estimates for SARS-CoV-2 RNA. In the linear regression models for MA and ME, the ambient temperature was a significant covariate (Supplemental Table 3). Similarly, the ambient temperature was a significant predictor in the generalized additive model, Poisson, and negative binomial model types for MA and ME. The temperature was not a significant predictor in any of the models for CT (though CT data represented only three months, so temperatures were less variable). Flow rate was a significant predictor in Poisson models for MA and ME and in the negative binomial model for MA only. The facility was a significant predictor in the generalized additive (MA), Poisson (MA and ME), and negative binomial (MA and ME) models.

Table 3 presents a comparison of the model fit statistics for models relating SARS-CoV-2 RNA to COVID-19 cases in each region. In terms of R^2 values in each of the three states, the multivariable generalized additive model performed best for MA

Table 1 | Population-level characteristics[a] of the areas served by the three locations[b] over the study periods

	Massachusetts (8/19/2020–1/19/202)	Connecticut (3/1/2020–6/1/2020)	Maine (9/1/2020–3/2/2021)
Case counts per day	427 (180, 1265)	45.5 (26.0, 78.8)	12.8 (1.64, 26.9)
Daily new cases/10,000 people	1.79 (0.76, 5.30)	3.49 (2.00, 6.05)	1.24 (0.16, 2.60)
Copies of SARS-CoV-2 (copies/mL)	327 (85.0, 858)	70997 (39917, 115719)	21.1 (8.84, 57.2)
Flow rate (millions of gallons/day)	138 (114, 182)	–	7.87 (6.58, 8.94)
Temperature (degrees Celsius)	10.6 (3.33, 16.7)	9.22 (6.89, 13.2)	5.84 (−0.49, 15.3)

[a]Values presented are median (25th percentile, 75th percentile).

[b]Values in Massachusetts and Maine were computed as population size-weighted averages among facilities.

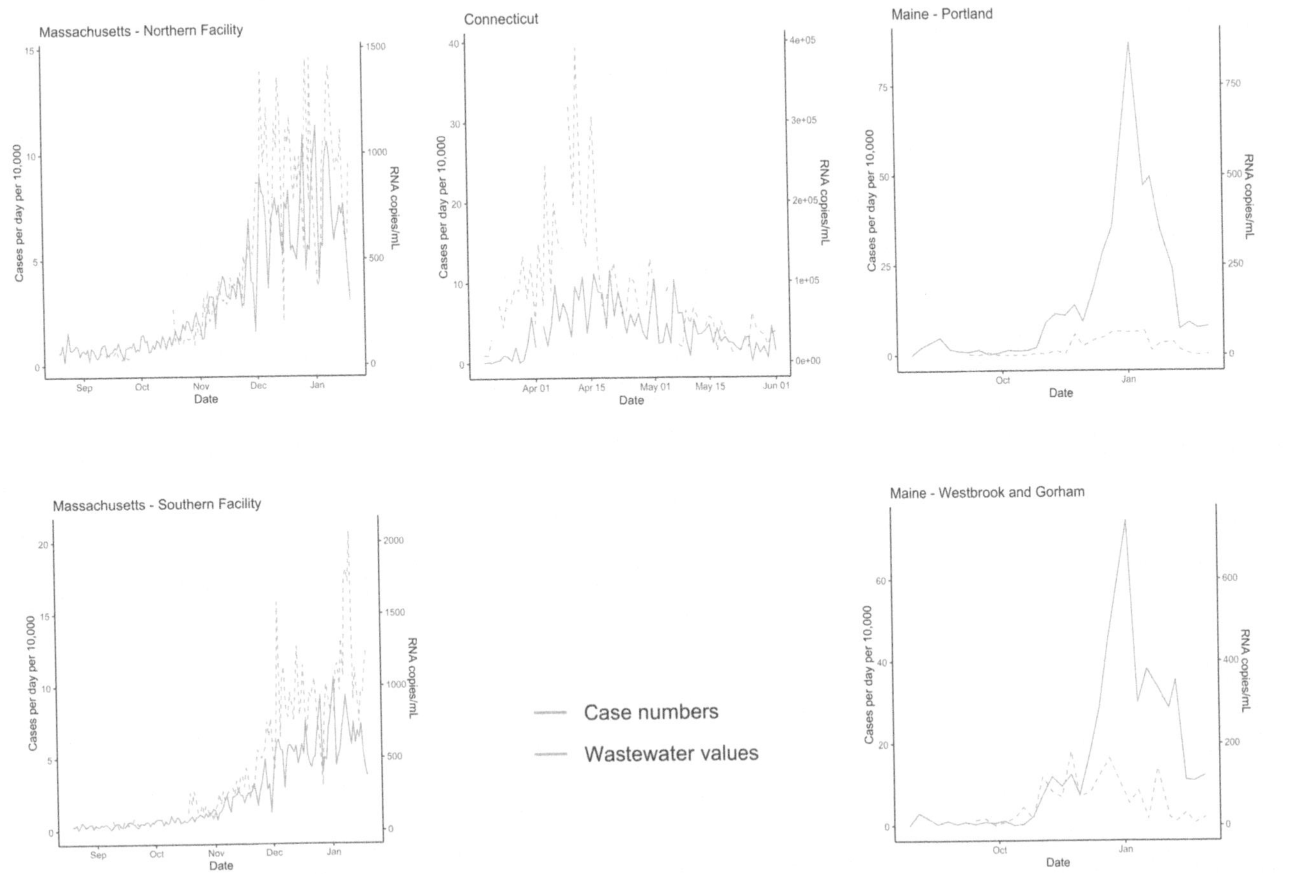

Figure 2 | Distributions of SARS-CoV-2 RNA in wastewater and sludge. The data collection time periods were 8/19/2020–1/19/2021 in Massachusetts, 3/1/2020–6/1/2020 in CT, and 9/1/2020–3/2/2021 in Maine. The vertical axis scales vary by facility.

Table 2 | Unadjusted and adjusted[a] results of statistical models relating natural log-transformed SARS-CoV-2 RNA and daily COVID-19 cases[b]

	Massachusetts	**Connecticut**	**Maine**
Unadjusted linear model; β (95% CI)	0.76 (0.71, 0.80)	0.65 (0.30, 1.00)	0.61 (0.40, 0.83)
Adjusted linear model; β (95% CI)	0.62 (0.55, 0.69)	0.71 (0.33, 1.08)	0.28 (0.05, 0.51)
Unadjusted generalized additive model; *p*-value	<0.01	<0.01	<0.01
Adjusted generalized additive model; *p*-value	<0.01	<0.01	0.03
Unadjusted Poisson model; β (95% CI)	0.67 (0.67, 0.68)	0.47 (0.43, 0.51)	0.37 (0.31, 0.44)
Adjusted Poisson model; β (95% CI)	0.59 (0.58, 0.60)	0.47 (0.43, 0.52)	0.39 (0.31, 0.47)
Unadjusted negative binomial model; β (95% CI)	0.71 (0.64, 0.77)	0.47 (0.22, 0.70)	0.49 (0.29, 0.70)
Adjusted negative binomial model; β (95% CI)	0.61 (0.55, 0.68)	0.46 (0.20, 0.72)	0.34 (0.16, 0.52)

[a]Adjusted models are adjusted for average daily temperature (degrees Celsius), flow rate (millions of gallons per day; Massachusetts and Maine only), and facility (Massachusetts and Maine only).
[b]Linear and generalized additive models estimate population-adjusted case counts. Poisson and negative binomial models estimate case numbers.

(adjusted R^2=0.86) and ME (adjusted R^2=0.65), although results were similar between generalized additive models and linear models. The multivariable Poisson model performed best for CT (pseudo-R^2=0.21 for Poisson model, adjusted R^2=0.16 for generalized additive and linear models) and also performed well for MA (pseudo-R^2=0.84). In all three states, the negative binomial model performed worst. In terms of the RMSE, the Poisson model seemed to better fit the data than the other model types for CT and ME. The RMSE values for MA were similar among the linear, generalized additive, and Poisson models.

Table 3 | Comparison of model fit for statistical models relating natural log-transformed SARS-CoV-2 RNA and daily COVID-19 cases

	Massachusetts	Connecticut	Maine
Unadjusted linear[a] model			
Adjusted R^2	0.80	0.16	0.40
Root-mean-square error	0.47	1.15	1.31
Adjusted[b] linear model			
Adjusted R^2	0.85	0.16	0.63
Root-mean-square error	0.41	1.14	0.99
Unadjusted generalized additive model			
Adjusted R^2	0.82	0.16	0.40
Root-mean-square error	0.45	1.15	1.31
Adjusted generalized additive model			
Adjusted R^2	0.86	0.16	0.65
Root-mean-square error	0.39	1.14	0.97
Unadjusted Poisson model			
Pseudo R^2	0.58	0.21	0.20
Root-mean-square error	0.66	0.67	1.02
Adjusted Poisson model			
Pseudo R^2	0.84	0.21	0.52
Root-mean-square error	0.39	0.67	0.79
Unadjusted negative binomial model			
Pseudo R^2	0.07	0.02	0.05
Root-mean-square error	0.67	0.67	1.03
Adjusted negative binomial model			
Pseudo R^2	0.15	0.02	0.16
Root-mean-square error	0.39	0.67	0.81

[a]Linear and generalized additive models estimate population-adjusted case counts. Poisson and negative binomial models estimate case numbers.
[b]Models are adjusted for average daily temperature (degrees Celsius), flow rate (millions of gallons per day; Massachusetts and Maine only), and facility (Massachusetts and Maine only).

In a sensitivity analysis using weekly case numbers for ME rather than daily numbers, results remained generally the same (Supplemental Table 4). Similar to the daily results, the adjusted generalized additive model had the highest (adjusted) R^2 (0.63), and the adjusted Poisson model had the lowest RMSE (0.78). When accounting for time lags in reporting of cases in MA, results remained similar at four-, five-, and six-day lags (Supplemental Table 5).

DISCUSSION

In this multi-state analysis of wastewater and sludge-measured SARS-CoV-2 RNA concentrations, linear, generalized additive, Poisson, and negative binomial models all indicated significant associations between SARS-CoV-2 RNA concentration and daily reported COVID-19 cases. The adjusted linear, generalized additive, and Poisson models resulted in higher R^2 and lower RMSEs across the three New England metropolitan regions. This is among the first studies to compare model fit for different types of statistical models relating SARS-CoV-2 RNA and COVID-19 cases across several locations. It is also one of the first to consider SARS-CoV-2 RNA measured in sludge. Our results could inform ongoing and future wastewater monitoring efforts to aid in pandemic preparedness and response.

Since SARS-CoV-2 was first detected in wastewater, researchers and public health practitioners have used the concentration of viral RNA to estimate COVID-19 cases in sewersheds (Ahmed *et al.* 2020). Previous research found nonparametric and parametric regression models to be useful tools when interpreting the concentration of viral particles in wastewater and sludge (McMahan *et al.* 2020; Vallejo *et al.* 2020). A study conducted by Wu *et al.* leveraged MWRA wastewater data to suggest the number of prevalent COVID-19 cases was higher than anticipated based on individual testing, with asymptomatic cases possibly explaining the difference between clinically documented cases and observed SARS-CoV-2 shedding in wastewater (Wu *et al.* 2020). The current study builds upon these findings by using uniform modeling techniques to compare associations with reported case numbers in communities in three New England states. Our results indicate that using a linear model versus a generalized additive model versus a Poisson model does not make a large difference in terms of model fit. For example, in MA both the R^2 and the RMSEs in each of the three models only differed by a maximum of 0.02. Additionally, as the significant covariates differed by a municipality, our analysis suggests that local context (including characteristics such as flow rate) should be considered when fitting models estimating COVID-19 cases from wastewater and sludge.

The strength of association between viral RNA in wastewater or sludge and COVID-19 case counts varied by geographic region. These differences may be explained by differences in the study time period (e.g., CT data were from spring 2020, whereas MA and ME data were collected from mid to late 2020 through early 2021), number of days included in the analysis (e.g., shorter time period in CT), testing capacity (especially if testing capacity impacts COVID-19 transmission and reported case counts), and community characteristics (e.g., occupational or policy differences affecting human mobility patterns). Collectively, these differences (along with differences in sample collection methods between sludge and wastewater) may help explain why model performance was weaker in CT than in MA or ME. Other differences may be due to how viral RNA is measured in wastewater and sludge (e.g., the limits of detection for the two wastewater sites were 4.80 copies/mL in MA and 0.76 copies/mL in ME). The higher limit of detection in MA implies that some concentrations of SARS-CoV-2 RNA could have been underestimated in these facilities; however, this detection limit was constant over time and would theoretically underestimate similar values each day. Nevertheless, given the ability to detect (with >99% probability) SARS-CoV-2 RNA if there were at least one case per 6,500 people in the MA area served and given the large population served, the somewhat higher detection limit was not likely to substantively affect the analysis. Furthermore, while the negative binomial regression relaxes the assumption of equal mean and variance, it is possible that the negative binomial model was consistently outperformed by the other models because it is more sensitive to smaller sample sizes than other similar models (e.g., Poisson regression), and our analysis was based on relatively small sample sizes (especially for CT) (Piegorsch 1990; Tang 2015).

Additionally, the performance of statistical models in the COVID-19 context may depend on factors related to viral shedding that we did not account for in our analysis. The viral shedding rate is still unclear, as are differences in viral shedding due to different COVID-19 variants and based on differences by vaccination status (Wölfel *et al.* 2020; Mallavarpu Ambrose *et al.* 2021). Some work suggests that the shedding rate may vary between 10^2 and 10^8 copies/g throughout a 14–21 day shedding period, but further research is still needed to determine the duration and amount of shedding throughout the disease course (Randazzo *et al.* 2020; Wölfel *et al.* 2020). Additionally, in interpreting the output of these statistical models, public health officials should remember that the models do not account for local outbreaks or clusters within a sewershed and should instead be used to estimate changes in the number of cases within the community for which the model is being fit.

Other factors that can affect the validity of the wastewater and sludge estimates for SARS-CoV-2 RNA concentration include precipitation and flow rate. All sewer systems where data were collected had combined sewer for at least half of their sewer systems; these combined systems carry stormwater and wastewater together (Hartford | The Clean Water Project n.d.; Wastewater Maintenance | Portland ME n.d.; MWRA – Combined Sewer Overflow Control Program 2022). In our analyses, precipitation was not a significant covariate (results with precipitation not shown). In ME, sampling dates were at times changed based on rain, so this could potentially explain the nonsignificant results in ME. Flow rate can partially account for precipitation; nonetheless, wastewater studies without flow rate data should consider precipitation rates (Wu *et al.* 2020).

WBE is a useful tool that, as evidenced by our findings, can be implemented through several straightforward statistical modeling types with data collected from either sludge or wastewater. Public health departments and organizations considering wastewater monitoring should work in partnership with local water and wastewater utilities that collect samples. We suggest that wastewater treatment facilities develop a sampling plan that includes 24-h composite sampling, as was done in CT and ME. We also encourage continued data sharing of results in a timely manner to advance public health goals of predicting and responding to pandemic conditions.

We found that incorporating a time lag did not impact the results greatly, so a time lag could be used depending on the average time from test to reporting but may only marginally affect results due to day-to-day similarities in RNA concentrations and cases. We found that including facility as a covariate affected the results and interpretation, suggesting that models may need to consider local context. Accounting for local context explicitly within models will also help to inform local public health measures, such as mask mandates, mobile testing centers, and vaccination clinic targeting, and prediction and forecasting efforts by health departments. These types of applications of WBE have been implemented globally. For example, South Africa has been using WBE to surveille the spread of COVID-19 and track new variants (Aguiar-Oliveira *et al.* 2020; Kumar *et al.* 2020; Fuschi *et al.* 2021).

There were several limitations of this study. First, the SARS-CoV-2 wastewater concentrations and number of cases in ME were measured weekly rather than daily in MA and CT. For consistency with the MA and CT data, we calculated daily values for ME by dividing the reported weekly cases by seven, assuming that there were no temporal trends over the course of the week. Our sensitivity analysis suggests that this assumption was reasonable, and future applications may prefer 7-day moving averages to smooth out the noise of individual day outliers (e.g., holiday and weekend testing differences). There could also have been differences in testing capabilities among the locations, which could have resulted in certain cities or towns having higher reported case numbers than others compared to the number of true cases. This limitation highlights the importance of WBE as a method of disease surveillance, as it can capture the total case burden, including those cases that may not be diagnosed. Second, we estimated population size for each sewershed without accounting for time-varying differences in population due to employment, education, or other trends. This could underestimate the strength of association if, for example, many more people work in the sewershed than reside there (thus contributing to the wastewater influent but not case counts). Relatedly, we assumed that everyone within the study communities contributed to public wastewater, though in certain areas some of the population may use private septic systems. Third, the timing, sampling, and analytic methods for wastewater and sludge collection and analysis varied by municipality. This could result in variable data quality and model performance. Fourth, daily viral RNA concentrations measured at the different facilities within MA and ME may not be independent of the other facility; if they were not independent, the standard errors may be artificially inflated. Relatedly, we did not account for temporal autocorrelation in any modeling strategy. Other modeling approaches, such as auto-regressive models, could address this unsatisfied assumption and could be considered in the future (Akaike 1969). Fifth, average daily temperature was, in some cases, highly correlated with SARS-CoV-2 copies/mL, which raises some concern for multicollinearity. We assessed multicollinearity using the variance inflation factor and did not find strong evidence; however, we cannot rule out that this assumption was violated. Given that temperature is an important factor in social distancing practices and evident case numbers, we felt it was important to incorporate it into the model. Sixth, we only assessed the effect of a time lag in MA, as we had data on the largest number of days in this location. CT, while sampled daily, had data on a shorter time period, and ME data were collected weekly. Nonetheless, we did not find a strong difference in the unlagged versus four-, five-, or six-day lagged models in MA.

Our study also had several strengths. We used consistent data management and statistical approaches to assess the relationship between viral RNA concentration and COVID-19 case counts in multiple metropolitan New England areas. We were able to adjust for relevant covariates such as flow rate, ambient temperature, and facility. Finally, we were able to compare four frequently used statistical modeling approaches. Our work could thus inform future community-level COVID-19 surveillance efforts leveraging wastewater-based modeling. Future studies should assess whether the statistical modeling types and/ or specific model parameters we estimated vary in their ability to predict COVID-19 trends prospectively and whether the model parameters need to be tuned to local conditions for communities outside of New England.

CONCLUSIONS

We observed that SARS-CoV-2 RNA concentrations in wastewater and sludge were significantly associated with local COVID-19 case numbers in MA, CT, and ME. The linear, generalized additive, and Poisson modeling strategies outperformed the negative binomial models. Public health surveillance efforts can confidently use any one of multiple statistical approaches to estimate COVID-19 burden in communities.

ACKNOWLEDGEMENTS

We would like to thank Eilidh Sidaway, Brianna Shelley, Bailey Gryskwicz for laboratory analyses at Saint Joseph's College of Maine, Dr Jordan Peccia for providing the data from New Haven, and the Tufts Data Lab for assistance with figure generation.

FUNDING STATEMENT

Support for this project was provided by NEIWPCC (www.neiwpcc.org), a regional commission that helps the states of the Northeast preserve and advance water quality. The support was in-kind, consisting of the staff time of Dr Christina Stringer. Dr Stringer served in an advisory capacity, overseeing the work conducted by the primary authors, Elyssa Anneser and Emily Riseberg. This included guidance and oversight of the study design, as well as analysis and interpretation of data, and playing a minor role in report writing and article submission. Laura Corlin was supported by Eunice Kennedy Shriver National Institute of Child Health & Human Development (NICHD) grant number K12HD092535 and by the Tufts University/Tufts Medical Center Rapid Response Seed Funding Program. Portland Water District funded the project in Portland, Westbrook, and Gorham, Maine.

CONFLICTS OF INTEREST

Y. Brooks received some material supplies from IDEXX. E. Anneser, E. Riseberg, L. Corlin, and C. Stringer declare no conflicts of interest.

DATA AVAILABILITY STATEMENT

All relevant data are available from an online repository or repositories https://github.com/emilyriseberg/COVID_WBE.

REFERENCES

Aguiar-Oliveira, M. d. L., Campos, A. R., Matos, A., Rigotto, C., Sotero-Martins, A., Teixeira, P. F. P. & Siqueira, M. M. 2020 Wastewater-based epidemiology (WBE) and viral detection in polluted surface water: a valuable tool for COVID-19 surveillance – a brief review. *International Journal of Environmental Research and Public Health* **17** (24), 9251. https://doi.org/10.3390/ijerph17249251

Ahmed, W., Angel, N., Edson, J., Bibby, K., Bivins, A., O'Brien, J. W., Choi, P. M., Kitajima, M., Simpson, S. L., Li, J., Tscharke, B., Verhagen, R., Smith, W. J. M., Zaugg, J., Dierens, L., Hugenholtz, P., Thomas, K. V. & Mueller, J. F. 2020 First confirmed detection of SARS-CoV-2 in.untreated wastewater in Australia: a proof of concept for the wastewater surveillance of COVID-19 in.the community. *Science of the Total Environment* **728**, 138764. https://doi.org/10.1016/j.scitotenv.2020.138764.

Akaike, H. 1969 Fitting autoregressive models for prediction. *Annals of the Institute of Statistical Mathematics* **21** (1), 243–247. https://doi.org/10.1007/BF02532251.

Archive of COVID-19 Cases in Massachusetts 2021 Mass.Gov. Available from: https://www.mass.gov/info-details/archive-of-covid-19-cases-in-massachusetts

Brooks, Y. M., Gryskwicz, B., Sheehan, S., Piers, S., Mahale, P., McNeil, S., Chase, J., Webber, D., Borys, D., Hilton, M., Robinson, D., Sears, S., Smith, E., Lesher, E. K., Wilson, R., Goodwin, M. & Pardales, M. 2021 Detection of SARS-CoV-2 in wastewater at Residential College, Maine, USA, August–November 2020. *Emerg Infect Dis J – CDC* **27** (12). https://doi.org/10.3201/eid2712.211199.

CDC 2020 *Research Use Only 2019-Novel Coronavirus (2019-NCoV) Real-Time RT-PCR Primers and Probes*. Centers for Disease Control and Prevention. Available from: http://www.cdc.gov/coronavirus/2019-ncov/lab/rt-pcr-panel-primer-probes.html

CDC 2021 *Symptoms of COVID-19*. Centers for Disease Control and Prevention. Available from: https://www.cdc.gov/coronavirus/2019-ncov/symptoms-testing/symptoms.html

Coronavirus Disease 2019 (COVID-19) 2021 Division of Disease Surveillance | Maine.Gov. Available from: https://www.maine.gov/dhhs/mecdc/infectious-disease/epi/airborne/coronavirus/index.shtml

COVID-19 Daily Report 2020. Available from: https://data.ct.gov/stories/s/COVID-19-Daily-Report/q5as-kyim/

COVID-19 Response Reporting n.d. Mass.Gov. Available from: https://www.mass.gov/info-details/covid-19-response-reporting (Accessed December 17 2021)

Feng, S., Roguet, A., McClary-Gutierrez, J. S., Newton, R. J., Kloczko, N., Meiman, J. G. & McLellan, S. L. 2021 Evaluation of sampling, analysis, and normalization methods for SARS-CoV-2 concentrations in wastewater to assess COVID-19 burdens in Wisconsin communities. *ACS ES&T Water* **1** (8), 1955–1965. https://doi.org/10.1021/acsestwater.1c00160.

Fuschi, C., Pu, H., Negri, M., Colwell, R. & Chen, J. 2021 Wastewater-based epidemiology for managing the COVID-19 pandemic. *ACS ES&T Water* **1** (6), 1352–1362. https://doi.org/10.1021/acsestwater.1c00050.

Goulding, N., Hickman, M., Reid, M., Amundsen, E. J., Baz-Lomba, J. A., O'Brien, J. W., Tscharke, B. J., de Voogt, P., Emke, E., Kuijpers, W., Hall, W. & Jones, H. E. 2020 A comparison of trends in wastewater-based data and traditional epidemiological indicators of stimulant consumption in three locations. *Addiction (Abingdon, England)* **115** (3), 462–472. https://doi.org/10.1111/add.14852.

Hartford | The Clean Water Project n.d. Available from: https://www.thecleanwaterproject.com/in-your-town/hartford (Accessed March 19 2022).

He, J., Guo, Y., Mao, R. & Zhang, J. 2021 Proportion of asymptomatic coronavirus disease 2019: a systematic review and meta-analysis. *Journal of Medical Virology* **93** (2), 820–830. https://doi.org/10.1002/jmv.26326.

Kumar, M., Patel, A. K., Shah, A. V., Raval, J., Rajpara, N., Joshi, M. & Joshi, C. G. 2020 First proof of the capability of wastewater surveillance for COVID-19 in India through detection of genetic material of SARS-CoV-2. *Science of the Total Environment* **746**, 141326. https://doi.org/10.1016/j.scitotenv.2020.141326.

La Rosa, G., Iaconelli, M., Mancini, P., Bonanno Ferraro, G., Veneri, C., Bonadonna, L., Lucentini, L. & Suffredini, E. 2020 First detection of SARS-CoV-2 in untreated wastewaters in Italy. *The Science of the Total Environment* **736**, 139652. https://doi.org/10.1016/j.scitotenv.2020.139652.

Mallavarpu Ambrose, J., Priya Veeraraghavan, V., Kullappan, M., Chellapandiyan, P., Krishna Mohan, S. & Manivel, V. A. 2021 Comparison of immunological profiles of SARS-CoV-2 variants in the COVID-19 pandemic trends: an immunoinformatics approach. *Antibiotics* **10** (5), 535. https://doi.org/10.3390/antibiotics10050535.

Mandal, P., Gupta, A. K. & Dubey, B. K. 2020 A review on presence, survival, disinfection/removal methods of coronavirus in wastewater and progress of wastewater-based epidemiology. *Journal of Environmental Chemical Engineering* **8** (5), 104317. https://doi.org/10.1016/j.jece.2020.104317.

McMahan, C. S., Self, S., Rennert, L., Kalbaugh, C., Kriebel, D., Graves, D., Deaver, J. A., Popat, S., Karanfil, T. & Freedman, D. L. 2020 *COVID-19 Wastewater Epidemiology: A Model to Estimate Infected Populations*. https://doi.org/10.1101/2020.11.05.20226738.

Medema, G., Heijnen, L., Elsinga, G., Italiaander, R. & Brouwer, A. 2020 Presence of SARS-Coronavirus-2 RNA in sewage and correlation with reported COVID-19 prevalence in the early stage of the epidemic in The Netherlands. *Environmental Science & Technology Letters* **7** (7), 511–516. https://doi.org/10.1021/acs.estlett.0c00357.

MWRA – Combined Sewer Overflow Control Program 2022. Available from: https://www.mwra.com/03sewer/html/sewcso.htm

MWRA – Wastewater COVID-19 Tracking n.d. Available from: https://www.mwra.com/biobot/biobotdata.htm (Accessed March 30 2022).

National Centers for Environmental Information 2021 *National Oceanic and Atmospheric Administration (NOAA)*. Available from: https://www.ncdc.noaa.gov/

Omori, R., Mizumoto, K. & Chowell, G. 2020 Changes in testing rates could mask the novel coronavirus disease (COVID-19) growth rate. *International Journal of Infectious Diseases* **94**, 116–118. https://doi.org/10.1016/j.ijid.2020.04.021.

Peccia, J., Zulli, A., Brackney, D. E., Grubaugh, N. D., Kaplan, E. H., Casanovas-Massana, A., Ko, A. I., Malik, A. A., Wang, D., Wang, M., Warren, J. L., Weinberger, D. M., Arnold, W. & Omer, S. B. 2020 Measurement of SARS-CoV-2 RNA in wastewater tracks community infection dynamics. *Nature Biotechnology* **38** (10), 1164–1167. https://doi.org/10.1038/s41587-020-0684-z.

Piegorsch, W. W. 1990 Maximum likelihood estimation for the negative binomial dispersion parameter. *Biometrics* **46** (3), 863–867. https://doi.org/10.2307/2532104.

Polo, D., Quintela-Baluja, M., Corbishley, A., Jones, D. L., Singer, A. C., Graham, D. W. & Romalde, J. L. 2020 Making waves: wastewater-based epidemiology for COVID-19 – approaches and challenges for surveillance and prediction. *Water Research* **186**, 116404. https://doi.org/10.1016/j.watres.2020.116404.

QuickFacts n.d. *United States Census Bureau*. Available from: https://www.census.gov/quickfacts/fact/table/MA,chelseacitymassachusetts/PST045219 (Accessed February 9 2021).

Randazzo, W., Cuevas-Ferrando, E., Sanjuán, R., Domingo-Calap, P. & Sánchez, G. 2020 Metropolitan wastewater analysis for COVID-19 epidemiological surveillance. *International Journal of Hygiene and Environmental Health* **230**, 113621. https://doi.org/10.1016/j.ijheh.2020.113621.

Shah, S., Gwee, S. X. W., Ng, J. Q. X., Lau, N., Koh, J. & Pang, J. 2022 Wastewater surveillance to infer COVID-19 transmission: a systematic review. *The Science of the Total Environment* **804**, 150060. https://doi.org/10.1016/j.scitotenv.2021.150060.

Sinclair, R. G., Choi, C. Y., Riley, M. R. & Gerba, C. P. 2008 Pathogen surveillance through monitoring of sewer systems. *Advances in Applied Microbiology* **65**, 249–269. https://doi.org/10.1016/S0065-2164(08)00609-6.

Tang, Y. 2015 Sample size estimation for negative binomial regression comparing rates of recurrent events with unequal follow-up time. *Journal of Biopharmaceutical Statistics* **25** (5), 1100–1113. https://doi.org/10.1080/10543406.2014.971167.

Vallejo, J. A., Rumbo-Feal, S., Conde-Pérez, K., López-Oriona, Á., Tarrío-Saavedra, J., Reif, R., Ladra, S., Rodiño-Janeiro, B. K., Nasser, M., Cid, Á., Veiga, M. C., Acevedo, A., Lamora, C., Bou, G., Cao, R. & Poza, M. 2020 Predicting the number of people infected with SARS-COV-2 in a population using statistical models based on wastewater viral load. *MedRxiv*. https://doi.org/10.1101/2020.07.02.20144865.

Vogels, C. B. F., Brito, A. F., Wyllie, A. L., Fauver, J. R., Ott, I. M., Kalinich, C. C., Petrone, M. E., Casanovas-Massana, A., Muenker, C., Moore, A. J., Klein, J., Lu, P., Lu-Culligan, A., Jiang, X., Kim, D. J., Kudo, E., Mao, T., Moriyama, M., Oh, J. I., Park, A., Silva, J., Song, E., Takahashi, T., Taura, M., Tokuyama, M., Venkataraman, A., Weizman, O-E., Wong, P., Yang, Y., Cheemarla, N. R., White, E. B., Lapidus, S., Earnest, R., Geng, B., Vijayakumar, P., Odio, C., Fournier, J., Bermejo, S., Farhadian, S., Dela Cruz, C. S., Iwasaki, A., Ko, A. I., Landry, M. L., Foxman, E. F. & Grubaugh, N. D. 2020 Analytical sensitivity and efficiency comparisons of SARS-CoV-2 RT–qPCR primer–probe sets. *Nature Microbiology* **5** (10), 1299–1305. https://doi.org/10.1038/s41564-020-0761-6.

Wang, J., Li, W., Yang, B., Cheng, X., Tian, Z. & Guo, H. 2021 Impact of hydrological factors on the dynamic of COVID-19 epidemic: a multi-region study in China. *Environmental Research* **198**, 110474. https://doi.org/10.1016/j.envres.2020.110474.

Wastewater Maintenance | Portland, ME n.d. Available from: https://www.portlandmaine.gov/489/Wastewater-Maintenance (Accessed March 19 2022).

Weidhaas, J., Aanderud, Z. T., Roper, D. K., VanDerslice, J., Gaddis, E. B., Ostermiller, J., Hoffman, K., Jamal, R., Heck, P., Zhang, Y., Torgersen, K., Laan, J. V. & LaCross, N. 2021 Correlation of SARS-CoV-2 RNA in wastewater with COVID-19 disease burden in sewersheds. *The Science of the Total Environment* **775**, 145790. https://doi.org/10.1016/j.scitotenv.2021.145790.

Wölfel, R., Corman, V. M., Guggemos, W., Seilmaier, M., Zange, S., Müller, M. A., Niemeyer, D., Jones, T. C., Vollmar, P., Rothe, C., Hoelscher, M., Bleicker, T., Brünink, S., Schneider, J., Ehmann, R., Zwirglmaier, K., Drosten, C. & Wendtner, C. 2020 Virological assessment of hospitalized patients with COVID-2019. *Nature* **581** (7809), 465–469. https://doi.org/10.1038/s41586-020-2196-x.

Wu, F., Zhang, J., Xiao, A., Gu, X., Lee, W. L., Armas, F., Kauffman, K., Hanage, W., Matus, M., Ghaeli, N., Endo, N., Duvallet, C., Poyet, M., Moniz, K., Washburne, A. D., Erickson, T. B., Chai, P. R., Thompson, J. & Alm, E. J. 2020 SARS-CoV-2 titers in wastewater are higher than expected from clinically confirmed cases. *MSystems* **5** (4). https://doi.org/10.1128/mSystems.00614-20.

Zarei, S., Salimi, Y., Repo, E., Daglioglu, N., Safaei, Z., Güzel, E. & Asadi, A. 2020 A global systematic review and meta-analysis on illicit drug consumption rate through wastewater-based epidemiology. *Environmental Science and Pollution Research International* **27** (29), 36037–36051. https://doi.org/10.1007/s11356-020-09818-6.

First received 14 January 2022; accepted in revised form 15 April 2022. Available online 26 April 2022

doi: 10.2166/wh.2023.238

A case study of a community-organized wastewater surveillance in a small community: correlating weekly reported COVID-19 cases with SARS-CoV-2 RNA concentrations during fall 2020 to summer 2021 in Yarmouth, ME

Yolanda M. Brooks [a,*], Bailey Gryskwicz[a], Eilidh Sidaway[a], Brianna Shelley[a], Laura Coroi[b], Margaret Downing[b], Tom Downing[b], Sharon McDonnell[b], Dan Ostrye[b], Katrina Hoop[c] and Gib Parrish[b]

[a] Department of Sciences, St. Joseph's College of Maine, 278 White's Bridge Rd, Standish, ME 04084, USA
[b] Wastewater Testing Team, Yarmouth Community Coronavirus Task Force, C/O Yarmouth Town Hall, 200 Main St., Yarmouth, ME 04096, USA
[c] Department of Social Sciences, University of Maine at Augusta, 46 University Drive, Augusta, ME 04330, USA
*Corresponding author. E-mail: yolandambrooks@gmail.com; ybrooks@sjcme.edu

YMB, 0000-0003-0638-3244

ABSTRACT

Wastewater surveillance offers a rapid evaluation of SARS-CoV-2 transmission in a community. We describe how a community group, the Yarmouth Wastewater Testing Team (YWTT), in Yarmouth, Maine, (population 8,990) utilized an asset-based community design framework to organize and manage a program to monitor SARS-CoV-2 RNA concentrations. From September 22, 2020 through June 8, 2021, the YWTT disseminated weekly reports of the wastewater results and reported COVID-19 cases within the Yarmouth postal code. After high and increasing SARS-CoV-2 RNA concentrations, the YWTT issued two community advisories to encourage extra care to reduce exposure. Correlations between SARS-CoV-2 RNA concentrations and COVID-19 cases were stronger the week after sampling, and the average of the COVID-19 cases during the week of sampling and the following week, indicating that surveillance provided advance notice of cases. A 10% increase in SARS-CoV-2 RNA concentrations was associated with a 13.29% increase in the average number of weekly reported cases of COVID-19 during the week of sampling and the following week ($R2 = 0.42$; $p < 0.001$). Adjusting for viral recovery (December 21, 2020 through June 8, 2021), improved R2 from 0.60 to 0.68. Wastewater surveillance was an effective tool for the YWTT to quickly respond to viral transmission.

Key words: community partnerships, COVID-19, RT-PCR, SARS-CoV-2, wastewater surveillance

HIGHLIGHTS

- Community organizations can successfully design wastewater surveillance programs.
- Weekly wastewater surveillance correlates to weekly reported COVID-19 cases.
- Severe acute respiratory syndrome coronavirus 2 (SARS-CoV-2) RNA concentrations are best correlated to cases the same week and the following week.
- Adjusting for viral recovery increased the fit of linear regression by 8%.
- Community-based initiatives can effectively respond to an increased viral transmission.

GRAPHICAL ABSTRACT

During COVID-19 pandemic, Yarmouth, Maine, monitored its residents for SARS-CoV-2 infections using wastewater testing and Health Department case reports to inform community response to the pandemic.

Due to presence of symptomatic and asymptomatic infections, relying on symptomatic cases alone as a measure of COVID-19 infection did not provide an accurate picture of rate of infection.

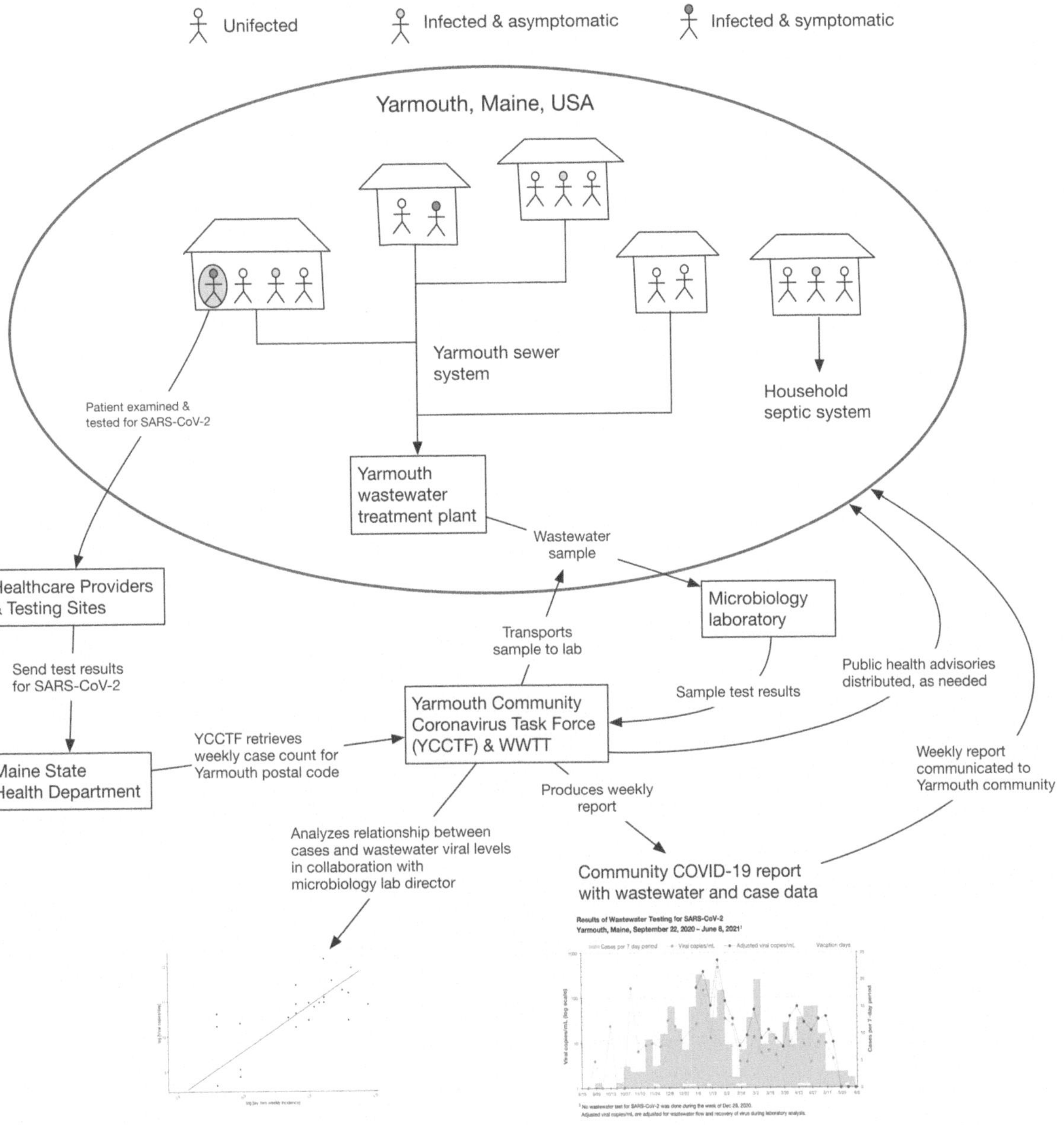

BACKGROUND

Wastewater-based epidemiology has been used as a public health tool to evaluate trends in the transmission of various infectious viruses such as hepatitis A, polio, adenovirus, norovirus GII, sapovirus, human herpesvirus 6, human herpesvirus 8, and influenza A (Lago *et al.* 2003; McCall *et al.* 2020; Bi *et al.* 2021). Trends in severe acute respiratory syndrome coronavirus 2 (SARS-CoV-2) RNA concentrations have been tracked in wastewater treatment plant (WWTP) influent in many communities across the world (Ahmed *et al.* 2020; Kumar *et al.* 2020; Medema *et al.* 2020; Sherchan *et al.* 2020; Prado *et al.* 2021; Street *et al.* 2021). Many of these studies have investigated wastewater surveillance in urban communities with populations ≥380,000. These systems are a mixture of industrial and residential wastewater and can have less microbial variability than municipal wastewater systems in small communities (Gudra *et al.* 2022).

Few studies have investigated the potential of wastewater surveillance in smaller municipal wastewater systems to detect and track trends in SARS-CoV-2 transmission. In a rural community of <5,000 people in Eastern Ontario, SARS-CoV-2 N1 and N2 markers were more readily detected from 24-h composite samples ($n = 5$ sampling dates) from pumping stations than those in the downstream lagoon (D'Aoust *et al.* 2021). Significant correlations between reported coronavirus disease (COVID-19) cases and concentrations of SARS-CoV-2 N2 markers ($n = 20$ or 21) were found in only one of the four WWTPs that served <25,000 people in Finland (Tiwari *et al.* 2022). Additionally, in a WWTP serving approximately 17,000 connections in Bexar County, Texas, trends in SARS-CoV-2 RNA concentrations from composite samples of wastewater influent ($n = 13$) were significantly correlated with weekly COVID-19 infection rates (Al-Duroobi *et al.* 2021). These studies demonstrate that wastewater-based surveillance can be used in smaller communities to detect SARS-CoV-2 RNA and correlate the concentrations to reported COVID-19 cases over short periods. More data are needed to demonstrate how wastewater-based surveillance can be used as a public health tool over longer periods in smaller communities and how local communities can meaningfully utilize these data to respond to the pandemic.

Traditional approaches to solving social problems, such as a public health crisis, view communities as having deficits and lacking resources. In contrast, asset-based community development (ABCD) begins with the assumption that all communities regardless of their privilege have people who can, and do, serve as 'connectors' to work collaboratively when strengths, interests, and passions are made visible (Kretzmann & McKnight 1993). ABCD includes individuals who identify their skills and interests and collaborate with existing institutions such as schools and public agencies (Hopes *et al.* 2015). During the COVID-19 pandemic, citizens in many large and small localities came together, illustrating the principles of ABCD to do the following activities: provide housing and home-based resources, run screenings, offer public health information for immigrant groups and other marginalized communities, prepare and deliver food for isolated individuals, and coordinate early childhood education (Hogan 2020; Ishisaka 2020; Michener *et al.* 2021). There are few studies, however, that have documented community-organized wastewater surveillance programs.

Many municipal wastewater surveillance programs are collaborations between local health departments, municipalities, utility companies, and testing laboratories; most of these programs have been housed in health departments, corporations, or local universities. Few wastewater surveillance programs have integrated ABCD into their approach. In this study, we describe a unique community, public utility, and collegiate partnership to monitor SARS-CoV-2 transmission in Yarmouth, ME, a small coastal community (population ~8,990), and to inform public health recommendations. The specific objectives of this study were to:

1. monitor SARS-CoV-2 transmission using wastewater influent from the Yarmouth municipal WWTP independent of weekly reported COVID-19 cases by the Maine Center for Disease Control (ME CDC);
2. qualitatively describe the ABCD initiatives that combat the transmission of SARS-CoV-2 in the Yarmouth community; and
3. mathematically evaluate the relationship between the concentrations of SARS-CoV-2 RNA in wastewater influent and weekly reported COVID-19 cases in a small community, Yarmouth, ME with linear regression analyses and Spearman's correlation coefficients. We also evaluate how lags and leads in SARS-CoV-2 RNA concentrations as well as viral recovery affect the correlation of SARS-CoV-2 RNA concentrations to weekly reported COVID-19 cases.

METHODS

Study area and the establishment of a community-based initiative to curb SARS-CoV-2 transmission in Yarmouth, ME

The Town of Yarmouth is a coastal Maine community with a population of 8,990 (U.S. Census Bureau 2020), which is principally a residential community approximately 15 min from Maine's largest municipality, Portland. The Town covers the same geographic area as the Yarmouth zip code, and about 70% (~6,300) of its residents are served by the Yarmouth WWTP. The remaining residents (~2,700) use residential septic systems. The wastewater system contains 30 pump stations and does not have any permitted industrial discharges. The Yarmouth WWTP is a sanitary sewage system with a capacity of 5,910 m^3/day of influent, and the mean flow during sample collection was 2,815 m^3/day (s.d. = 814 m^3/day).

In February 2020, Yarmouth community members became concerned about the arrival of COVID-19 in the United States and, after discussions with Town officials, formed the Yarmouth Community Coronavirus Task Force (YCCTF). This group was comprised of over 50 core individuals from a range of backgrounds that first gathered in March 2020 for a presentation by a local epidemiologist on the science of pandemics. Citizens volunteered to serve on various subcommittees that performed a wide range of duties. For example, while some sewed and promoted face masks, others served as 'neighborhood communicators' whose job it was to connect community members with updates and important information regarding COVID-19 prevention.

First-hand accounts of establishing a wastewater surveillance program from members of the YCCTF

In July 2020, several YCCTF members learned that some communities were measuring SARS-CoV2 RNA concentrations in wastewater to evaluate viral transmission (Biobot Analytics 2020; Gallatin City-County Health Department 2020). They also learned of the possible establishment of a national wastewater testing program for COVID-19 (Mehrotra *et al.* 2020). After discussions with Town of Yarmouth officials and staff at the Yarmouth WWTP, the YCCTF established the Yarmouth Wastewater Testing Team (YWTT) consisting of six Yarmouth residents with various backgrounds.

The YWTT worked with the Yarmouth Town Manager to identify initial funding for a wastewater surveillance program for 6 months (mid-September 2020–mid-March 2021) through the CARES Act's *Keep Maine Healthy Grant – Waste Water*. An additional 3 months of testing from mid-March 2021 to early June 2021 was funded jointly by the Town of Yarmouth and individual contributions. The YWTT volunteers managed the Yarmouth wastewater surveillance program, including locating a local laboratory, organizing the logistics of coordinating sample collection and transportation, and communicating results to the public.

Wastewater collection, the concentration of viral nucleic acids, and the quantification of SARS-CoV-2 N1 and N2

From September 22, 2020 to June 8, 2021, 24-h composite samples of untreated influent were aseptically collected once a week from the Yarmouth WWTP, typically starting on Monday at 7:00 am and finishing on Tuesday at 7:00 am; 300 ml of influent was collected between every 15,000 and 30,000 gal using a Hach AS 950 (Hach Company, Loveland, CO) with a refrigeration unit that stored the composite sample at 4 °C. The total volume of each composite sample was ~2 L. The average pH of the composite was 7.6 (s.d. = 0.1). The average concentrations of suspended solids and BOD were 231 mg/L (s.d. = 55 mg/L) and 276 mg/L (s.d. = 336 mg/L), respectively, and were consistently measured in samples collected 1 day after the composite samples. No sample was collected and tested during the week of December 28, 2020. The samples were placed in a cooler with ice and transported to the laboratory. At the laboratory, samples were stored at 4 °C and analyzed within 72 h after the sampling event. For the samples collected from December 21, 2020, to June 8, 2021, within 72 h of sample collection, 131.4 µl of 1:10 dilution of bovine respiratory syncytial virus (BRSV, Zoetis, Parsipanny, NJ) was added to 105 ml of the wastewater influent sample prior to the concentration step. Using the Water SARS-CoV-2 reverse transcription polymerase chain (RT-PCR) test (IDEXX Laboratories, Westbrook, ME), the biological material was concentrated with two centrifugation steps: 4,700 × *g* for 30 min followed by the addition of 10.5 g of polyethylene glycol 8000 and 2.364 g of NaCl per sample and a centrifugation step of 12,000 × *g* for 2 h. Prior to the extraction of nucleic acids, the IDEXX Water Internal Control (IDEXX Laboratories, Westbrook, ME) was added to each sample, positive control, and negative extraction control (sterile dH_2O). The nucleic acids were extracted using the IDEXX Water DNA/RNA Magnetic Bead kit. The quantitative PCRs (qPCRs) were measured using absolute concentrations of the N1 and N2 markers in a single reaction using the QuantStudio 3 (Applied BioSytems, Waltham, MA). Sample volumes of 5 µl were added to the RT-qPCR run for all samples, non-template controls, positive controls, and all steps of the standard curve. Each RT-qPCR run contained duplicates of the following: each sample, non-template control, one step of the standard curve (2,000 or 20,000 copies/5 µl), negative extraction control (nuclease-free H_2O), and positive control of undiluted BRSV (starting on December 21, 2021) and 20 µl of 1:100 dilution of SARS-CoV-2 culture fluid (Isolate: USA-WA1/2020; ZeptoMetrix, Buffalo, NY). The 2019-nCoV_N_Positive Control plasmid (Integrated DNA Technologies Inc., Coralville, IA) was the quantitative standard, and the standard curve included five steps. A new standard curve was made every four qPCR runs, and each step was run in triplicate in the qPCR along with a negative control. The standard curves had a limit of quantification at either 2 copies or 20 copies per reaction (average C_t values = 39.36 and 36.19, respectively) combined for both the N1 and N2 markers. Samples with detectable but not quantifiable concentrations of SARS-Cov-2 N1/N2 markers were reported at the experimental limit of quantification, either 2 copies/rxn or 20 copies/rxn. All samples that contained undetectable SARS-CoV-2 N1/N2 markers were reported at half of the theoretical limit of detection, either 1 copy/rxn or 10 copies/rxn, as previously recommended (CDC n.d.).

Collection of reported cases, data analyses, and statistical analyses

In all samples, concentrations of SARS-CoV-2 RNA were adjusted for the 24 h flow of the wastewater; and from December 21, 2020, to June 8, 2021, the SARS-CoV-2 RNA concentrations were also adjusted for viral recovery using BRSV spiked into each sample.

The number of COVID-19 cases each week for the Yarmouth, Maine, zip code consisted of cases reported to and compiled by the Maine Center for Disease Control & Prevention (Maine CDC). The weekly reported cases spanned from Monday to Sunday each week and were publicly available on the subsequent Wednesday (Supplementary Material, Figure S1).

All statistical analyses were performed in R Version 4.1.3 using the base and stats package (R Core Team 2022). The alpha value for all statistical tests was 0.05. All graphical plots were created using the ggplot2, ggpubr, and ggpmisc packages. Linear regression was used to analyze associations between the number of reported COVID-19 cases and concentrations of SARS-CoV-2 RNA, as this type of regression analysis provided a good fit in wastewater surveillance in medium and larger communities (Anneser *et al.* 2022). The following equation was used:

$$\log(S_n) = m_n * \log(I_n) + b \tag{1}$$

S_n was the concentration of SARS-CoV-2 RNA per day normalized with the flow rate where:
S_1 included samples collected from September 22, 2020, to June 8, 2021 but did not account for the viral recovery;
S_2 included samples collected from December 21, 2020 to June 8, 2021 but did not account for viral recovery; and
S_3 included samples collected from December 21, 2020 to June 8, 2021 and accounted for viral recovery with BRSV.

Within each variation of S_n, five variations of the independent variable, I_n, were used to evaluate the various lags and leads in the number of weekly reported COVID-19 cases for individuals residing in the Yarmouth zip code (Supplementary Material, Figure S1).

I_1 was the reported cases for the week before wastewater sampling (1–7 days prior to wastewater sampling);
I_2 was the average reported cases for the same week as wastewater sampling and the week prior to wastewater sampling;
I_3 was the reported cases for the week of sampling (0–6 days after wastewater sampling);
I_4 was the average of the reported cases for the same week as wastewater sampling and the week after wastewater sampling; and
I_5 was the reported cases for the week after sampling (7–13 days after sampling).

Within each variation of S_n, the best-fit equation from the variations of I_n was the equation with the smallest Akaike Information Criterion (AIC) value. Comparisons of best-fitted equations among S_{1-3} were analyzed using their R^2 values.

Non-parametric Spearman's rho correlation coefficients also evaluated the relationships between SARS-CoV-2 RNA concentrations (S_n) to the various lags and leads in reported COVID-19 cases (I_n). The non-parametric test was chosen as the Shapiro–Wilk test demonstrated that the data were not normally distributed.

RESULTS

Summary of wastewater samples and SARS-CoV-2 RNA concentrations

In total, 37 weekly samples from 24-h composites of wastewater influent were collected from the Yarmouth WWTP from September 22, 2020 to June 8, 2020. No sampling occurred on December 28, 2020. A summary of the concentrations of SARS-CoV-2 RNA adjusted for the flow rate and recovery, as well as the weekly reported cases of COVID-19 in Yarmouth, are available in the Supplementary Material, Example report to stakeholders. The detection limit (and the limit of quantification) of the concentration of SARS-CoV-2 RNA was generally <log 9.1 copies/day (and between log 9.9 and 10.6 copies/day; Figure 1(a)) and was dependent on the flow rate (and the viral recovery for samples collected from December 21 to June 8, 2021). Seven samples (19%) had undetectable concentrations of SARS-CoV-2 RNA: September 22, 2020; October 6, 2020; October 20, 2020; October 27, 2020; May 25, 2021; June 1, 2021; and June 8, 2021 (Figure 1(a)). Two samples (5%) had detectable but not quantifiable concentrations of SARS-CoV-2 RNA (November 10, 2020 and February 22, 2021). The average concentration of SARS-CoV-2 RNA adjusted for the flow rate only (and recovery) was log 10.16 copies/day (log 10.68 copies/day) and ranged from <log 8.51 to log 12.01 copies/day (<log 8.60 to log 12.19 copies/day; Figure 1(a)). For the recovery-adjusted samples, the recovery of BRSV averaged 40%, with a range of 18–96%, which was >10% recovery, which was the

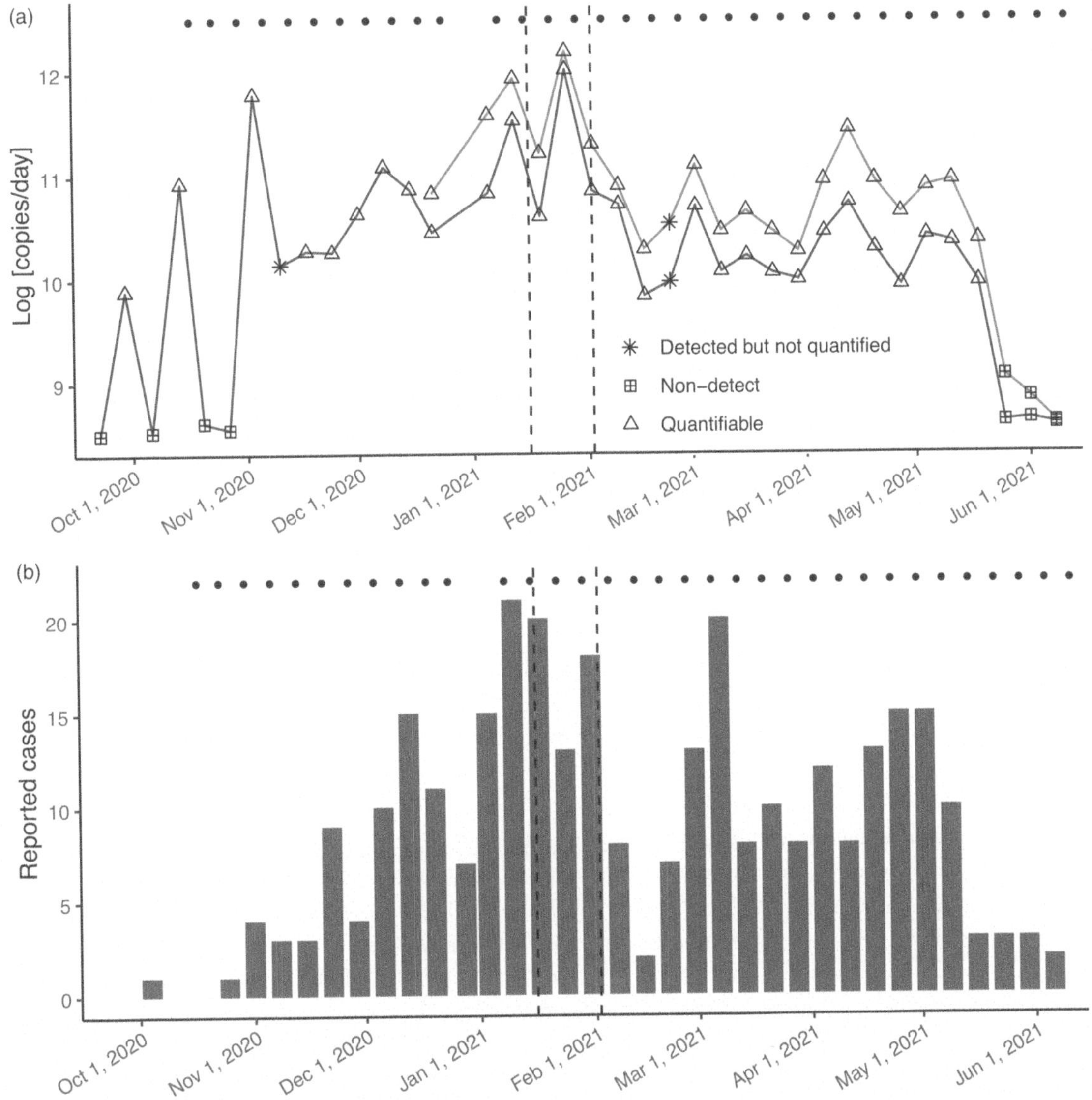

Figure 1 | (a) Log concentrations of SARS-CoV-2 RNA (copies/day) in the wastewater influent from the Yarmouth WWTP from September 22, 2020 to June 8, 2021. Values were adjusted for the flow rate only (blue) and for the flow rate and recovery (red). No wastewater sample was analyzed during the week of December 28, 2020. Samples with undetected concentrations of SARS-CoV-2 RNA are represented as the log of ½ value of the detection limit and adjusted for the flow rate (and recovery if applicable). Concentrations of SARS-CoV-2 RNA that were detected but not quantifiable are represented as the log-transformed limit of quantification and were adjusted for the flow rate (and recovery if applicable). (b) Number of cases of COVID-19 within the zip code for Yarmouth, ME as reported weekly by the Maine CDC. The dashed line represents the date of public advisories of high SARS-CoV-2 RNA concentrations in RNA. The black dots represent the dates that weekly reports of SARS-CoV-2 RNA were announced to the public. Please refer to the online version of this paper to see this figure in colour: https://dx.doi.org/10.2166/wh.2023.238.

threshold to re-run samples in a study of SARS-CoV-2 RNA measured from settled solids in untreated wastewater (Yu *et al.* 2021).

Description of the trends in SARS-CoV-2 RNA concentrations in wastewater and weekly reported COVID-19 cases in Yarmouth, ME between September 22, 2020 and June 8, 2021

From September 22 to October 27, 2020, SARS-CoV-2 RNA concentrations fluctuated each week between non-detected and $\geq$1 log above the detection limit (Figure 1(a)). Concentrations of SARS-CoV-2 RNA were detected from November 3, 2020 to May 18, 2021. Many peaks in SARS-CoV-2 RNA concentrations, as defined by $>$1 log change in detectable concentrations adjusted for the flow rate, occurred within 2 weeks after holidays and Yarmouth public school vacations, such as October

14, 2020 (2 days after Indigenous People's Day/Columbus Day); January 12, 2021 (8 days after the return from Christmas/ New Year's break); and January 26, 2021 (8 days after Martin Luther King Day) (Figure 1(a)).

There were <5 weekly reported COVID-19 cases from residents in Yarmouth during the weeks starting on September 22, 2020 to November 15, 2020, the week of February 8, 2021, and May 10, 2021 to June 6, 2021 (Figure 1(b)). As shown in Figure 1(b), weekly cases rose during fall 2020 and peaked in early January 2021. Additional peaks occurred in early March and late April 2021.

Comparing the SARS-CoV-2 RNA concentrations and reported COVID-19 cases in Yarmouth, ME

Overall, linear regression analyses demonstrated a significant temporal correlation between the log concentrations of SARS-CoV-2 RNA and the log number of weekly reported COVID-19 cases even when including a lag or a lead up to 1 week ($p < 0.05$; Figure 2(a)–2(o)). The best-fit regressions for each dependent variable of S_{1-3} and the log concentration of SARS-CoV-2 RNA vs. the log of the number of reported COVID-19 cases were with I_4, which represents the average of the reported COVID-19 cases for the same week and the week after the wastewater sample was collected (Figure 2(j)–2(l)). From September 22, 2020 to June 8, 2021, a 13.3% increase in I_4 was associated with a 10% increase in the concentration of SARS-CoV-2 RNA, S_1 (Figure 2(j); AIC = 77.82; $N = 35$). For the period from December 21, 2020 to June 8, 2021, the log concentrations of SARS-CoV-2 RNA were adjusted for the flow rate only, S_2, or the flow rate and viral recovery, S_3. Adjusting for viral recovery in addition to flow rate resulted in a 2.7% increase of COVID-19 cases associated with a 10% increase in SARS-CoV-2 RNA concentrations. Specifically, the best-fit equation for S_2, indicated an 18.9% increase in I_4 was associated with a 10% increase in S_2 (Figure 2(k); AIC = 40.85; $N = 24$), while the best-fit equation for S3 indicated a 21.6% increase in I_4 was associated with a 10% increase in S_3 (Figure 2(l); AIC = 38.89; $N = 24$).

We compared the R^2 values of the three best-fit equations from S_{1-3} to evaluate which linear equation best represented the relationship between the SARS-CoV-2 RNA concentrations and the number of reported COVID-19 cases, I_4. The smallest R^2 value, 0.4 (Figure 2(a)–2(o)), was from the S_1 equation, which explained 42% of the variation in I_4. The largest R^2 was from S_3, 0.68, which explained 68% of the variation in I_4. The R^2 value of the S_2 equation was 0.60. Thus, adjusting for viral recovery in addtion to flow rate improved the fit of the linear regression model by 8%.

Overall, Spearman's rho correlation coefficients of the concentrations of SARS-CoV-2 RNA, S_n, and the reported cases of COVID-19, I_n, were positively and significantly correlated with all leads and lags in the reported clinical cases ($p < 0.05$) except for three instances: S_2 vs. I_1, S_2 vs. I_2 and S_3 vs. I_1 (Table 1). The largest Spearman's rho correlation coefficients for the concentrations of SARS-CoV-2 RNA adjusted for flow rates only ($S_1$and S_2) were with the number of reported COVID-19 cases 1 week after the wastewater sample, I_5 ($\text{rho}_{S_1-I_5} = 0.744$ and $\text{rho}_{S_2-I_5} = 0.613$, respectively; Table 1). When viral recovery was considered along with flow rate (S_3) the largest Spearman's rho correlation coefficient of the concentrations of SARS-CoV-2 RNA was with average of the number of COVID-19 cases reported for the week of, and 1 week after, wastewater sampling, I_4 ($\text{rho}_{S_3-I_4} = 0.706$). Reported COVID-19 cases measured 1 week after sampling had a similar rho of 0.700 (Table 1).

Qualitative description of the formation of the YCCTF, a community-based organization

Qualitative feedback from members of the YCCTF illustrated that the ABCD model can be an effective approach for understanding local organizing. As one team member explained, getting 'well-known' residents involved in ABCD from the beginning was critical in establishing buy-in among members of the public. As volunteers identified duties and roles that often fit with their professional line of work or personal experiences, the Yarmouth community focused its efforts on the needs of seniors, K-12 students, and business owners. A member of the YWTT who has lived in the community for 36 years and has been involved in both social service management and political campaigns explained: 'I know how to organize people to do good stuff.' Echoing other members' sentiments, she explained that she and community members 'began to create capacity to meet the needs of citizens through ordinary channels.' The Task Force was established as a grassroots, loose collective that was non-hierarchical. In addition, she emphasized that the response to the pandemic was a collective effort, separate from the Town's government. She added that we were 'doing what we think will benefit the Yarmouth people.' Another member, an environmental scientist who has lived in Yarmouth for 30 years described the process to identify the strengths of community members as 'organic.' His introduction to the YCCTF's work occurred through the Rotary Club. He first helped with grocery shopping and later joined the YWTT, given his experience with wastewater treatment. He explained: 'I figured out what it would take to do this work that I understood.' While he identified his specific skills

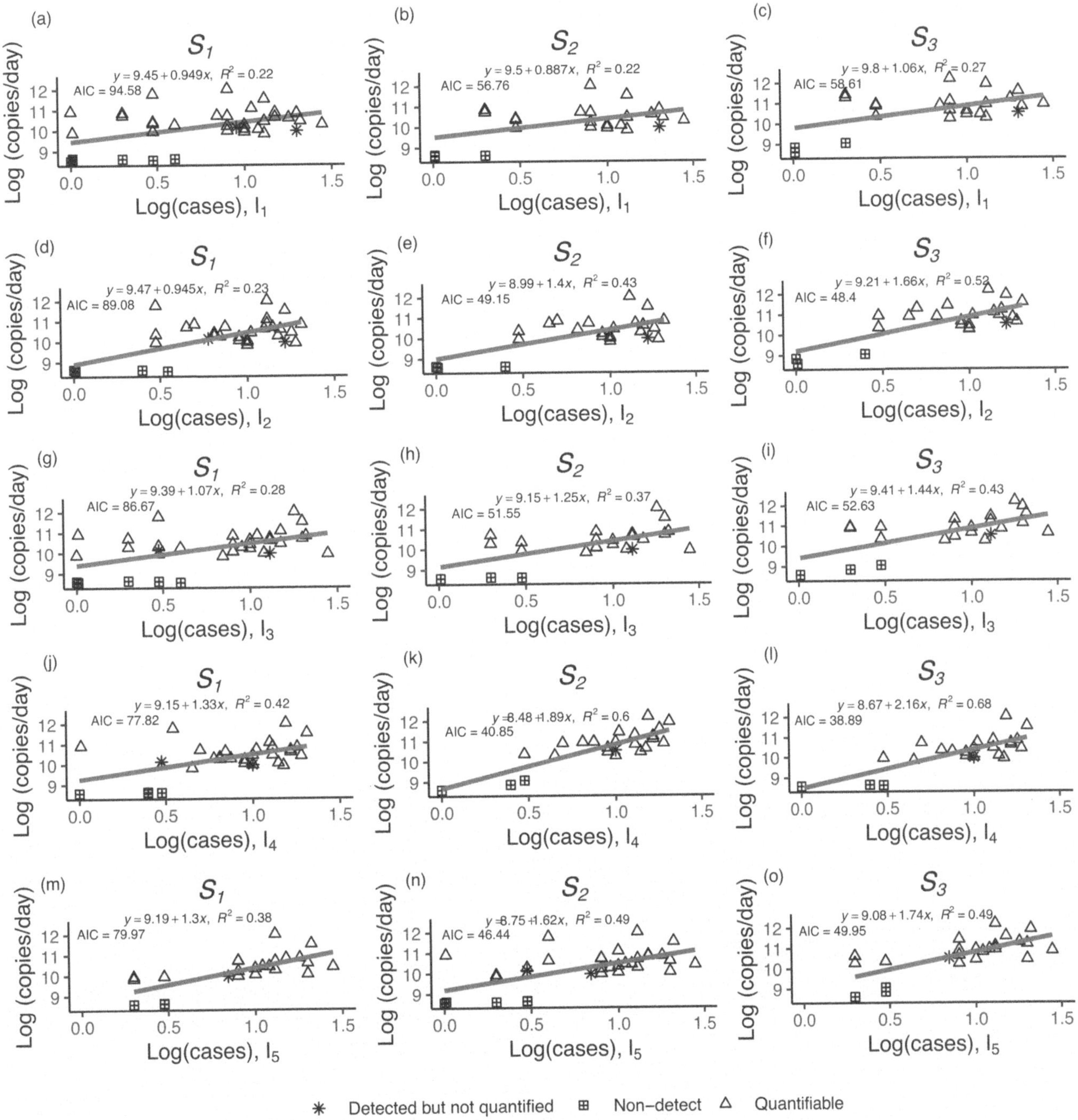

Figure 2 | (a–o) Evaluation of the association of the log-transformed of the number of weekly reported COVID-19 cases within the zip code for Yarmouth with leads and lags of up to 1 week from the sample collection, I_{1-5}, to the log-transformed concentrations of SARS-CoV-2 RNA (copies/day) adjusted for the flow rate, S_1 (a, d, g, j, m), during September 22, 2020 to June 8, 2021; S_2 (b, e, h, k, n), during December 21, 2020 to June 8, 2021; and S_3 (c, f, i, l, o), adjusted for the flow rate and viral recovery during December 21, 2020 to June 8, 2021. I_1 was the number of weekly reported cases 1 week prior to sample collection. I_2 was the average of the weekly reported cases for 1 week before and the same week as the wastewater was collected. I_3 was the number of weekly reported cases for the same week as the wastewater was collected. I_4 was the average of the weekly reported cases for the same week and 1 week after the wastewater was collected. I_5 was the number of weekly reported cases 1 week after the wastewater was collected. The blue lines represent the linear regression of each dataset with the corresponding parameters of each equation detailed in each graph. Samples with non-detected concentrations of SARS-CoV-2 are represented at the log-transformed ½ value of the detection limit and adjusted for the flow rate. Concentrations of SARS-CoV-2 that were detected but not quantifiable are represented at the log-transformed limit of quantification and were adjusted for the flow rate. Please refer to the online version of this paper to see this figure in colour: https://dx.doi.org/10.2166/wh.2023.238.

Table 1 | Spearman rho correlation coefficients of the log concentrations of SARS-CoV-2 RNA adjusted for flow rates, S_1 and S_2, during September 22, 2020 to June 8, 2021 and December 21, 2020 to June 8, 2021, respectively, and S_3, which evaluated viral recovery and flow rate during December 21, 2020 to June 8, 2021 to the log-transformed number of weekly reported COVID-19 cases (with variations in lead and lag times up to 1 week from the wastewater sample collection, I_{1-5}) in the Town of Yarmouth, ME, I_n

Weekly reported cases of COVID-19, log I_n	N	Spearman's rho
S_1 Dates of samples – September 22, 2020–June 8, 2021, viral recovery not measured		
I_1 (1 week before the wastewater sample)	37	0.331*
I_2 (average of the same week as sampling and the week before)	36	0.406*
I_3 (same week as the wastewater sample)	36	0.472**
I_4 (average of the same week as sampling and the week after)	35	0.561***
I_5 (1 week after the wastewater sample)	35	0.613***
S_2 Dates of samples – December 21, 2020–June 8, 2021, viral recovery not measured		
I_1 (1 week before the wastewater sample)	24	0.219
I_2 (average of the same week as sampling and the week before)	24	0.383
I_3 (same week as the wastewater sample)	24	0.490*
I_4 (average of the same week as sampling and the week after)	24	0.692***
I_5 (1 week after wastewater sample)	24	0.744***
S_3 Dates of samples – December 21, 2020–June 8, 2021, viral recovery measured		
I_1 (1 week before the wastewater sample)	24	0.302
I_2 (average of the same week as sampling and the week before)	24	0.485*
I_3 (same week as the wastewater sample)	24	0.532**
I_4 (average of the same week as sampling and the week after)	24	0.706***
I_5 (1 week after the wastewater sample)	24	0.700***

p-values: * < 0.05; ** < 0.01; *** < 0.001.

and expertise, he also was quick to highlight the egalitarian way others on the YCCTF functioned, adding that 'nobody cared about who got credit for [the] work.' Merging personal interests and professional expertise, the YWTT members collaborated with the community's water management facility to obtain wastewater samples, identified a college to conduct laboratory analyses, and continuously reassessed their goals and the implications of their work to the community. Members of the YWTT team spoke about the challenge of justifying the expenses associated with testing the wastewater. Members did not want the testing to be simply a 'scientific exercise,' but rather something that could be translated into action should the testing yield unfavorable results. As one member explained, the YWTT saw testing as 'another way to view information [about the pandemic] and as a tool to inform neighbors.' Additionally, a member emphasized that 'Too often these projects are undertaken by people/organizations removed from the local community and there is not the support or use of the data by the community contributing the samples or meant to act on it.' Similar to challenges identified by others studying ABCD, the YWTT continuously engaged in discussion about the purpose, duties, and costs of wastewater surveillance.

Dissemination of wastewater surveillance data to the community and the YCCTF's response to changing concentrations of SARS-CoV-2 RNA and reported COVID-19 cases

The YWTT waited until October 17, 2020, after 4 initial weeks of wastewater testing, to begin preparing and disseminating weekly wastewater testing reports to the Yarmouth Town Manager, the Superintendent of the Yarmouth WWTP, Yarmouth Community Services, and Yarmouth residents (Figure 1(a) and 1(b)). The YWTT issued subsequent reports for every week that testing was performed. An example of the reports is provided in the Supplementary Material. The first communications began as emails with a report that included a brief overview of wastewater testing for SARS-CoV-2 RNA, a graph and table with historical and current wastewater data, a note on the wastewater sampling methodology, and frequently asked questions. Weekly COVID-19 cases reported by the Maine CDC for the Yarmouth zip code were added in early November 2020. Eventually, concentrations of SARS-CoV-2 RNA that were adjusted for the flow rate and viral recovery were added. The reports were posted each week on 'Be Well Yarmouth,' the Town of Yarmouth's COVID-19 website (Yarmouth Community

Coronavirus Task Force 2022). Beginning on December 13, 2020, the report was also sent to the Maine Department of Health and Human Services and a larger local audience; an email summary and interpretation of each report's findings were added to facilitate communication. Beginning with the first report, neighborhood communicators informed their neighbors of the recent wastewater results and weekly reported COVID-19 cases in Yarmouth. Additionally, they relayed pertinent information regarding the COVID-19 response in the community, such as community testing events, vaccination events, and other resources available to community members. On January 16, 2021 and February 2, 2021, when concentrations of SARS-CoV-2 RNA were higher and increasing from the previous weeks, the YWTT prepared and the YCCTF issued community advisories alongside the wastewater reports to encourage extra care to reduce exposure in the coming weeks (Figure 1(a) and 1(b)). The extra care emphasized in the reports included actions such as wearing masks and following US CDC and Maine CDC recommendations regarding when to get tested for COVID-19, quarantining, isolation, and locations of testing facilities. The advisories were sent based on the recommendations of two or three members of the YCCTF who regularly reviewed the clinical and wastewater data to look for increasing trends in cases in the previous 4 weeks. At the time of the dissemination of the advisories, the existing literature did not recommend specific quantitative thresholds for genetic markers measured from wastewater or clinical cases to trigger public health action to mitigate the transmission of SARS-CoV-2.

Retrospective analysis of the clinical and wastewater data revealed that the advisories were disseminated when (1) significant increases in unadjusted wastewater SARS-CoV-2 levels (at least 3× the concentrations in the previous 2–3 weeks and concentrations >100,000 copies/L, as compared to an unadjusted 'baseline' of about 5,000–10,000 copies/L) and (2) concurrent increases or high numbers of COVID cases in the Yarmouth zip code area, as reported by Maine CDC, or in Yarmouth schools over a 2–4-week period. (We did not have SARS-CoV-2 RNA concentrations adjusted for viral recovery for the test results on which the first advisory was based.) There was no other period from September 2020 to June 2021 when these conditions were met. There were increases in SARS-CoV-2 RNA concentrations and reported cases from February 22 to March 7 and during April 2021, but the RNA concentrations never reached the levels seen in January 2021.

DISCUSSION

Wastewater surveillance is a popular public health tool that can help monitor the transmission of SARS-CoV-2 and other pathogens. The purpose of this study was to demonstrate the performance of a wastewater surveillance program using an ABCD framework in a small community, Yarmouth, ME. Specifically, the YCCTF and the Town of Yarmouth, ME, established and funded a community-based wastewater surveillance program, which monitored the presence and concentrations of SARS-CoV-2 RNA in the majority of the Yarmouth community for ~$400 per sample for 37 weeks. To our knowledge, this is one of the few documented examples of community-based wastewater surveillance in a small community.

We described the results of longitudinal wastewater surveillance and concurrent weekly reported COVID-19 cases in the Town of Yarmouth. We also evaluated the relationship between the concentrations of SARS-CoV-2 RNA and various lags and leads in weekly reported COVID-19 cases. Temporal trends in Yarmouth's wastewater and clinical data followed general trends reported in the Northeast United States. Specifically, in Yarmouth and metropolitan Boston, MA, clinical cases increased from September 23, 2020 until mid-January 2021, followed by another spike in mid-April 2021 (Xiao *et al.* 2022). Overall, our analyses demonstrated that there was a significant and positive temporal correlation between the number of weekly reported COVID-19 cases in Yarmouth (I_n) and the concentrations of SARS-CoV-2 RNA (S_{1-3}) in the municipal wastewater with both linear regression and Spearman's rho correlations (Table 1 and Figure 2(a)–2(o)). Similarly, a wastewater surveillance study of a WWTP in metropolitan Charlotte, NC, USA (population = 68,895) demonstrated that linear regressions and Spearman's correlations of RT-qPCR measurements of SARS-CoV-2 N1 and N2 markers with reported COVID-19 cases on the same day and up to 14 days later were significantly and positively correlated (Barua *et al.* 2022). Additionally, Spearman's rho and linear regression analyses provided a positive and significant correlation between the concentrations of N2 markers measured from grab samples in WWTPs in metropolitan Reno-Sparks, NV with reported COVID-19 cases on the same day as sampling as well as a 7-day lead in cases ($p < 0.001$; Li *et al.* 2022). The significant association between the SARS-CoV-2 RNA concentrations and lags in reported cases could be the result of viral shedding by infected individuals in wastewater prior to the onset of symptoms, delays in seeking diagnostic testing, or lags in reporting positive COVID-19 cases to health officials.

Specifically, we compared the linear relationship between SARS-CoV-2 RNA concentrations, S_{1-3}, in Yarmouth wastewater to various leads and lags in the reported COVID-19 cases (I_{1-5}). Overall, the best-fit of the equations, as evaluated by the lowest AIC value, was from the average number of COVID-19 cases reported from the week of sampling and the following week (I_4; Table 1). Similarly, a study of municipal wastewater surveillance in metropolitan Boston, MA found better correlations between daily measurements of SARS-CoV-2 RNA concentrations 4–5 days prior to the daily COVID-19 case counts (Wu *et al.* 2022). Wastewater surveillance in three municipal WWTPs in the Charlotte, NC metropolitan area demonstrated the best correlation of daily measurements of SARS-CoV-2 N1 and N2 markers in 24 h composite samples to the 7-day rolling average of reported COVID-19 cases with a wastewater sample lead of 5–11 days (Barua *et al.* 2022). Our results provide further evidence that reported cases of COVID-19 in sewersheds servicing small and large communities were best correlated with retrospective concentrations of SARS-CoV-2 RNA measured with daily and weekly sampling schedules. Therefore, wastewater surveillance can be used as an early warning system in both large and small communities.

Spearman's correlations also demonstrated that the strongest correlations of SARS-CoV-2 RNA concentrations, S_{1-3}, were with reported COVID-19 cases averaged from the same week and the week after sampling, I_4, or the week after sampling, I_5. Spearman's correlation and linear regressions in SARS-CoV-2 RNA concentrations measured between December 21, 2020 and June 8, 2021 adjusted for the flow rate only, S_2 differed in which the type of lag in reported COVID-19 cases had the strongest relationship, I_4 for linear regression and I_5 for Spearman's correlation. Specifically, Spearman's correlation (and R^2 values) with I_4 had a rho value of 0.052 larger (or 0.105 smaller) than I_5. This difference in the ranks of the strongest relationship of the variables measured with Spearman's correlation coefficients and linear regression analyses could be attributed to the increased variability of an evaluation of a subset of our entire dataset. Although a previous study demonstrated that linear regression analyses can adequately explain the relationship between SARS-CoV-2 RNA concentrations and reported COVID-19 cases (Anneser *et al.* 2022), such an analysis assumes a constant slope between the variables, which is not required to calculate Spearman's correlation coefficients.

In our study, there were 10 instances (26% of samples) in which the concentration of SARS-CoV-2 RNA was either below detection or detected but not quantifiable. In the best-fit model for S_1, 60% of the undetected ($n = 4$) and detected but not quantifiable samples ($n = 6$) occurred when $I_4 \leq 3$ cases. This indicates that the correlation between S_1 and $I_4$was still meaningful when S_1 was nondetectable.

The adjustment of the SARS-CoV-2 RNA concentrations for viral recovery from December 21, 2020 to June 8, 2021 improved the goodness-of-fit measure, R^2, from 0.60 to 0.68 when plotted against the average reported COVID-19 cases during the week of and the week after sampling, I_4 (Figure 2(k) and 2(l)). The significance of the correlation of I_4 and S_{2-3} did not change. Although previous studies demonstrated that normalizing viral RNA concentrations for viral recovery improved correlations to case data in the samples gathered from municipal WWTPs servicing smaller communities, the correlations were still statistically significant without the normalization (Feng *et al.* 2021; Nagarkar *et al.* 2022; Vadde *et al.* 2022). Thus, our data, and data from previous studies, indicate that the added time and cost required to assess viral recovery in smaller municipal systems may not be necessary to measure SARS-CoV-2 RNA concentrations from wastewater influent.

On November 3, 2020, the SARS-CoV-2 RNA concentration increased to log 11.78 copies/day, but during the week of sampling and the following week, there were 5 reported COVID-19 cases (Figure 1(a) and 1(b)). The sampling occurred the week that Yarmouth's high school switched to remote learning due to an outbreak at the school. It is possible that the COVID-19 infections were underreported among high-school students due to asymptomatic infections and symptomatic students who did not have sought care. Similarly, college students living in dormitories were less likely to have symptomatic infections (Poletti *et al.* 2021; Schmitz *et al.* 2021) and may not have sought testing. Wastewater surveillance of SARS-CoV-2 in a southern Nevada community (population of 16,399) estimated 3.3 times more COVID-19 infections than the number of confirmed cases from March 2020 to April 2021 (Vo *et al.* 2022). Additionally, wastewater surveillance of SARS-CoV-2 in metropolitan Boston, MA estimated at least 3.8 times more infections than reported cases during March 18–23, 2020 (Wu *et al.* 2020). These studies indicate that reported cases underrepresent the total COVID-19 cases in a sewershed, and wastewater-based epidemiology can provide a better estimate of the total caseload.

Our study supports the theory that community-based organizations, such as the YCCTF, can quickly engage community members and provide them with information and local resources to mitigate COVID-19 transmission (Korfmacher & Harris-Lovett 2022). Alongside the weekly wastewater testing reports, the YCCTF also provided approachable explanations of the testing results to the local community. Neighborhood leaders sought out and interacted with members of harder-to-reach populations. After viral peaks in the wastewater, neighborhood leaders encouraged residents to be more careful, which may have helped reduce SARS-CoV-2 RNA concentrations in the subsequent weeks. Additionally, the wastewater surveillance program provided reassurance to the Yarmouth community that COVID-19 was being monitored and addressed at the local level. The wastewater program's timely information about SARS-CoV-2 transmission in the community was viewed as being beneficial enough to the community's COVID-19 monitoring and response activities that the Town of Yarmouth and individual residents contributed funds to extend the initial 6-month pilot program for 3 months and, after a 3-month break, to resume the program in September 2021 for another 6 months.

The wastewater surveillance program in Yarmouth had some limitations. Although we were able to establish significant correlations between weekly measurements of SARS-CoV-2 RNA and reported COVID-19 cases, previous studies demonstrated that once weekly wastewater sampling did not identify significant trends in SARS-CoV-2 transmission in the sewersheds of larger communities (population >200,000) where COVID-19 cases are reported daily (Feng *et al.* 2021; Xie *et al.* 2021). Another limitation was ~4 days turnaround time from sample collection to receiving results. Quicker availability of testing results would allow timelier assessment and reporting of SARS-CoV-2 levels in the community and faster mobilization of community support in response to increased viral levels. Because the Yarmouth sewershed only covers about 70% of Yarmouth residents, the wastewater surveillance program provided an incomplete picture of viral transmission in Yarmouth, including the presence of non-resident workers and the absence of some residents during part of the weekly 24-h sampling period, which may have also contributed to a possibly unrepresentative picture of viral levels in the community. Lastly, the correlation of SARS-CoV-2 RNA concentrations with reported COIVD-19 cases is dependent on the availability and use of individual testing in the community.

Overall, this study demonstrates the feasibility and usefulness of weekly wastewater testing in a small community. Our study also documents a community-based initiative to respond to the COVID-19 pandemic. The study of the Yarmouth wastewater surveillance program can be used as a framework to help other small communities establish wastewater surveillance, interpret surveillance data, and disseminate testing results to the community. Future studies could evaluate the cost–benefit analysis of a wastewater surveillance program in a small community.

ACKNOWLEDGEMENTS

The wastewater surveillance program and the study were funded by the CARES Act's *Keep Maine Healthy Grant - Waste Water*, the Town of Yarmouth, ME, and individual contributions. We thank the staff of the Yarmouth Municipal Wastewater Treatment Plant for help with wastewater sample collection. We also thank the community members of the Yarmouth Community Coronavirus Task Force for their help in mobilizing community support for this project. We thank Dr Brian Swalla for his critical editing of the manuscript.

DATA AVAILABILITY STATEMENT

All relevant data are included in the paper or its Supplementary Information.

CONFLICT OF INTEREST

Y.M.B. received in-kind materials from IDEXX Laboratories, Inc.

REFERENCES

Ahmed, W., Tscharke, B., Bertsch, P. M., Bibby, K., Bivins, A., Choi, P., Clarke, L., Dwyer, J., Edson, J., Nguyen, T. M. H., O'Brien, J. W., Simpson, S. L., Sherman, P., Thomas, K. v., Verhagen, R., Zaugg, J. & Mueller, J. F. 2020 SARS-CoV-2 RNA monitoring in wastewater as an early warning system for COVID-19 transmission in the community: a temporal case study. *Science of the Total Environment* 144216. https://doi.org/10.1016/j.scitotenv.2020.144216.

Al-Duroobi, H., Moghadam, S. v., Phan, D. C., Jafarzadeh, A., Matta, A. & Kapoor, V. 2021 Wastewater surveillance of SARS-CoV-2 corroborates heightened community infection during the initial peak of COVID-19 in Bexar County, Texas. **2**, 15. https://doi.org/10.1093/femsmc/xtab015.

Anneser, E., Riseberg, E., Brooks, Y. M., Corlin, L. & Stringer, C. 2022 Modeling the relationship between SARS-CoV-2 RNA in wastewater or sludge and COVID-19 cases in three New England regions. *Journal of Water and Health* **20** (5), 816–828. https://doi.org/10.2166/wh.2022.013.

Barua, V. B., Juel, M. A. I., Blackwood, A. D., Clerkin, T., Ciesielski, M., Sorinolu, A. J., Holcomb, D. A., Young, I., Kimble, G., Sypolt, S., Engel, L. S., Noble, R. T. & Munir, M. 2022 Tracking the temporal variation of COVID-19 surges through wastewater-based epidemiology during the peak of the pandemic: a six-month long study in Charlotte, North Carolina. *Science of the Total Environment* **814**. https://doi.org/10.1016/J.SCITOTENV.2021.152503.

Bi, C., Ramos-Mandujano, G., Tian, Y., Hala, S., Xu, J., Mfarrej, S., Esteban, C. R., Delicado, E. N., Alofi, F. S., Khogeer, A., Hashem, A. M., Almontashiri, N. A. M., Pain, A., Izpisua Belmonte, J. C. & Li, M. 2021 Simultaneous detection and mutation surveillance of SARS-CoV-2 and multiple respiratory viruses by rapid field-deployable sequencing. *Med* **2** (6), 689–700.e4. https://doi.org/10.1016/J.MEDJ.2021.03.015.

Biobot Analytics 2020 *Data – Biobot*. Available from: https://biobot.io/data/.

CDC n.d. *Wastewater Surveillance Data Reporting and Analytics*. Available from: https://www.cdc.gov/healthywater/surveillance/wastewater-surveillance/data-reporting-analytics.html (accessed 9 July 2022).

D'Aoust, P. M., Towhid, S. T., Mercier, É., Hegazy, N., Tian, X., Bhatnagar, K., Zhang, Z., Naughton, C. C., MacKenzie, A. E., Graber, T. E. & Delatolla, R. 2021 COVID-19 wastewater surveillance in rural communities: comparison of lagoon and pumping station samples. *Science of the Total Environment* **801**, 149618. https://doi.org/10.1016/j.scitotenv.2021.149618.

Feng, S., Roguet, A., Mcclary-Gutierrez, J. S., Newton, R. J., Kloczko, N., Meiman, J. G. & Mclellan, S. L. 2021 Evaluation of sampling, analysis, and normalization methods for SARS-CoV-2 concentrations in wastewater to assess COVID-19 burdens in Wisconsin communities. https://doi.org/10.1021/acsestwater.1c00160.

Gallatin City-County Health Department 2020 *Wastewater Data: Wastewater Testing and COVID-19. Healthy Gallatin*. Available from: https://www.healthygallatin.org/coronavirus-covid-19/wastewater-data/.

Gudra, D., Dejus, S., Bartkevics, V., Roga, A., Kalnina, I., Strods, M., Rayan, A., Kokina, K., Zajakina, A., Dumpis, U., Ikkere, L. E., Arhipova, I., Berzins, G., Erglis, A., Binde, J., Ansonska, E., Berzins, A., Juhna, T. & Fridmanis, D. 2022 Detection of SARS-CoV-2 RNA in wastewater and importance of population size assessment in smaller cities: an exploratory case study from two municipalities in Latvia. *Science of the Total Environment* **823**, 153775. https://doi.org/10.1016/J.SCITOTENV.2022.153775.

Hogan, G. 2020 *Neighbors Mobilize To Help Each Other During Coronavirus Pandemic. Gothamist*. Available from: https://gothamist.com/news/neighbors-mobilize-help-each-other-during-coronavirus-pandemic.

Hopes, K., Mcknight, J. & Lawrence, H. 2015 *Asset Based Neighborhood Organizing: The Method of the Abundant Community Initiative in Edmonton, Canada*.

Ishisaka, N. 2020 *Coronavirus Sparks an Epidemic of People Helping People in Seattle. Seattle Times*. Available from: https://www.seattletimes.com/seattle-news/coronavirus-sparks-an-epidemic-of-people-helping-people-in-seattle/?fbclid=IwAR04uxMeSQXV2WX5lYjhyVQEtEWwb189GWXzuTSL16ILg8Yu3v99FUovbpE.

Korfmacher, K. S. & Harris-Lovett, S. 2022 Invited perspective: implementation of wastewater-based surveillance requires collaboration, integration, and community engagement. *Environmental Health Perspectives* **130** (5). https://doi.org/10.1289/EHP11191.

Kretzmann, J. P. & McKnight, J. L. 1993 *Building Communities From the Inside Out: A Path Toward Finding and Mobilizing a Community's Assets*, 1st edn. ACTA Publications, Evanston, IL.

Kumar, M., Patel, A. K., Shah, A. v., Raval, J., Rajpara, N., Joshi, M. & Joshi, C. G. 2020 First proof of the capability of wastewater surveillance for COVID-19 in India through detection of genetic material of SARS-CoV-2. *Science of The Total Environment* **746**, 141326. https://doi.org/10.1016/J.SCITOTENV.2020.141326.

Lago, P. M., Gary, H. E., Pérez, L. S., Cáceres, V., Olivera, J. B., Puentes, R. P., Corredor, M. B., Jímenez, P., Pallansch, M. A. & Cruz, R. G. 2003 Poliovirus detection in wastewater and stools following an immunization campaign in Havana, Cuba. *International Journal of Epidemiology* **32** (5), 772–777. https://doi.org/10.1093/ije/dyg185.

Li, L., Mazurowski, L., Dewan, A., Carine, M., Haak, L., Guarin, T. C., Dastjerdi, N. G., Gerrity, D., Mentzer, C. & Pagilla, K. R. 2022 Longitudinal monitoring of SARS-CoV-2 in wastewater using viral genetic markers and the estimation of unconfirmed COVID-19 cases. *Science of The Total Environment* **817**, 152958. https://doi.org/10.1016/J.SCITOTENV.2022.152958.

McCall, C., Wu, H., Miyani, B. & Xagoraraki, I. 2020 Identification of multiple potential viral diseases in a large urban center using wastewater surveillance. *Water Research* **184**. https://doi.org/10.1016/j.watres.2020.116160.

Medema, G., Heijnen, L., Elsinga, G., Italiaander, R. & Brouwer, A. 2020 Presence of SARS-Coronavirus-2 RNA in sewage and correlation with reported COVID-19 prevalence in the early stage of the epidemic in The Netherlands. *Environmental Science & Technology Letters* **7** (7), 511–516. https://doi.org/10.1021/acs.estlett.0c00357.

Mehrotra, A., Larsen, D. A. & Jha, A. K. 2020 *It's Time to Begin a National Wastewater Testing Program for Covid-19*. STAT. Available from: https://www.statnews.com/2020/07/09/wastewater-testing-early-warning-covid-19-infection-communities/.

Michener, L., Aguilar-Gaxiola, S., Alberti, P. M., Castaneda, M. J., Castrucci, B. C., Harrison, L. M., Hughes, L. S., Richmond, A. & Wallerstein, N. 2021 Engaging with communities – lessons (re)learned from COVID-19. *Preventing Chronic Disease* **17**. https://doi.org/10.5888/PCD17.200250.

Nagarkar, M., Keely, S. P., Jahne, M., Wheaton, E., Hart, C., Smith, B., Garland, J., Varughese, E. A., Braam, A., Wiechman, B., Morris, B. & Brinkman, N. E. 2022 SARS-CoV-2 monitoring at three sewersheds of different scales and complexity demonstrates distinctive relationships between wastewater measurements and COVID-19 case data. *Science of the Total Environment* **816**. https://doi.org/10.1016/J.SCITOTENV.2021.151534.

Poletti, P., Tirani, M., Cereda, D., Trentini, F., Guzzetta, G., Sabatino, G., Marziano, V., Castrofino, A., Grosso, F., Gabriele, C. D., Piccarreta, R., Andreassi, A., Melegaro, A., Gramegna, M., Ajelli, M. & Merler, S. 2021 Association of age with likelihood of developing symptoms and critical disease among close contacts exposed to patients with confirmed SARS-CoV-2 infection in Italy + supplemental content. *JAMA Network Open* **4** (3), 211085. https://doi.org/10.1001/jamanetworkopen.2021.1085.

Prado, T., Fumian, T. M., Mannarino, C. F., Resende, P. C., Motta, F. C., Eppinghaus, A. L. F., Vale, C. d., Braz, V. H., de Andrade, R. M. S., da S, J., Maranhão, R., & Miagostovich, A. G. & P, M. 2021 Wastewater-based epidemiology as a useful tool to track SARS-CoV-2 and support public health policies at municipal level in Brazil. *Water Research* **191**, 116810. https://doi.org/10.1016/J.WATRES.2021.116810.

R Core Team 2022 R: A language and environment for statistical computing. R Foundation for Statistical Computing, Vienna, Austria. URL https://www.R-project.org/.

Schmitz, B. W., Innes, G. K., Prasek, S. M., Betancourt, W. Q., Stark, E. R., Foster, A. R., Abraham, A. G., Gerba, C. P. & Pepper, I. L. 2021 Enumerating asymptomatic COVID-19 cases and estimating SARS-CoV-2 fecal shedding rates via wastewater-based epidemiology. *Science of The Total Environment* **801**, 149794. https://doi.org/10.1016/J.SCITOTENV.2021.149794.

Sherchan, S. P., Shahin, S., Ward, L. M., Tandukar, S., Aw, T. G., Schmitz, B., Ahmed, W. & Kitajima, M. 2020 First detection of SARS-CoV-2 RNA in wastewater in North America: a study in Louisiana, USA. *Science of The Total Environment*, 140621. https://doi.org/10.1016/j.scitotenv.2020.140621.

Street, R., Mathee, A., Mangwana, N., Dias, S., Sharma, J. R., Ramharack, P., Louw, J., Reddy, T., Brocker, L., Surujlal-Naicker, S., Berkowitz, N., Malema, M. S., Nkambule, S., Webster, C., Mahlangeni, N., Gelderblom, H., Mdhluli, M., Gray, G., Muller, C. & Johnson, R. 2021 Spatial and temporal trends of SARS-CoV-2 RNA from wastewater treatment plants over 6 weeks in Cape Town, South Africa. *International Journal of Environmental Research and Public Health* **18** (22), 12085. https://doi.org/10.3390/IJERPH182212085.

Tiwari, A., Lipponen, A., Hokajärvi, A. M., Luomala, O., Sarekoski, A., Rytkönen, A., Österlund, P., Al-Hello, H., Juutinen, A., Miettinen, I. T., Savolainen-Kopra, C. & Pitkänen, T. 2022 Detection and quantification of SARS-CoV-2 RNA in wastewater influent in relation to reported COVID-19 incidence in Finland. *Water Research* **215**, 118220. https://doi.org/10.1016/J.WATRES.2022.118220.

U.S. Census Bureau 2020 *QuickFacts. 2020 Census.*

Vadde, K. K., Al-Duroobi, H., Phan, D. C., Jafarzadeh, A., Moghadam, S. v., Matta, A. & Kapoor, V. 2022 Assessment of concentration, recovery, and normalization of SARS-CoV-2 RNA from two wastewater treatment plants in Texas and correlation with COVID-19 cases in the community. *ACS ES&T Water*. https://doi.org/10.1021/ACSESTWATER.2C00054.

Vo, V., Tillett, R. L., Papp, K., Shen, S., Gu, R., Gorzalski, A., Siao, D., Markland, R., Chang, C.-L., Baker, H., Chen, J., Schiller, M., Betancourt, W. Q., Buttery, E., Pandori, M., Picker, M. A., Gerrity, D. & Oh, E. C. 2022 Use of wastewater surveillance for early detection of Alpha and Epsilon SARS-CoV-2 variants of concern and estimation of overall COVID-19 infection burden. *Science of The Total Environment* **835**, 155410. https://doi.org/10.1016/j.scitotenv.2022.155410.

Wu, F., Zhang, J., Xiao, A., Gu, X., Lee, W. L., Armas, F., Kauffman, K., Hanage, W., Matus, M., Ghaeli, N., Endo, N., Duvallet, C., Poyet, M., Moniz, K., Washburne, A. D., Erickson, T. B., Chai, P. R., Thompson, J. & Alm, E. J. 2020 SARS-CoV-2 titers in wastewater Are higher than expected from clinically confirmed cases. *MSystems* **5** (4). https://doi.org/10.1128/mSystems.00614-20.

Wu, F., Xiao, A., Zhang, J., Moniz, K., Endo, N., Armas, F., Bonneau, R., Brown, M. A., Bushman, M., Chai, P. R., Duvallet, C., Erickson, T. B., Foppe, K., Ghaeli, N., Gu, X., Hanage, W. P., Huang, K. H., Lee, W. L., Matus, M., McElroy, K. A., Nagler, J., Rhode, S. F., Santillana, M., Tucker, J. A., Wuertz, S., Zhao, S., Thompson, J. & Alm, E. J. 2022 SARS-CoV-2 RNA concentrations in wastewater foreshadow dynamics and clinical presentation of new COVID-19 cases. *Science of The Total Environment* **805**, 150121. https://doi.org/10.1016/j.scitotenv.2021.150121.

Xiao, A., Wu, F., Bushman, M., Zhang, J., Imakaev, M., Chai, P. R., Duvallet, C., Endo, N., Erickson, T. B., Armas, F., Arnold, B., Chen, H., Chandra, F., Ghaeli, N., Gu, X., Hanage, W. P., Lee, W. L., Matus, M., McElroy, K. A., Moniz, K., Rhode, S. F., Thompson, J. & Alm, E. J. 2022 Metrics to relate COVID-19 wastewater data to clinical testing dynamics. *Water Research* **212**. https://doi.org/10.1016/J.WATRES.2022.118070.

Xie, Y., Challis, J. K., Oloye, F. F., Asadi, M., Cantin, J., Brinkmann, M., McPhedran, K. N., Hogan, N., Sadowski, M., Jones, P. D., Landgraff, C., Mangat, C., Servos, M. R. & Giesy, J. P. 2021 RNA in municipal wastewater reveals magnitudes of COVID-19 outbreaks across four waves driven by SARS-CoV-2 variants of concern. *ACS Environmental Science and Technology Water*. https://doi.org/10.1021/ACSESTWATER.1C00349/ASSET/IMAGES/LARGE/EW1C00349_0004.JPEG.

Yarmouth Community Coronavirus Task Force 2022 *Be Well Yarmouth*. https://www.bewellyarmouth.org/.

Yu, A., Wolfe, M., Leon, T., Duong, D., Kennedy, L., Ravuri, S., Wigginton, K. R. & Boehm, A. 2021 Estimating relative abundance of two SARS-CoV-2 variants through wastewater surveillance at two large metropolitan sites. https://doi.org/10.21203/RS.3.RS-1083575/V1.

First received 16 September 2022; accepted in revised form 14 February 2023. Available online 27 February 2023

doi: 10.2166/wh.2022.020

Challenges in realising the potential of wastewater-based epidemiology to quantitatively monitor and predict the spread of disease

Julian Faraway[a,*], James Boxall-Clasby[b], Edward J. Feil[c], Marjorie J. Gibbon[c], Oliver Hatfield[d], Barbara Kasprzyk-Hordern IWA [b] and Theresa Smith[a]

[a] Department of Mathematical Sciences, University of Bath, Bath BA2 7AY, UK
[b] Department of Chemistry, University of Bath, Bath BA2 7AY, UK
[c] Department of Biology and Biochemistry, University of Bath, Bath BA2 7AY, UK
[d] Institute for Mathematical Innovation, University of Bath, Bath BA2 7AY, UK
*Corresponding author. E-mail: jjf23@bath.ac.uk

JF, 0000-0002-7308-3084; BK-H, 0000-0002-6809-2875; TS, 0000-0002-7085-3864

ABSTRACT

Researchers around the world have demonstrated correlations between measurements of SARS-CoV-2 RNA in wastewater (WW) and case rates of COVID-19 derived from direct testing of individuals. This has raised concerns that wastewater-based epidemiology (WBE) methods might be used to quantify the spread of this and other diseases, perhaps faster than direct testing, and with less expense and intrusion. We illustrate, using data from Scotland and the USA, the issues regarding the construction of effective predictive models for disease case rates. We discuss the effects of variation in, and the problem of aligning, public health (PH) reporting and WW measurements. We investigate time-varying effects in PH-reported case rates and their relationship to WW measurements. We show the lack of proportionality of WW measurements to case rates with associated spatial heterogeneity. We illustrate how the precision of predictions is affected by the level of aggregation chosen. We determine whether PH or WW measurements are the leading indicators of disease and how they may be used in conjunction to produce predictive models. The prospects of using WW-based predictive models with or without ongoing PH data are discussed.

Key words: COVID-19, prediction models, public health surveillance, SARS-CoV-2, sewage

HIGHLIGHTS

- This study is a worked example of predicting the spread of disease using wastewater-based epidemiology (WBE).
- It demonstrates non-linear relationships between wastewater and public health measures of the disease.
- It demonstrates time-varying relationships in WBE.
- It investigates time lags in the prediction of WBE.
- It makes recommendations for building predictive models in WBE.

INTRODUCTION

The utility of wastewater-based epidemiology (WBE) as an efficient approach to monitoring the health of populations and the environment has been increasingly recognised over the last decade. WBE has been used to estimate illicit drug use trends (Thomas *et al.* 2012; Ort *et al.* 2014) but has also been applied to estimate public exposure to alcohol (Reid *et al.* 2011; Baz-Lomba *et al.* 2016), tobacco (Castiglioni *et al.* 2015), counterfeit medicines (Venhuis *et al.* 2014; Causanilles *et al.* 2018), as well as levels of stress biomarkers, such as isoprostanes (Ryu *et al.* 2016; O'Brien *et al.* 2019), and usage of pharmaceuticals (Escolà Casas *et al.* 2021; Kasprzyk-Hordern *et al.* 2021). Interest in this approach dramatically increased because of the COVID-19 pandemic, resulting in rapid conceptual and methodological advances. In our examples, we examine the application of WBE to monitoring SARS-CoV-2 and highlight some of the uncertainties and challenges that remain in modelling the data. These considerations will apply to many other pathogens of ongoing interest and are relevant for planning responses to future pandemics.

If it were possible to obtain real-time reliable estimates of disease prevalence in a community, there would be little need for WBE. In practice, it is expensive, time-consuming, and intrusive to obtain good public health (PH) data because this requires direct sampling or recording of individuals. In contrast, wastewater (WW) can be sampled at a lower cost, processed more quickly, and carried out with much lower privacy concerns. WBE is already being used around the world to monitor SARS-CoV-2 RNA in WW in various ways. We identify three increasingly challenging modes of use:

1. *Presence or absence of RNA in WW.* In countries where a so-called zero-COVID policy has been attempted, this can identify the presence of infection in the community that might prompt lockdown measures.
2. *Qualitative level of RNA in WW.* For example, this might be reported as absent, low, medium, or high. This gives an indication of the general level of COVID-19 infection in the community without attempting a numeric estimation. It also provides an indication of change over time in levels of infection. Public dashboards for WW RNA reports often use this mode (see Scotland example below).
3. Numeric estimates of the numbers of individuals currently infected with COVID-19 coupled with a prediction of near-future numbers of infected along with realistic expression of uncertainty in the estimates and predictions.

WBE has largely succeeded in the first two of these modes, but the third goal of numerical estimation and prediction remains a challenge. A discussion of this challenge is the topic of this article. The first two modes of operation do not require detailed PH data, while the third mode requires at least some quality PH data (in our example, we wish laboratory-confirmed COVID-19 cases) and a model for the relationship between PH and WW data. For WW data to provide an added value beyond PH data, it needs to produce satisfactory predictions in the absence of PH data after some initial (or training) PH data have been used to produce a predictive model. Once a WW-based model has been established, it might compensate for future shortcomings in PH data. For example, there have been failures in PH testing laboratories and the exceedance of test capacity. In the longer term, one might allow that the WW model would suffice without any PH data.

Many recent papers (Ai *et al.* 2021; Feng *et al.* 2021; Greenwald *et al.* 2021; Melvin *et al.* 2021; Zulli *et al.* 2021; Wurtzer *et al.* 2022) show a correlation between the daily or weekly amount of COVID-19 RNA in WW and the numbers of COVID-19 infections reported to the PH authorities, but some do not go beyond showing a correlation. Certainly, the presence of a correlation indicates the potential for a prediction but substantial additional work is necessary before an operational model that provides numerical predictions can be produced. There is some debate about how much WW data can anticipate PH data, which relates to the goal of WW models providing an 'early warning system'.

The quality of predictions depends on the quality of WW measurements. Improving these measurements is a major focus of WBE research. There are many sources of variation that affect these measurements that are discussed in Alhama *et al.* (2021) and Wade *et al.* (2022). We accept that these measurements are subject to some variability. Our interest lies in the consequences of this variability.

We start with a description of the WW and PH data used in our running Scotland example. We discuss how WW and PH data are aligned and the issues with the use and quality of PH data. We look at the form of the relationship between PH and WW data and how this can vary over time. We consider the relative advantages of predicting with WW and PH data. We conclude with a discussion of how predictive models might continue to perform into the future.

DATA

The data for the main example come from Scotland from the beginning of September 2020 to the end of August 2021. Public Health Scotland has publicly provided daily reports of laboratory-confirmed COVID-19 cases broken down by the 14 Health Area Boards that cover all of Scotland – see Public Health Scotland (2021). The Scottish Environmental Protection Agency has collated and publicly provided data – see Scottish Environmental Protection Agency (2021) – from up to 110 WW treatment plants (WWTPs) in our data. Composite samples were drawn over 24-h periods. Samples were processed and analysed using RT-qPCR methods to produce a measure of SARS-CoV-2 RNA in gene copies per litre. Some WWTPs were sampled frequently but other WWTPs less densely and not for the whole period of this study. Further details will be provided in the analysis to follow.

Our purpose here is not to conclude anything about Scotland specifically, but to use this data as a representative of the type of data that might be collected anywhere. Data of this nature are subject to many uncertainties that will be present in any location. Our interest lies in the modelling process and the general lessons that may be learned.

In practice, one might also allow additional data such as flow and water quality data from WWTPs. Such data might improve the quality of the predictions but will not always be available in all locations. In this study, we restrict attention to only the case and RNA count data to focus on the modelling aspects.

We also consider data from the USA provided publicly by Biobot (2021) (see Duvallet *et al.* (2021) for an analysis partially based on this data). These data differ in that only weekly COVID-19 case data are available, and the WWTPs within each of the 30 counties available have already been combined. We use this data to confirm that the general conclusions drawn from the Scotland data apply elsewhere.

Registering WW and PH data

In our example data from Scotland, there are 14 PH areas reporting daily cases. These areas cover Scotland completely. We also have WW measures from 110 WWTPs. These WWTPs serve catchments that are usually (but not always) within a single PH area. Not all WW flows to a WWTP, particularly in rural areas, and not all WWTPs (particularly smaller ones) are sampled. Three sparsely populated areas have only one WWTP sampled, while one area has 17 WWTPs. In Glasgow, the largest city, almost all the population is served by a sampled WWTP, while as little as a third of the population in the more sparsely populated areas is served by a sampled WWTP. Overall, about 70% of the population is served by a WWTP included in the data. A WWTP serving a large population might be sampled several times a week across the whole period of the study, while smaller WWTPs might be sampled less frequently and only in the later part of the study. Although reporting and measurement practices vary across the world, the challenges seen in our example data are likely to arise in many locations. In less wealthy countries, less information is likely to be available.

We have taken a simple approach to aligning the WW to the PH data.

1. For each WWTP, during the period for which observations are available, we use LOCF (last observation carried forward, i.e., the estimate does not change until a new observation becomes available) to create a daily estimated WW measure. We avoid interpolation because, when making predictions, we will know only past values and not the future value.
2. For each day within each area, we computed a weighted average of the WWTPs where the weights are equal to the population served by the WWTP.
3. Reporting of cases is sensitive to the day of the week. We use a rolling weekly mean to avoid day-of-the-week effects. The rolling mean uses only the past seven values and no future values.
4. In processing both the WW and PH data, we have taken care to avoid using data from the future, but this also results in some small lag to both estimates.

One might reasonably suggest improvements to this procedure. We might use Geographic Information Systems data to better allocate WWTPs across PH areas. We might use other methods of integrating data from the recent past. But the point is that uncertainty will unavoidably be added, and the nature of the data will limit the localisation and confidence in our predictions.

There are time lags in both the WW and PH data processes. The amount of RNA deposited in WW varies over time within an individual as determined by the faecal shedding profile and between individuals – see Hoffmann & Alsing (2021). When considering data from a narrow locality such as a school or hospital, it is necessary to model these effects if we wish to estimate the number of infected individuals. For the larger areas we are considering, these effects are averaged over many individuals, and it is not feasible to deconvolve the individual profiles. We do not consider the faecal shedding profile in the models considered later for this reason, but there will be a time lag between infection and measurable RNA appearing at the WWTP. Similarly, PH reporting will experience time lags due to the delay in the appearance of symptoms and the processing and reporting of PCR test results. We have not modelled either of these lags in the models to follow.

Issues with PH data

WW and PH data both measure the quantity of disease present in the population. We have acknowledged the variability in WW measurements, but PH measures have their own set of problems.

Although WW measures prevalence, many studies report the correlation between RNA measures and incidence. Incidence relates to new cases, whereas prevalence relates the number of currently infected individuals. RNA concentrations measure prevalence but imperfectly. Governments tend to be more interested in incidence, particularly large increases that might motivate interventions, so there is impetus to make incidence the target of predictive modelling. One can attempt to estimate the prevalence by integrating the incidence with an appropriate weighting function that takes into account the length of infection and other factors. By the same reasoning, incidence can be regarded as a form of derivative of the prevalence function and is inherently more difficult to predict. For the purposes of this study, we use the 7-day rolling mean of the daily incidence. This has the advantage of resolving the day-of-the-week variation and the use of the mean results in a measure that is somewhere between incidence and prevalence. This will be the target of our predictive model and we will call it the *case rate*.

One major goal of WBE is to replace more expensive and intrusive PH measures of SARS-CoV-2 (or other diseases) with less-expensive WW measures. To build a predictive model for case rates, we need a training set with reliable PH measures. This may be difficult to obtain, particularly when our goal is to avoid the necessity of collected PH measures in the longer term.

In the UK, symptomatic individuals and, sometimes, close contacts are strongly encouraged to take a PCR test and report the results to the PH authorities. Unfortunately, from a modelling perspective, some individuals will be asymptomatic and others will not be compliant. Furthermore, there have been problems with some commercial PCR testing laboratories. Independently, the UK's Office of National Statistics (ONS) takes weekly random samples from the population to estimate the proportion who are currently infected. This survey is small in scale and can only produce reliable estimates over large areas – in this case, all of Scotland. It cannot be used alone to provide the localised information we require for our prediction model. Even so, we can use it to check the PH data.

We compute the ratio of the weekly number of cases across all Scotland to the weekly estimated number from the ONS infection survey. This ratio is shown in Figure 1 as it varies over time. We see that the ratio reaches a maximum value of about 0.35. We expect that the ratio will be less than one because infected individuals will only report at most once in the PH data but will test positive over more than 1 week for infection survey purposes. Furthermore, the infection survey will also capture asymptomatic cases. We could develop a method to estimate the prevalence from the PH data, but there is a more serious problem. The ratio is not constant over time. A plausible explanation is that PH reporting improved over the time period, but other explanations are possible.

Any proposed prediction model will need to address this variable bias in the PH measure. In this location, we have the benefit of two independent measures of the case rate, but this will not be available in general (it is not present in our secondary USA example).

Case rates are not proportional to RNA measures

We start with a focus on the data from the largest city in Scotland, Glasgow, which has the densest data across the available areas as seen in Figure 2. Both the PH and the WW measures are caused by the true number of infected individuals. The PH

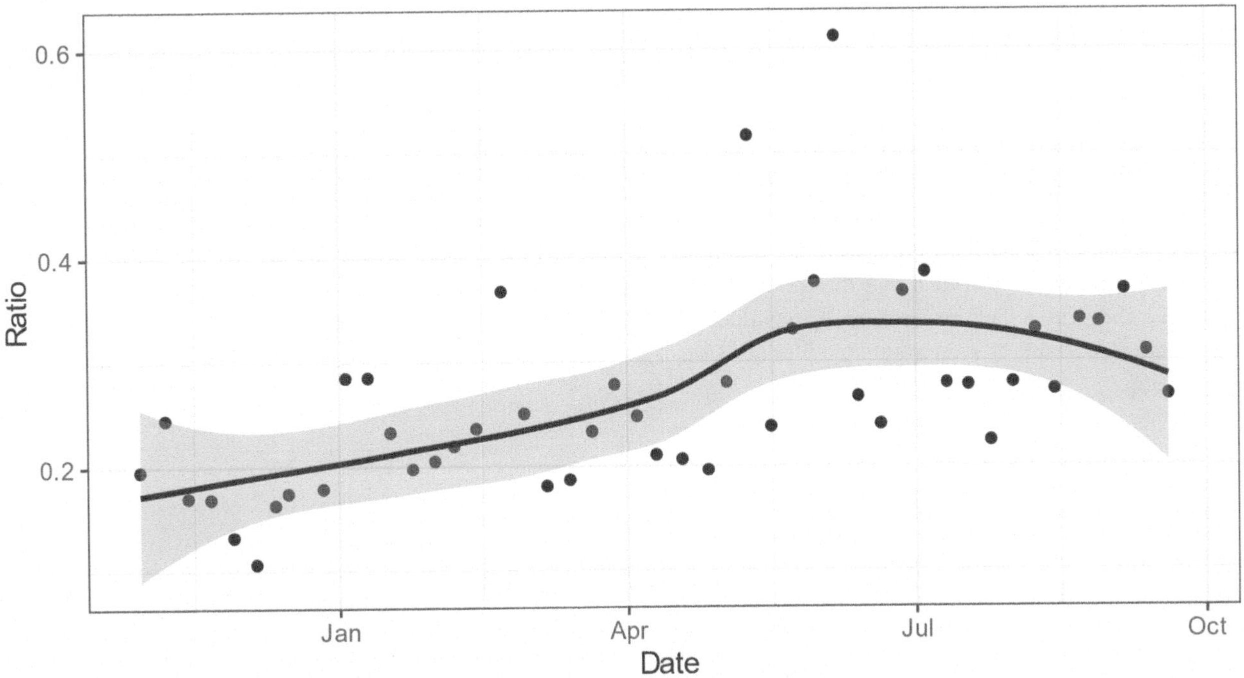

Figure 1 | Ratio of reported COVID-19 cases to cases estimated by the ONS infection survey.

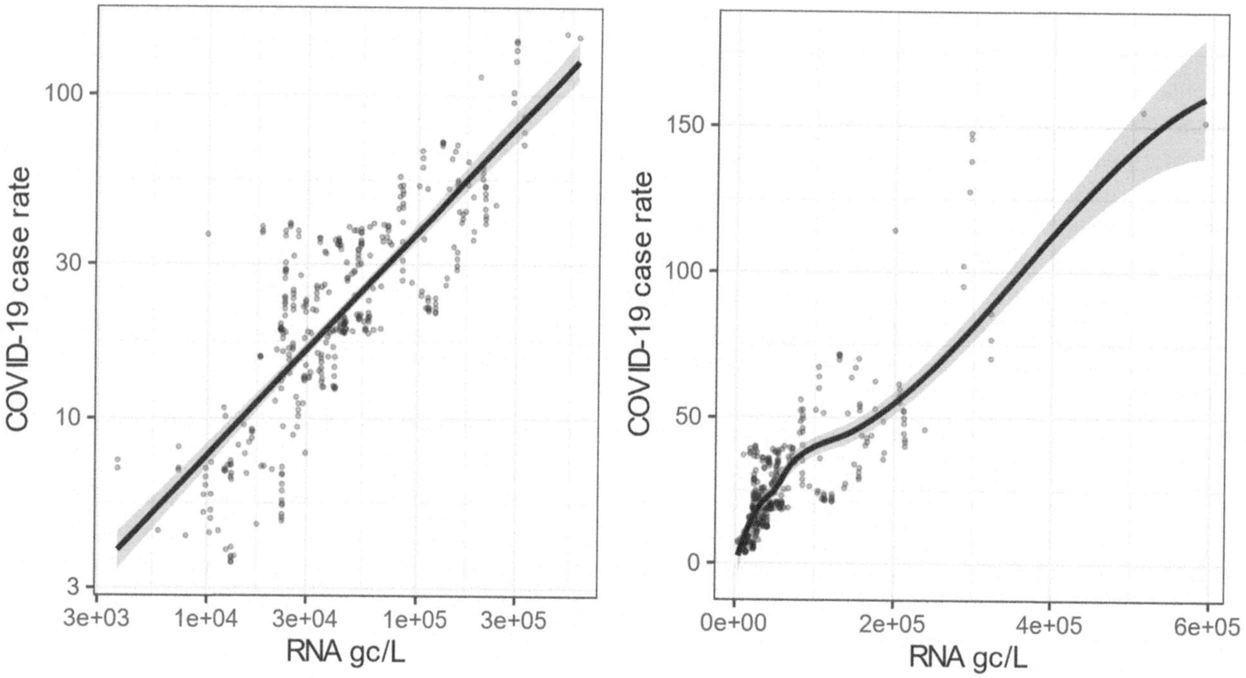

Figure 2 | SARS-CoV2 RNA and COVID-19 case rates in Glasgow. The case rate is the number infected per 100,000 population (as defined earlier). A log-scale is used on the left with a linear fit and 95% confidence band. On the right, the same data are shown untransformed and with smoothed LOESS fit.

measure is less variable but more expensive to obtain, while the WW measure is more variable but less expensive. The interest lies in predicting the PH measure from the WW measure because we envision a future use where only the WW measure is available. Furthermore, the PH measure is in the units we need, while the WW measure is an RNA concentration, which is not directly useful. For these reasons, our models treat the RNA measure as the predictor and the case rate as the response.

There are strong biological reasons to expect that cases should be proportional to measured RNA. Let c be cases (per 100,000 population) and g be gene copies per litre. By the proportionality assumption, these are linked by:

$$c = pg \tag{1}$$

Ideally, one would determine the constant p by some calibration process that would depend on the locality. We can then measure g and directly estimate c. Naturally, we expect some measurement errors. We expect that other variables, such as flow rates or water quality indicators, will add variation. If we wish to use WBE to estimate disease prevalence when PH data are not available, it is critical that this underlying linear relationship holds.

Because c and g often vary over orders of magnitude, it is understandably common to plot them on a log-scale (the base of the log is not important) as seen in the first plot of Figure 2. An approximate linear relationship is seen, as is typical in such plots, like so:

$$\log c = a + b \log g \tag{2}$$

This linearity on the log-scale becomes $c = pg^b$ on the untransformed scale where $a = \log p$. Unless $b=1$ (which we would prefer), the relationship is non-linear. In the Glasgow data, we can estimate b from the slope as 0.69 with a standard error (SE) of 0.026. We see that $b<1$, so that a convex curvilinear relationship is fitted. Of all the 14 areas in this data, this is the highest estimate of b. Furthermore, when we fit a smoothed curve to the untransformed data as seen in the second plot of the figure, we do not even see the expected power curve relationship. We will see later that this is typical and not unique to Glasgow.

How can we address this situation? As we shall see later, there is a time-varying relationship that may explain at least part of the problem. Furthermore, if we had additional auxiliary data, we might be able to normalise (or correct) the measurements to restore a $c = pg$ relationship.

An additional concern is the classic errors-in-variables or measurement error problem as described in Fuller (1987). Both c and g are subject to a substantial measurement error, which leads to some complications with the estimation of p. There are well-established methods for working with this, which we do not pursue further here.

There remains a possibility that even after these corrections and adjustments, the proportional relationship is not restored. This would make predictive modelling substantially more difficult. We return to this issue in the discussion.

For very low or zero counts, there are difficulties in using the log transformation. There are also complications when the RNA measure approaches the limits of quantification and detection. Some modification to the modelling approach is necessary, but these complications do not arise with the Glasgow data due to the high population and prevalence of COVID-19.

Locality of the relationship between case rates and RNA measures

Figure 3 shows the same information as in the right panel of Figure 2 but for four representative areas. We see that the shape of the relationship between case rates and RNA measures varies by locality. The weakest relationship is seen in Orkney, which is the least populated area shown. The heterogeneity in the relationship carries over to the areas not shown. The relationship varies in ways that would be difficult to parameterise. We can make a prediction for these locations, but we would find this difficult in new locations without substantial PH data.

Let us suppose that, with appropriate normalisation and improvements in measurement, we can get closer to linearity on a log-scale. As is, the data already show a roughly approximate linear relationship on the log-scale. Consider a linear model on this log-scale, where the parameters vary by area i:

$$\log c = a_i + b_i \log g \tag{3}$$

In fitting this model, we find that the values of a and b vary significantly by i. The estimated values of the slope b are always well less than 1, with more populated areas having $b \approx 0.65$, while the less populated areas show lower values indicating a weaker relationship in those areas. In these areas, the WWTP coverage is lower, fewer samples are taken, and the infection rates tend to be lower due to the sparser population. The intercept a_i represents the amount RNA measured per infected

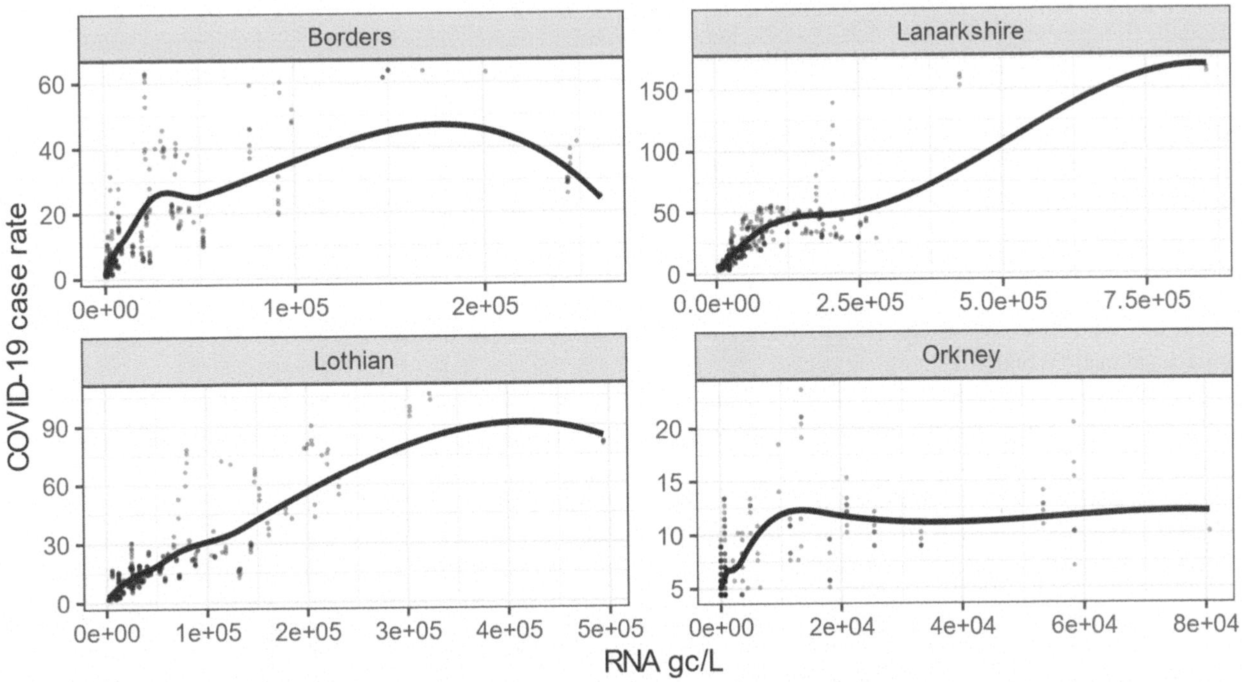

Figure 3 | Relationship between SARS-CoV2 RNA and COVID-19 case rates in four areas of representative areas of Scotland.

individual, but these show substantial variation. There are several plausible reasons for this variation. The amount of time RNA spends in the sewerage will vary according to the locality and other environmental factors that will affect the rate of decay.

Provided these parameters remain constant in time, we might use training data to build a model to predict future case rates based on future WW measures. A more problematic situation would arise where we introduce WBE to a new location where we lack the training data we have in this example. Can these parameters be determined without data? How would this work for a new virus or other target? We might attempt some physical modelling of the amount of RNA (or other output) to be expected per infected person that would be sampled at a given WWTP, given knowable characteristics of the catchment and necessary environmental information. This was attempted in Morvan *et al.* (2021), but without much success. Progress in this area is necessary to cope with situations where little or no PH information is available, or the target is new.

Aggregation and prediction uncertainty

Suppose we fit the linear model to the logged Glasgow-only data as described previously. We use this model to predict the case rate based on the average RNA measure observed across all of Scotland for the entire period, which is 56,300 gc/L. Predicting the case rate for this central value of RNA is an easier task than a prediction for a very low or very high value of RNA. The predicted number of cases is 25.3 per 100,000 with a 95% prediction interval of (10.6, 60.4). The width of the interval reflects the uncertainty. For decision-making purposes, we would prefer this to be narrower.

We can reduce the prediction uncertainty by aggregating over all of Scotland and use this aggregate data to fit the linear model. We compute population-weighted averages of both PH and WW data over all the areas and fit the same model. Under this model, the predicted case rate for the same RNA measurement of 56,300 gc/L is 21.3, with a 95% prediction interval of (11.6, 39.1). Thus, fitting a model based on aggregated data reduces the variation, allowing us to obtain a much narrower prediction interval, relatively speaking.

In contrast, should we wish to predict cases within the catchment of a single WWTP and even if we were able to specify suitable parameters for the catchment area of that WWTP, we would obtain a prediction interval considerably wider than the Glasgow interval.

We have used a simple but transparent modelling approach. We might improve the prediction uncertainty with some more sophisticated modelling, but the natural variation in the data, as seen in the figures, limits the precision of the predictions. The use of auxiliary data not available to us here might also improve the predictions. The acceptable level of uncertainty would depend on the intended use of these predictions.

Time-varying relationships

We have already seen that the PH reporting of cases varies over time. Now we look at how the relationship between PH and WW measures changes over time. Consider the varying coefficient model:

$$c(t) = a_i(t) + b(t)g(t) \tag{4}$$

where the intercept coefficient functions $a_i(t)$ vary with area i, but the slope coefficient $b(t)$ is the same across areas. Variations on this choice are reasonable, but the essential conclusion would be the same. These models can be fit using the mgcv package in R (Wood 2017).

In Figure 4, we can see that the $a(t)$ parameters change over time. We show only four of the areas in this plot, but the other 10 are qualitatively similar. The larger the value, the less RNA is measured per infected individual and so smaller values are preferable. We can see that there were large declines in the early part of 2021 indicating a greater recovery of RNA, but this improvement was reversed later in the year.

The estimated slope coefficient $b(t)$ is seen in Figure 5. We see that this varies over time and does not approach the desired value of one. This indicates that the non-proportionality continues throughout the study period while varying in its form.

Why do we see this variability in the relationship? We saw earlier that PH reporting improved during the period of the study, but this works in opposition to the trends seen in the $a(t)$. Several explanations are plausible.

In Figure 6, we see that the number of samples drawn per month increased sharply in early 2021 and fell off again later in the year. This is one plausible explanation for why coefficient function varies in the manner seen previously. More samples

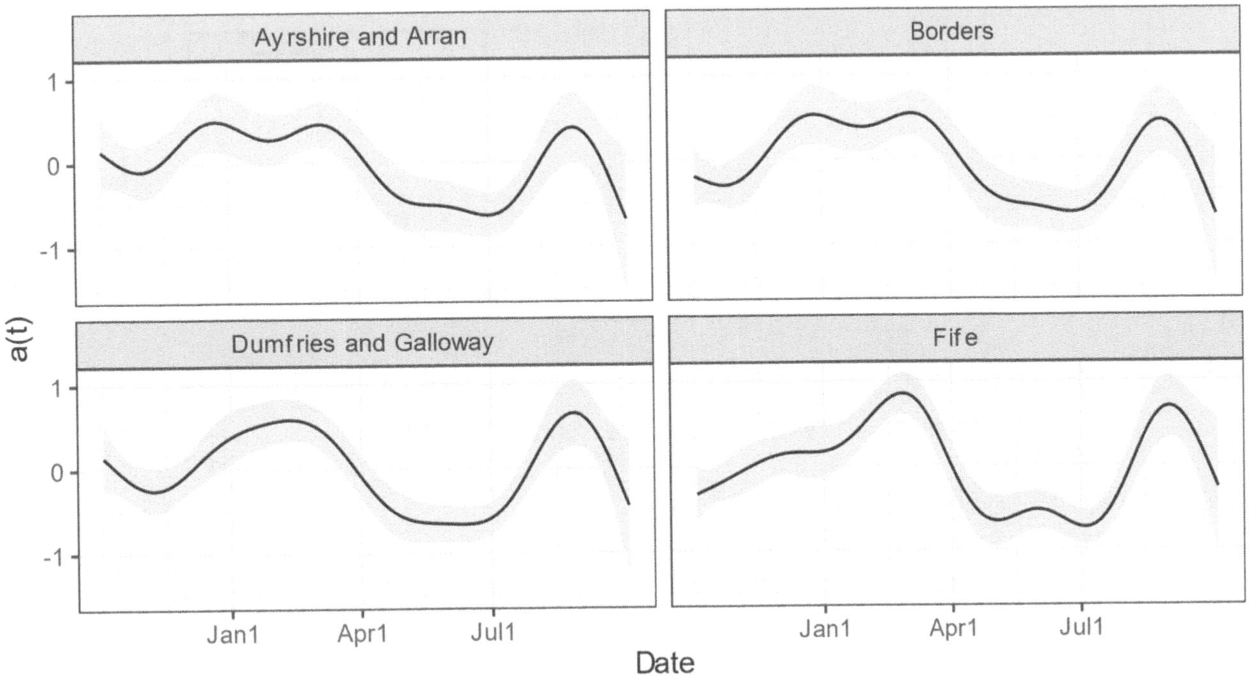

Figure 4 | Estimated *a*(*t*) for four selected areas.

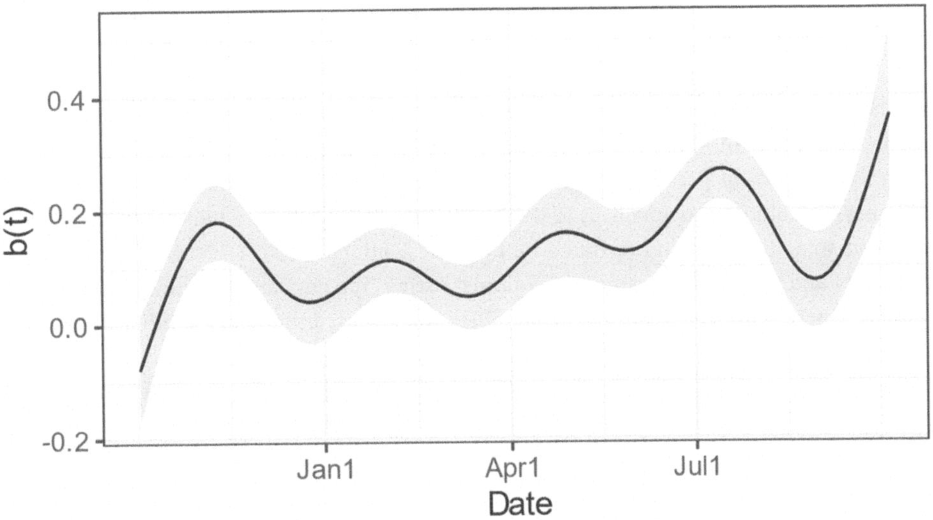

Figure 5 | Estimated *b*(*t*) for the time-varying model.

would have resulted in more accurate and less variable estimates of the area of RNA concentrations, which would have strengthened the correlation with observed cases. However, there are several other reasons why these functions might vary:

- Variations in PH reporting (as seen in this and the USA data);
- Variations in laboratory methods (these have improved over the period);
- Variations in faecal shedding possibly due to increasing vaccination or SARS-CoV-2 variants;
- Environmental variations such as agricultural runoff or weather, for example.

The time-varying relationship means that it will be difficult to build predictive WBE systems for the long term, and continuing updates, requiring the collection of further PH data, may be necessary.

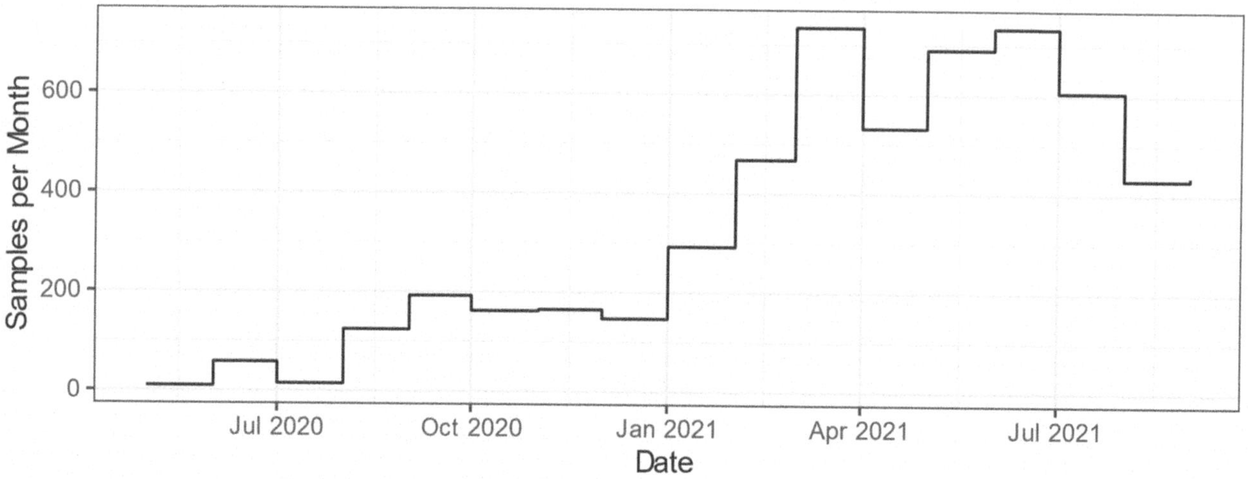

Figure 6 | Number of WW samples taken each month.

Predictive ability of WW measurements in comparison with PH measures

How well does WW data predict PH case rates? First, we consider how well WW data can predict future PH data. Consider lagged models of the form:

$$\log c(t) = a_i + b_i \log g(t - j) \quad \text{for} \quad j = 0, 1, \ldots 7 \tag{5}$$

We report R^2 and SE of the models in the table below. These are in-sample estimates, but the model class is quite simple. We would expect the out-of-sample performance to be a little worse.

In Table 1, we see that the fit of the model is best at lag=0, concurrent prediction and that the greater the lag, the worse the fit of the model. This indicates that although we can use WW to predict cases further ahead in time, the quality of these predictions is worse than the concurrent prediction. There is some variation in the findings in published studies. Some claim improved prediction ahead of time by as much as 2 weeks (Melvin *et al.* 2021), while others make more modest claims (Morvan *et al.* 2021), and some (like us) do not find a look-ahead (Zulli *et al.* 2021). New infections take varying amounts of time to appear in the PH and WW systems, and as mentioned earlier, the data analysis induces some lag on both variables. With adequate resourcing, there is no strong evidence that either system can anticipate the other by any significant number of days.

Now consider using PH data to predict PH data in the future. If we fit a model of the form where we predict 1 week ahead:

$$\log c(t) = a_i + b_i \log c(t - 7) \tag{6}$$

Table 1 | Performance of lagged prediction models

Lag in days	R^2	Standard error
0	0.715	0.526
1	0.702	0.536
2	0.687	0.548
3	0.671	0.561
4	0.653	0.575
5	0.634	0.589
6	0.615	0.603
7	0.596	0.617

we find an R^2 of 0.845 and an SE of 0.383. This is better than even the concurrent-based WW prediction model. This model is very simplistic in that we are essentially using last week's cases to predict this week's cases. Certainly, we could use forecasting techniques to do better than this.

Do WW measurements improve this last model? We consider the model:

$$\log c(t) = a_i + b_i \log c(t-7) + d_i \log g(t-7) \tag{7}$$

This results in an R^2 of 0.851 and an SE of 0.375. This is a negligible improvement in predictive ability.

These results suggest that if you have good PH data, you do not gain anything with additional WW data. If you do not have PH data, you can use WW data to predict PH data, but it works best concurrently.

Our model framing with the PH measure as the response and the WW predictor carries the assumption that the PH measure is more reliable. If this assumption is correct and the PH data continue to be available, WW measures add little. But if PH measures are not reliable or do not continue to be available, the WW measures become valuable. Also, there is the possibility that PH measures are not consistently reliable due to testing failure (BBC 2021), in which case the WW measures provide an alternative estimate of disease prevalence.

USA comparison

To assess the generalisability of our findings to other contexts, we repeated the analysis data from the USA provided publicly by Biobot (2021). Similar issues were seen in the USA data.

As seen in the top two plots of Figure 7, the relationship between cases and RNA appears approximately linear on logged scales, but the non-linearity is apparent on the untransformed scale. We show only Indiana county, Pennsylvania, here, but this is typical of the 30 counties in the data. There is greater heterogeneity in the amount of RNA observed per case in the USA data compared to Scotland, which reflects the greater geographical spread in the USA, differences in WW infrastructure, and variation in PH testing and reporting.

Time-varying relationships were also seen when the same model was fit to the USA as can be seen in the lower two plots of Figure 7. Again, greater heterogeneity was seen in the USA with no consistent patterns seen in the $a(t)$.

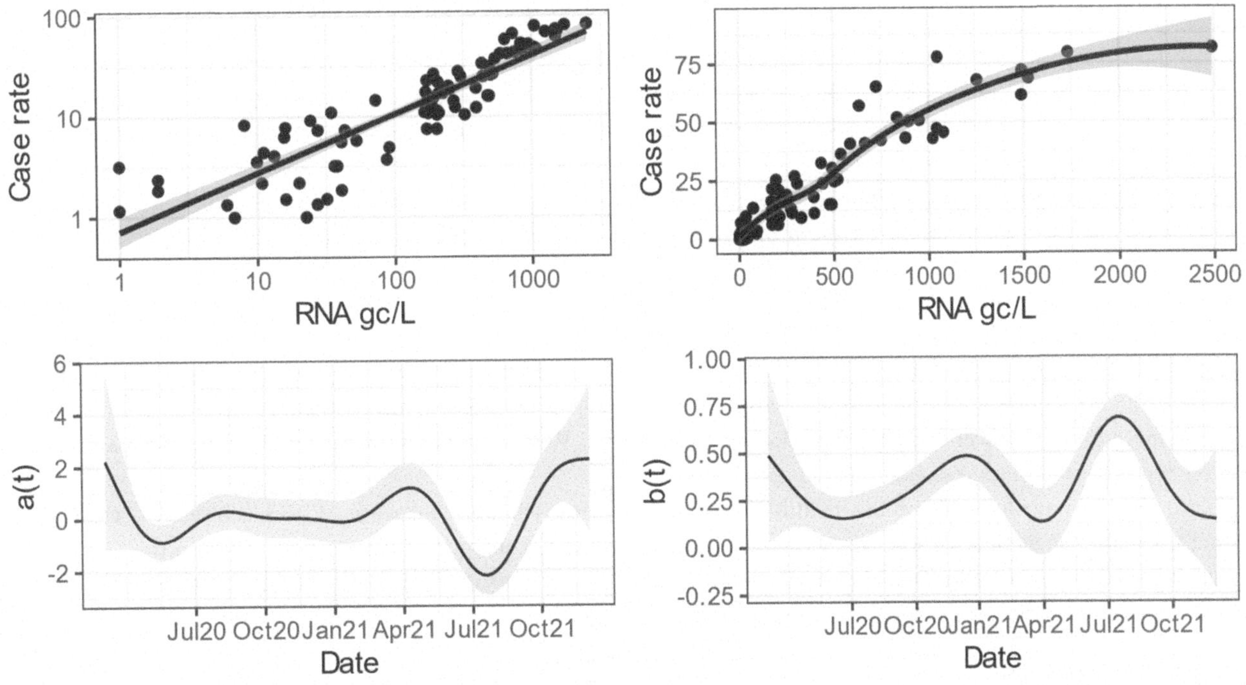

Figure 7 | Top two plots show relationship between cases and RNA for Indiana county, PA, first on a logged scale and then on the untransformed scale. The bottom two plots show estimates from the time-varying model for the USA data. The lower left plot is the estimate $a(t)$ for Indiana county, while the lower right plots show the estimated $b(t)$.

There was less scope to try lagged models with the USA data because of the weekly data. For a concurrent model predicting cases using RNA as in Equation (5), the value of R^2 was 0.681 and the SE was 0.587, while for the week-lagged model, the value of R^2 was 0.470 and the SE was 0.755. As with the Scotland data, concurrent prediction was best. The fit was just slightly worse for the USA data. Again, the WW data did not help a PH-only forecasting model.

DISCUSSION

When building a predictive model, it is not sufficient to fit the present data well. It is necessary to consider how well the model will perform in the future. We consider three scenarios.

1. Large amounts of PH data continue to be available. We will be able to update our models continuously. We will be free to use complex models that require substantial PH data to estimate. This will be particularly valuable when a time-varying relationship persists. Some recommendations for the updating of prediction models may be found in Jenkins *et al.* (2021). With both PH- and WW-based estimates of disease prevalence, we will have some insurance against anomalies that may arise in either reporting system. This is the most expensive scenario since we must resource two systems of measurement.
2. A small amount of occasional PH data is available. Suppose we can normalise or adjust our WW measurements, such that our RNA concentration g follows the proportional relationship $c=pg$, where c is the case rate as discussed in Equation (1). A single observation of c obtained from a brief infection survey would suffice to estimate p. This scenario does not require that we maintain extensive PH testing infrastructure.
3. No PH data are available. This scenario would arise for a disease or location for which we have not been collecting data or is new. We would need to match this disease with a similar disease for which we know p. For example, if we have SARS-CoV-3, we can use our knowledge of SARS-CoV-2.

How can we improve the process? We can improve the WW measurement process in how the samples are drawn and analysed. With current levels of research, we can expect reductions in variation, which will result in more precise predictions. We can also increase sampling which will also reduce variation through smoothing and averaging, but this will also increase the cost. Beyond these desirable measurement process improvements, there is a strategic modelling choice.

We can build complex statistical or machine learning models that require large amounts of WW and PH data to setup. Along with helpful auxiliary environmental variables, we may achieve strongly predictive models. The drawback of this approach is that it requires substantial PH data. An alternative is to build a stronger understanding of the sewerage system by which individuals deposit quantities of RNA (or other substances), while lesser quantities of RNA are sampled at the WWTP. This requires physical models that describe the effects of environmental variables such as water quality or flow rates. Better methods of normalisation using biochemical markers would be valuable. Although promising, crAssphage has not been a clear success – see Ai *et al.* (2021) and Greenwald *et al.* (2021). The USA data discussed here used pepper mild mottle virus for normalisation, but the results were not better than the unnormalised Scottish data. If we could obtain reliable measures of g that follow the proportional $c=pg$ relationship, calibrating the relationship to case rates will not require much PH data.

We have focussed on SARS-CoV-2, which is a relatively difficult target but it has generated much useful data. Some diseases are endemic and vary slowly – these targets are easily monitored using WBE. Other diseases, such as influenza, are mostly present at low levels but have occasional larger outbreaks. WBE can be useful here, particularly in the absence of much PH reporting. Reductions in the variation in WW measurement will help quickly distinguish real large changes in case rates from noise. The monitoring of novel disease outbreaks using established WBE infrastructure, before PH reporting can be widely deployed, requires some progress as discussed above.

In this article, we have focussed on the quantitative monitoring or prediction of disease. WBE has already been successful for other uses and has achieved good progress with the quantification problem. Further improvements in the measurement process and advanced understanding of how RNA and other substances progress through the sewerage system will enable more precise and less-expensive estimates and predictions of the disease.

ACKNOWLEDGEMENTS

We are grateful to the Scottish government and Biobot for making the data available. Funding from the UKRI (EP/V028499/1) is greatly appreciated.

DATA AVAILABILITY STATEMENT

Code and links to data are available from https://github.com/julianfaraway/wbequant.

CONFLICT OF INTEREST STATEMENT

The authors declare there is no conflict.

REFERENCES

Ai, Y., Davis, A., Jones, D., Lemeshow, S., Tu, H., He, F., Ru, P., Pan, X., Bohrerova, Z. & Lee, J. 2021 Wastewater-based epidemiology for tracking COVID-19 trend and variants of concern in Ohio, United States. medRxiv: 2021.2006.2008.21258421. Preprint.

Alhama, J., Maestre, J. P., Martín, M. Á. & Michán, C. 2021 Monitoring COVID-19 through SARS-CoV-2 quantification in wastewater: progress, challenges and prospects. *Microbial Biotechnology* **15** (6), 1719–1728.

Baz-Lomba, J. A., Salvatore, S., Gracia-Lor, E., Bade, R., Castiglioni, S., Castrignanò, E., Causanilles, A., Hernandez, F., Kasprzyk-Hordern, B. & Kinyua, J. 2016 Comparison of pharmaceutical, illicit drug, alcohol, nicotine and caffeine levels in wastewater with sale, seizure and consumption data for 8 European cities. *BMC Public Health* **16** (1), 1–11.

BBC 2021 *Covid Test Lab in Wolverhampton Suspended Over Wrong Results*. Available from: https://www.bbc.co.uk/news/health-54923641. (Accessed 13 June 2022).

Biobot 2021. Available from: https://biobot.io/data/. (Accessed 13 June 2022).

Castiglioni, S., Senta, I., Borsotti, A., Davoli, E. & Zuccate, E. 2015 A novel approach for monitoring tobacco use in local communities by wastewater analysis. *Tobacco Control* **24** (1), 38–42.

Causanilles, A., Rojas Cantillano, D., Emke, E., Bade, R., Baz-Lomba, J. A., Castiglioni, S., Castrignanò, E., Gracia-Lor, E., Hernández, F., Kasprzyk-Hordern, B., Kinyua, J., McCall, A.-K., van Nuijs, A. L. N., Plósz, B. G., Ramin, P., Rousis, N. I., Ryu, Y., Thomas, K. V. & de Voogt, P. 2018 Comparison of phosphodiesterase type V inhibitors use in eight European cities through analysis of urban wastewater. *Environment International* **115**, 279–284.

Duvallet, C., Wu, F., McElroy, K. A., Imakaev, M., Endo, N., Xiao, A., Zhang, J., Floyd-O'Sullivan, R., Powell, M. M., Mendola, S., Wilson, S. T., Cruz, F., Melman, T., Sathyanarayana, C. L., Olesen, S. W., Erickson, T. B., Ghaeli, N., Chai, P., Alm, E. & Matus, M. 2021 Nationwide trends in COVID-19 cases and SARS-CoV-2 wastewater concentrations in the United States. medRxiv: 2021.2009.2008.21263283. Preprint.

Escolà Casas, M., Schröter, N. S., Zammit, I., Castaño-Trias, M., Rodriguez-Mozaz, S., Gago-Ferrero, P. & Corominas, L. 2021 Showcasing the potential of wastewater-based epidemiology to track pharmaceuticals consumption in cities: comparison against prescription data collected at fine spatial resolution. *Environment International* **150**, 106404.

Feng, S., Roguet, A., McClary-Gutierrez, J. S., Newton, R. J., Kloczko, N., Meiman, J. G. & McLellan, S. L. 2021 Evaluation of sampling, analysis, and normalization methods for SARS-CoV-2 concentrations in wastewater to assess COVID-19 burdens in Wisconsin Communities. *ACS ES&T Water* **1** (8), 1955–1965.

Fuller, W. A. 1987 *Measurement Error Models*. Wiley, New York.

Greenwald, H. D., Kennedy, L. C., Hinkle, A., Whitney, O. N., Fan, V. B., Crits-Christoph, A., Harris-Lovett, S., Flamholz, A. I., Al-Shayeb, B., Liao, L. D., Beyers, M., Brown, D., Chakrabarti, A. R., Dow, J., Frost, D., Koekemoer, M., Lynch, C., Sarkar, P., White, E., Kantor, R. & Nelson, K. L. 2021 Tools for interpretation of wastewater SARS-CoV-2 temporal and spatial trends demonstrated with data collected in the San Francisco Bay Area. *Water Research X* **12**, 100111.

Hoffmann, T. & Alsing, J. 2021 Faecal shedding models for SARS-CoV-2 RNA amongst hospitalised patients and implications for wastewater-based epidemiology. medRxiv: 2021.2003.2016.21253603. Preprint.

Jenkins, D. A., Martin, G. P., Sperrin, M., Riley, R. D., Debray, T. P. A., Collins, G. S. & Peek, N. 2021 Continual updating and monitoring of clinical prediction models: time for dynamic prediction systems? *Diagnostic and Prognostic Research* **5** (1), 1.

Kasprzyk-Hordern, B., Proctor, K., Jagadeesan, K., Lopardo, L., O'Daly, K. J., Standerwick, R. & Barden, R. 2021 Estimation of community-wide multi-chemical exposure via water-based chemical mining: key research gaps drawn from a comprehensive multi-biomarker multi-city dataset. *Environment International* **147**, 106331.

Melvin, R. G., Chaudhry, N., Georgewill, O., Freese, R. & Simmons, G. E. 2021 Predictive power of SARS-CoV-2 wastewater surveillance for diverse populations across a large geographical range. medRxiv: 2021.2001.2023.21250376. Preprint.

Morvan, M., Lo Jacomo, A., Souque, C., Wade, M. J., Hoffmann, T., Pouwels, K., Singer, A., Bunce, J. T., Engeli, A., Grimsley, J. M. S., O'Reilly, K. & Danon, L. 2021 Estimating SARS-CoV-2 prevalence from large-scale wastewater surveillance: insights from combined analysis of 44 sites in England. Preprint.

O'Brien, J. W., Choi, P. M., Li, J., Thai, P. K., Jiang, G., Tscharke, B. J., Mueller, J. F. & Thomas, K. V. 2019 Evaluating the stability of three oxidative stress biomarkers under sewer conditions and potential impact for use in wastewater-based epidemiology. *Water Research* **166**, 115068.

Ort, C., Banta-Green, C. J., Bijlsma, L., Castiglioni, S., Emke, E., Gartner, C., Kasprzyk-Hordern, B., Reid, M. J., Rieckermann, J. & Van Nuijs, A. L. 2014 Sewage-based epidemiology requires a truly transdisciplinary approach. *GAIA – Ecological Perspectives for Science and Society* **23** (3), 266–268.

Public Health Scotland 2021. Available from: https://www.opendata.nhs.scot/dataset/weekly-covid-19-statistical-data-in-scotland. (Accessed 13 June 2022).

Reid, M. J., Langford, K. H., Mørland, J. & Thomas, K. V. 2011 Analysis and interpretation of specific ethanol metabolites, ethyl sulfate, and ethyl glucuronide in sewage effluent for the quantitative measurement of regional alcohol consumption. *Alcoholism: Clinical and Experimental Research* **35** (9), 1593–1599.

Ryu, Y., Gracia-Lor, E., Bade, R., Baz-Lomba, J. A., Bramness, J. G., Castiglioni, S., Castrignanò, E., Causanilles, A., Covaci, A., de Voogt, P., Hernandez, F., Kasprzyk-Hordern, B., Kinyua, J., McCall, A.-K., Ort, C., Plósz, B. G., Ramin, P., Rousis, N. I., Reid, M. J. & Thomas, K. V. 2016 Increased levels of the oxidative stress biomarker 8-iso-prostaglandin F2α in wastewater associated with tobacco use. *Scientific Reports* **6** (1), 39055.

Scottish Environmental Protection Agency 2021. Available from: https://informatics.sepa.org.uk/RNAmonitoring/. (Accessed 13 June 2022).

Thomas, K. V., Bijlsma, L., Castiglioni, S., Covaci, A., Emke, E., Grabic, R., Hernandez, F., Karolak, S., Kasprzyk-Hordern, B., Lindberg, R. H., Lopez de Alda, M., Meierjohann, A., Ort, C., Pico, Y., Quintana, J. B., Reid, M., Rieckermann, J., Terzic, S., van Nuijs, A. L. & de Voogt, P. 2012 Comparing illicit drug use in 19 European cities through sewage analysis. *Science of the Total Environment* **432**, 432–439.

Venhuis, B. J., Zwaagstra, M. E., Keizers, P. H. & de Kaste, D. 2014 Dose-to-dose variations with single packages of counterfeit medicines and adulterated dietary supplements as a potential source of false negatives and inaccurate health risk assessments. *Journal of Pharmaceutical and Biomedical Analysis* **89**, 158–165.

Wade, M. J., Lo Jacomo, A., Armenise, E., Brown, M. R., Bunce, J. T., Cameron, G. J., Fang, Z., Farkas, K., Gilpin, D. F., Graham, D. W., Grimsley, J. M. S., Hart, A., Hoffmann, T., Jackson, K. J., Jones, D. L., Lilley, C. J., McGrath, J. W., McKinley, J. M., McSparron, C., Nejad, B. F., Morvan, M., Quintela-Baluja, M., Roberts, A. M. I., Singer, A. C., Souque, C., Speight, V. L., Sweetapple, C., Walkers, D., Watts, G., Weightman, A. & Kasprzyk-Hordern, B. 2022 Understanding and managing uncertainty and variability for wastewater monitoring beyond the pandemic: lessons learned from the United Kingdom national COVID-19 surveillance programmes. *Journal of Hazardous Materials* **424**, 127456.

Wood, S. N. 2017 *Generalized Additive Models: An Introduction with R*. CRC Press/Taylor & Francis Group, Boca Raton.

Wurtzer, S., Waldman, P., Levert, M., Cluzel, N., Almayrac, J. L., Charpentier, C., Masnada, S., Gillon-Ritz, M., Mouchel, J. M., Maday, Y. & Boni, M. 2022 SARS-CoV-2 genome quantification in wastewaters at regional and city scale allows precise monitoring of the whole outbreaks dynamics and variants spreading in the population. *Science of the Total Environment* **810**, 152213.

Zulli, A., Pan, A., Bart, S. M., Crawford, F. W., Kaplan, E. H., Cartter, M., Ko, A. I., Cozens, D., Sanchez, M., Brackney, D. E. & Peccia, J. 2021 Predicting daily COVID-19 case rates from SARS-CoV-2 RNA concentrations across a diversity of wastewater catchments. *FEMS Microbes* **2**, xtab022.

First received 14 January 2022; accepted in revised form 27 May 2022. Available online 11 June 2022

doi: 10.2166/wh.2022.223

Prevalence of enteric viruses in wastewater in Egypt after the COVID-19 pandemic

Dina Nadeem Abd-Elshafy [a,b,*], Rola Nadeem [b,c] and Mahmoud Mohamed Bahgat [b,c]

[a] Environmental Virology Laboratory, Department of Water Pollution Research, Institute of Environmental Research and Climate Changes, The National Research Centre, Giza, Egypt
[b] Research Group Immune- and Bio-markers for Infection, The Centre of Excellent for Advanced Science, The National Research Centre, Giza, Egypt
[c] Department of Therapeutic Chemistry, Institute of Pharmaceutical and Drug Industries Research, The National Research Centre, Dokki, Giza, Egypt
*Corresponding author. E-mail: dnanadeem@yahoo.com

DNA, 0000-0002-2777-2487; RN, 0000-0002-4093-2270; MMB, 0000-0002-6485-8722

ABSTRACT

People's hygienic habits greatly affect the spreading rate of enteric viruses. After the COVID-19 pandemic, many people followed announced precautions and improved their hygienic status to protect themselves from SARS-CoV-2 infection. Here, we studied if this indirectly affected the prevalence of enteric viruses in Egypt. A total of 21 samples (one sample per week) were collected from the Zenin wastewater treatment plant (WWTP) through the period between August 2021 and March 2022. Detection of adenovirus, hepatitis A virus (HAV), and rotavirus showed their presence in 66, 14.3, and 9.5% of the collected samples, respectively. Comparing those percentages to previously published data concerned with the detection of the same viruses from the same WWTP or others revealed a remarkable decrease in the prevalence of the three viruses after the COVID-19 pandemic. This allows the conclusion that safety precautions against SARS-CoV-2 lead indirectly to a reduction of adenovirus, HAV, and rotavirus prevalence rates.

Key words: adenovirus, COVID-19 pandemic, enteric viruses, Hepatitis A virus, rotavirus, wastewater

HIGHLIGHTS

- Safety precautions against SARS-CoV-2 lead indirectly to a remarkable reduction in the prevalence rates of enteric viruses.
- Prevalence of adenovirus decreased to 66% after the pandemic.
- Hepatitis A virus showed 14.3% positivity of tested samples.
- Rotavirus prevalence became 9.5% after the pandemic.
- All those prevalence percentages were remarkably less than detection percentage before the COVID-19 pandemic.

INTRODUCTION

Prevalence of enteric viruses is affected by hygienic status, therefore, their infection rates increase in countries with low hygienic levels (Payment 2001). According to the WHO, enteric viruses such as HAV, HEV, rotavirus, adenovirus, norovirus, and others are classified as pathogens having moderate to high harmful effects on human health (WHO 2011). Reports detected those viruses in Egypt (Shaheen *et al.* 2018; Elmahdy *et al.* 2020).

Usage of wastewater in crops irrigation makes them a source of enteric viral infection (Gerba & Choi 2006) and this was shown in different reports in Egypt with the presence of human adenovirus and rotavirus in green onion and lettuce (Shaheen *et al.* 2019), presence of HAV and human norovirus in watercress, strawberry, and coriander (Elmahdy *et al.* 2022).

Following hygienic habits such as carefully washing fruits and vegetables before eating, washing hands with soap and water, cleaning surfaces from dust, and using water filters in houses can help protect humans from enteric virus infections.

After the outbreak of the COVID-19 pandemic, many people tended to follow announced hygienic precautions to protect themselves from SARS-CoV-2 infection including not only the above-mentioned precautions but in addition many people soak fruits and vegetables in water containing vinegar before eating, give up buying street food and boil drinking water, which can all protect from enteric viruses transmission.

Thus, here we aimed to study if safety precautions against SARS-CoV-2 indirectly affected the prevalence of enteric viruses among Egyptians.

MATERIALS AND METHODS

Wastewater samples

One litre of wastewater was collected weekly through the period between August 2021 and March 2022 (21 samples in total) from the Zenin wastewater treatment plant (WWTP) which is one of the largest WWTPs in the Giza government, Egypt. It receives 275,000–550,000 m^3/day with an average of 385,000 m^3/day of wastewater flow (Latif 2022).

Adsorption-elution water concentration method

50 mL of 1 M aluminium chloride was added to each wastewater sample to give a final concentration of 0.05 M followed by adjusting pH to 3.5 with 1 N HCl. Samples were filtered through 0.45 μm nitrocellulose membrane then 200 mL 0.5 mM H_2SO_4, pH 3.0 passed after to ensure viral particle retention and eliminate all bio-solids. The membrane was removed from the holder and soaked in a Petri dish containing 11 mL of 1 mM NaOH; pH 10.5 for 10 min. Later, the upper surface of the membrane was scraped to elute the viruses. Volumes of 50 μL 50 mM H_2SO_4 and 50 μL 100X 1 nM Tris-EDTA, pH 8.0 were added to neutralize the elute and samples were stored at −20 °C until use (Katayama *et al.* 2002).

Enteric virus detection in concentrated wastewater samples

A volume of 140 μL of concentrated sample was subjected to viral RNA/DNA extraction (Qiagen, Germany) followed by reverse transcription (SuperScript III Reverse transcriptase, Invitrogen, USA) for RNA viruses. Quantitative RT-PCR was performed using SYBER green master mix (Luna, New England) and in-house designed primers (Table 1). Reaction mixture was subjected to the following temperature conditions, initial denaturation: 95 °C for 60 s, denaturation at 95 °C for 15 s, annealing: 55 °C for adenovirus and 57 °C for both HAV and rotavirus for 10 s, extension at 60 °C for 15 s for 40 cycles (Buston *et al.* 2009).

RESULTS AND DISCUSSION

Since the beginning of the COVID-19 pandemic, many people cared more about the hygienic status of their children, family, and houses and stopped many habits such as eating food from shops or street food, eating fruits and vegetables washed only with water and carefully washed their hands with soap and water and used disinfectants to sterilize many products before taking them inside their houses to protect themselves from SARS-CoV-2 infection. As the prevalence of enteric viruses in a given population depends greatly on hygienic status, we believe that protective precautions against SARS-CoV-2 must have an indirect effect on decreasing their infection rate.

To prove that, 21 samples were collected weekly through the period from August 2021 to March 2022 covering periods of increasing infection rates among Egyptians during the summer months and school season.

Samples were collected from Zenin WWTP, which is one of the largest WWTP in Egypt and receives a huge amount of wastewater per day, and thus detection rate of viruses in it can represent a wide sector of other WWTPs.

Adenovirus is one of the enteric viruses spreading in Egypt. Results in Figure 1 and Table 2 showed that 66% of samples were positive. Although this is a high prevalence, comparing it to pre-corona pandemic reports, it is lower than the recorded prevalence in the same WWTP in 2017 in untreated and treated wastewater which was 84 and 50%, respectively (Elmahdy *et al.* 2019). Another study also concerned with the same WWTP showed 91.7% positivity of adenovirus (El-Senousy *et al.*

Table 1 | Sets of in-house primers used for viral detection

Virus	Primer sequence	Annealing temperature
Adenovirus	F: CCCACGGTGGCGCCTAC R: AACCGCAGCGTCAAACGCT	55
HAV	F:TGATTAGCATGGAGCTGTAGG R: CAAAGCATCTCTTCATAGAAGTA	57
Rotavirus	F: TCAGTCTATTTTAAAGAGTACTCA R: TTTGATTCTCCCGATTGTTGATA	57

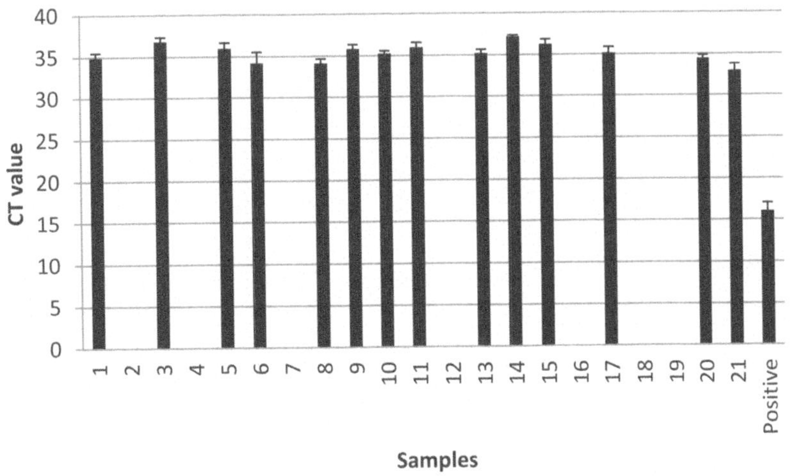

Figure 1 | CT values for adenovirus detection in wastewater samples showing 14 positive samples with prevalence percentage of 66%.

Table 2 | Number of adenovirus, HAV and rotavirus positive wastewater samples

	Viruses		
Months/number of samples	**Adenovirus**	**HAV**	**rotavirus**
August/1	1	0	0
September/3	2	1	0
October/3	2	2	0
November/2	1	0	0
December/2	1	0	0
January/4	3	0	1
February/4	2	0	0
March/2	2	0	1
Positive/total	14/21	3/21	2/21
Prevalence percentages	66%	14.3%	9.5%

2013). On the contrary, another study reporting on a different WWTP called Abou-Rawash showed the presence of adenovirus in 50% of wastewater samples (Elmahdy *et al.* 2020).

After a number of reported HAV outbreaks among European travellers, Egypt was considered to be an endemic country (Frank *et al.* 2007; MacDonald *et al.* 2013) due to the high prevalence rate of the virus (Divizia *et al.* 1999). Reports showed the presence of HAV in 83.3% of wastewater samples collected from Zenin (Hamza *et al.* 2017). Results in Figure 2 and Table 2 showed that only 14.3% of the samples were positive, indicating a sharp decrease in HAV spreading.

Finally, our results showed a prevalence rate of 9.5% to rotavirus (Figure 3; Table 2), which when compared with the pre-corona period indicates a remarkable decrease. The virus was detected in Egyptian wastewater with different percentages starting from 50% in Abou-Rawash WWTP (Elmahdy *et al.* 2020) and 25% in the same WWTP (Shaheen *et al.* 2018) to 21.9% in Zenin (Shaheen & Elmahdy 2019).

Figure 4 compared prevalence rates of the three detected viruses before and after the COVID-19 pandemic. Results showed that before the pandemic the mean percentages of detection of adenovirus, hepatitis A virus, and rotavirus were 75.2, 83.3, and 32.2%, respectively, while after the pandemic it was 66, 14.3, and 9.5%, respectively, confirming a real decrease in the prevalence of the three tested enteric viruses.

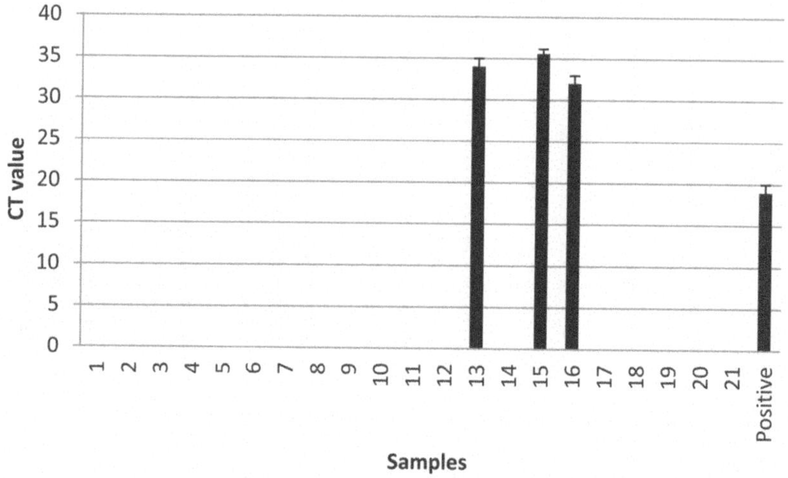

Figure 2 | CT value of hepatitis A virus detection in wastewater samples showing three positive samples with prevalence percentage of 14.3%.

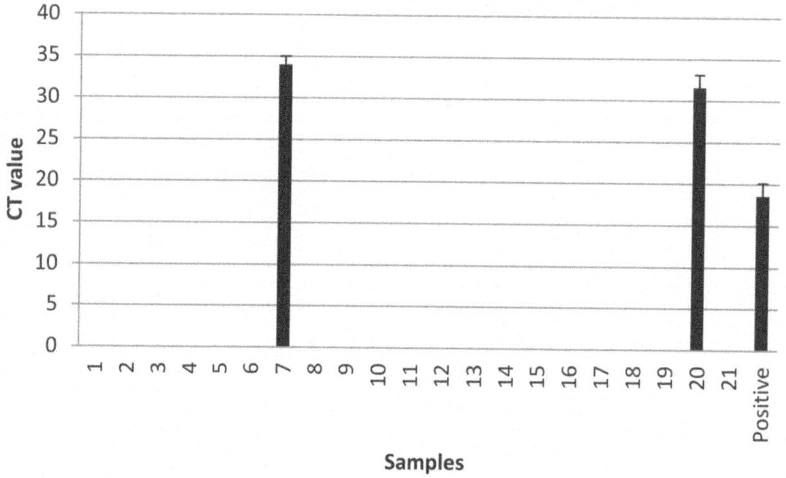

Figure 3 | CT value of rotavirus detection in wastewater samples showing two positive samples with prevalence percentage of 9.5%.

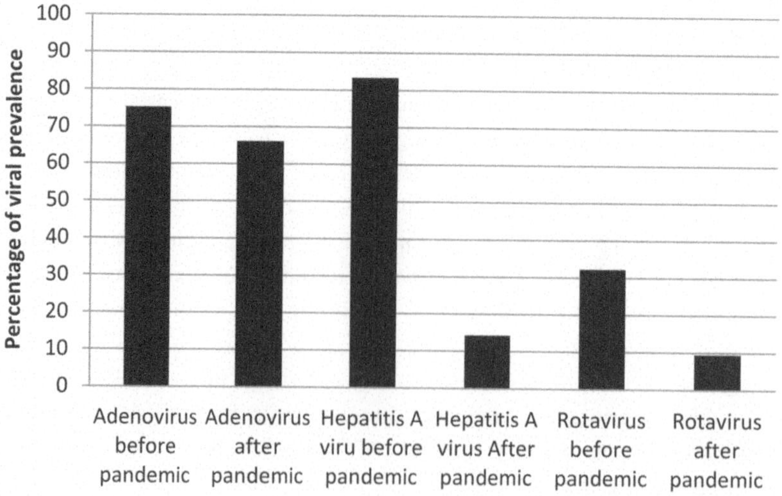

Figure 4 | Comparing prevalence rates of adenovirus, hepatitis A virus and rotavirus before and after the COVID-19 pandemic.

CONCLUSION

Safety precautions against SARS-CoV-2 lead indirectly to a remarkable reduction in the prevalence rates of adenovirus, HAV, and rotavirus prevalence in Egypt.

FUNDING

This work was funded by the National Research Center (Grant number: E-120806).

DATA AVAILABILITY STATEMENT

All relevant data are included in the paper or its Supplementary Information.

CONFLICT OF INTEREST

The authors declare there is no conflict.

REFERENCES

Buston, S. A., Benes, V., Garson, J. A., Hellemans, J., Huggett, J., Kubista, M., Mueller, R., Nolan, T., Pfaffl, M. W., Shipley, G. L., Vandesompele, J. & Wittwer, C. T. 2009 The MIQE guidelines: minimum information for publication of quantitative real-time PCR experiments. *Clin. Chem.* **55** (4), 611–622. doi:10.1373/clinchem.2008.112797.

Divizia, M., Gabrieli, R., Stefanoni, M. L., Renganathan, E., El Ghazzawi, E., Kader, O. A., Gamil, F., El Sawaf, G., El Sherbini, E., Saleh, E., Degener, A. M., Noce, A., Zaratti, L., Modesti, A. & Pana, A. 1999 HAV and HEV infection in hospitalised hepatitis patients in Alexandria, Egypt. *Eur. J. Epidemiol.* **15**, 603–609.

Elmahdy, E. M., Ahmed, N. I., Shaheen, M. N. F., Mohamed, E. B. & Loutfy, S. A. 2019 Molecular detection of human adenovirus in urban wastewater in Egypt and among children suffering from acute gastroenteritis. *J. Water Health* **17** (2), 287–294. https://doi.org/10.2166/wh.2019.303.

Elmahdy, E. M., Shaheen, M. N. F., Rizk, N. R. & Saad-Hussein, A. 2020 Quantitative detection of human adenovirus and human rotavirus group A in wastewater and El-Rahawy drainage canal influencing river Nile in the North of Giza, Egypt. *Food Environ. Virol.* **12**, 218–225. https://doi.org/10.1007/s12560-020-09429-x.

Elmahdy, M. E., Shaheen, M. N. F., Mahmoud, L. H. I., Hammad, I. A. & Soliman, E. R. S. 2022 Detection of norovirus and hepatitis A virus in strawberry and green leafy vegetables by using RT-qPCR in Egypt. *Food Environ. Virol.* **14** (2), 178–189.

El-Senousy, W. M., Barakat, A. B., Ghanem, H. E. & Kamel, M. A. 2013 Molecular epidemiology of human adenoviruses and rotaviruses as candidate viral indicators in the Egyptian sewage and water samples. *World Appl. Sci. J.* **27** (10), 1235–1247. doi:10.5829/idosi.wasj.2013.27.10.81200.

Frank, C., Walter, J., Muehlen, M., Jansen, A., van Treeck, U., Hauri, A. M., Zoellner, I., Rakha, M., Hoehne, M., Hamouda, O., Schreier, E. & Stark, K. 2007 Major outbreak of hepatitis A associated with orange juice among tourists, Egypt, 2004. *Emerging Infect. Dis.* **13**, 156–158.

Gerba, C. & Choi, C. 2006 Role of irrigation water in crop contamination by viruses. In *Viruses in foods* (S.M. Goyal, ed.). Springer, New York, NY, pp. 257–263.

Hamza, H., Abd-Elshafy, D. N., Fayed, S. A., Bahgat, M. M., El-Esnawy, N. A. & Abdel-Mobdy, E. 2017 Detection and characterization of hepatitis A virus circulating in Egypt. *Arch. Virol.* **162**, 1921–1931. doi:10.1007/s00705-017-3294-4.

Katayama, H., Shimasaki, A. & Ohgaki, S. 2002 Development of a virus concentration method and its application to detection of enterovirus and norwalk virus from coastal seawater. *Applied and Environmental Microbiology* **68**, 1033–1039.

Latif, E. F. 2022 Applying novel methods in conventional activated sludge plants to treat low-strength wastewater. *Environmental Monitoring and Assessment* **194**, 323.

MacDonald, E., Steens, A., Stene-Johansen, K., Gillesberg Lassen, S., Midgley, S., Lawrence, J., Crofts, J., Ngui, S. L., Balogun, K., Frank, C., Faber, M., Gertler, M., Verhoef, L., Koopmans, M., Sane, J., van Pelt, W., Sundqvist, L. & Vold, L. 2013 Increase in hepatitis A in tourists from Denmark, England, Germany, The Netherlands, Norway and Sweden returning from Egypt, November 2012 to March 2013. *Eurosurveillance* **18**, 20468.

Payment, P. 2001 Transmission of gastrointestinal diseases: hygiene as the final barrier. *Am. J. Infect. Control* **29** (4), 218–221.

Shaheen, M. N. F. & Elmahdy, E. M. 2019 Molecular detection of group C rotavirus in environmental samples in Giza, Egypt. *Asian J. Water Environ. Pollut.* **16** (4), 17–22.

Shaheen, M. N. F., Abd El-Daim, S. A., Ahmed, N. I. & Elmahdy, E. M. 2018 Molecular detection of three gastroenteritis viruses in urban sewage treatment plant and river water in Egypt. *Egypt. J. Aquat. Biol. Fish.* **22** (5), 615–627.

Shaheen, M. N. F., Elmahdy, E. M. & Chawla-sarkar, M. 2019 Quantitative PCR-based identification of enteric viruses contaminating fresh produce and surface water used for irrigation in Egypt. *Environ. Sci. Pollut. Res. Int.* **26** (21), 21619–21628.

WHO 2011 *World Health Organization: Guidelines for Drinking-Water Quality*, 4th edn. WHO Press, Geneva, Switzerland.

First received 5 August 2022; accepted in revised form 25 October 2022. Available online 7 November 2022

doi: 10.2166/wh.2022.021

Monitoring the exposure and emissions of antibiotic resistance: Co-occurrence of antibiotics and resistance genes in wastewater treatment plants

Ruud Steenbeek *,†, Peer H. A. Timmers†, Danielle van der Linde, Kay Hup, Luc Hornstra and Frederic Been

KWR Water Research, Groningenhaven 7, 3433 PE, Nieuwegein, The Netherlands

*Corresponding author. E-mail: ruud.steenbeek@kwrwater.nl

†Both authors contributed equally.

RS, 0000-0001-8101-3986

ABSTRACT

The COVID-19 pandemic has brought new momentum to 'wastewater-based epidemiology' (WBE). This approach can be applied to monitor the levels of antibiotic-resistant genes (ARGs), which in terms are used to make inferences about the burden of antimicrobial resistance (AMR) in human settlements. However, there is still little information about temporal variability in ARG levels measured in wastewater streams and how these influence the inferences made about the occurrence of AMR in communities. The goal of this study was hence to gain insights into the variability in ARG levels measured in the influent and effluent of two wastewater treatment plants in The Netherlands and link these to levels of antibiotic residues measured in the same samples. Eleven antibiotics were detected, together with all selected ARGs, except for *VanB*. Among the measured antibiotics, significant positive correlations ($p > 0.70$) with the corresponding resistance genes and some non-corresponding ARGs were found. Mass loads varied up to a factor of 35 between days and in concomitance with rainfall. Adequate sampling schemes need to be designed to ensure that conclusions are drawn from valid and representative data. Additionally, we advocate for the use of mass loads to interpret levels of AMR measured in wastewater.

Key words: antibiotic-resistant genes, antibiotics, between-day variability, human exposure and emission, wastewater-based epidemiology

HIGHLIGHTS

- Monitoring of antibiotics and antibiotic-resistant genes (ARGs) by wastewater-based epidemiology (WBE).
- Able to detect 11 antibiotics and 13 corresponding ARGs.
- Correlations found between antibiotics and ARGs in effluent.
- Important variability in ARG mass loads observed between sampling days.
- WBE needs to be implemented to design adequate sampling schemes to avoid misinterpreting the burden of antibiotic resistance.

GRAPHICAL ABSTRACT

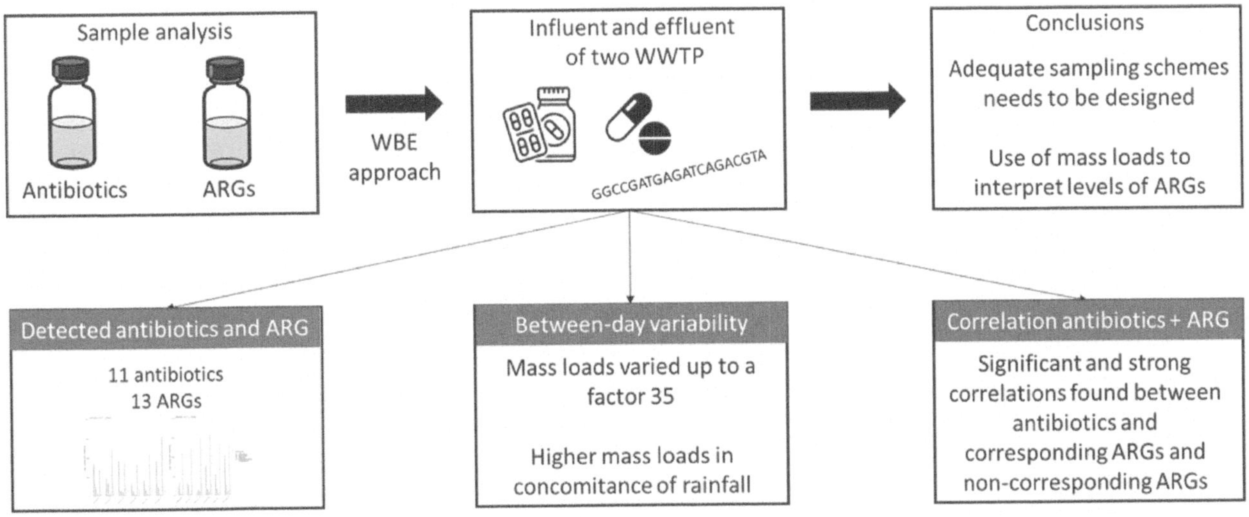

1. INTRODUCTION

The COVID-19 pandemic, which has been affecting people globally since the early 2020, has brought new momentum to a discipline known as 'wastewater-based epidemiology' (WBE) or 'sewage surveillance'. This approach has been used to monitor the circulation of infectious disease agents found in urine and feces since the 1980s (Asghar *et al.* 2014) and mid-2000s, to monitor (illicit) drug use and exposure to contaminants (Gracia-Lor *et al.* 2017). Additionally, this approach relies on the measurement of relevant parameters (e.g., RNA, DNA, chemicals and human metabolites) in wastewater samples collected at the influent of wastewater treatment plants (WWTPs). Levels of these parameters, measured at different locations and times, are then used as proxy to derive information relevant for public health, as is currently being done with SARS-CoV-2 (Medema *et al.* 2020a). Prior to the COVID-19 outbreak, one of the most important and promising applications of WBE has been the monitoring of antimicrobial resistance (AMR), and more specifically, antibiotic resistance (Aarestrup & Woolhouse 2020). The interest around sewage surveillance to monitor AMR raised from the acknowledgement that WWTPs are a major player in the spread of antibiotic resistance genes (ARGs) to the environment (Schlüter *et al.* 2007; Bouki *et al.* 2013; Gatica & Cytryn 2013; Rizzo *et al.* 2013). While these studies mainly focused on the emissions and risks of ARGs reaching surface waters for human health and the environment (Rizzo *et al.* 2013; Vaz-Moreira *et al.* 2014), researchers recently recognized that influents of WWTPs can be used to monitor the burden of AMR at the community level. In fact, there has been an increasing number of global studies which have monitored resistance at the population level by analysis of samples collected at the inlet of WWTPs (Hendriksen *et al.* 2019; Hutinel *et al.* 2019; Pärnänen *et al.* 2019). Following a WBE approach, these studies aim to provide valuable information about resistance prevalence, in a similar manner to traditional clinical surveillance. However, in the WBE approach, the information provided is at the community level and at different geographical and temporal resolutions (Huijbers *et al.* 2019). The available environmental data can then be linked to other relevant epidemiological and/or socio-economic factors and potentially identify significant covariates such as levels of sanitation and antibiotic use (Hendriksen *et al.* 2019; Hutinel *et al.* 2019; Pärnänen *et al.* 2019). Furthermore, the implementation of metagenomic sequencing provides the opportunity to characterize entire communities (i.e., the general population), which would otherwise not be monitored (Aarestrup & Woolhouse 2020). Similarly to ARGs, antibiotic residues have also been largely measured in WWTP influents and effluents (Wang *et al.* 2020). While most of these studies focused on assessing removal and emissions to the environment, recent reports focusing on estimating antibiotic use through wastewater analysis have also been reported (Castrignanò *et al.* 2020). Studies evaluating levels of both antibiotics and resistance genes in wastewater samples have also been reported, although to a lesser extent, limiting the strength of established relationships (Duarte *et al.* 2019). The combined analysis of antibiotic residues and ARGs in WWTPs could complement existing conventional surveillance systems in humans and animals which in developing countries, are affected by limited data and resources. Additionally, these studies can help to better understand the occurrence of AMR in non-pathogenic bacteria in large (healthy) communities, which is currently largely unknown (Manaia *et al.* 2018; Muloi *et al.* 2018).

However, before this approach can be routinely and reliably applied to monitor the occurrence of AMR and potentially assess the resistance burden in large communities across the world, additional research is necessary. Studies carried out so far, especially those whose goal was to map resistance in various countries through the wastewater analysis, focused less on short-temporal variability of gene counts, its impact on sampling strategies, and the derivation of (gene) mass loads (used to compare results across different catchments) (Hendriksen *et al.* 2019; Hutinel *et al.* 2019; Pärnänen *et al.* 2019). For example, some extensive studies reported on the short-term and long-term fluctuations of antibiotic mass loads at the influent of WWTPs and the impact on samplings and removal strategies (Coutu *et al.* 2013), but less is known about fluctuations in ARG counts and how these affect inferences made from wastewater analysis about the occurrence of AMR in investigated communities. The goal of this study was hence to gain insights into the between-day variability of ARG levels measured at the influent and effluent of two WWTPs in The Netherlands and link these to levels of antibiotic residues measured in the same samples. Between-day variability in levels of ARGs and antibiotics, and their relationships were then investigated to determine the impact that these have on sampling strategies and inferences drawn about occurrence of AMR in communities.

2. MATERIALS AND METHODS

2.1. Sampling

Samples were taken from WWTP 1 (52.09083 N°, 5.12222 E°, 300,000 inhabitants served, mean flow rate 20,700 m^3/day) and WWTP 2 (52.02 N°, 5.04306 E°, 96,000 inhabitants served, mean flow rate 54,400 m^3/day). A map with the sampling locations are shown in Supplementary Material, Figure S1 and the distance between the two WWTPs is 12.3 km. One liter of influent and effluent at the beginning of the WWTPs was sampled in six sampling rounds (the second and the last week of each month) in May, July and September 2019 to have three seasons covered in the sampling campaign. At the WWTPs, 24-h volume proportional composite samples were taken. In total, 12 influent and 12 effluent samples were collected in a large bucket made of high-density polyethylene (HDPE). Samples were immediately transported to the laboratory and separated in two aliquots. The first was immediately stored at −20 °C (for chemical analysis, see the following section), while the second was immediately processed according to the protocol described below (for biological analysis). Samples were collected with an iron spoon in HDPE bottles that were previously cleaned with methanol and ultrapure water. All glassware used for chemical and biological analyses was precleaned with acetone and petroleum ether.

2.2. Sample preparation

2.2.1. Microbiological sample preparation

Samples were transported directly to the laboratory and vacuum filtered through 0.22-μm pore-size polycarbonate track-etch filter membranes (Sartorius, Göttingen, Germany). The filter membranes were pre-treated with ammonium oxalate to remove iron, which diminish the quantitative polymerase chain reactions (qPCRs). Additionally, an internal control plasmid was added prior to DNA extraction (2.5×10^4 gene copies/μL) to allow quantification of the DNA loss caused by the extraction process, as done previously (Wullings *et al.* 2007). Samples were stored at −20 °C until DNA was extracted. DNA was subsequently extracted from the filters using DNeasy PowerSoil Kit (QIAGEN Benelux B.V., Venlo, the Netherlands). DNA extraction was performed according to the DNeasy PowerSoil Kit instructions. DNA extracts were stored at −20 °C until further analysis.

2.2.2. Chemical sample preparation

All reference standards were purchased from Sigma–Aldrich (St. Louis, MO, USA). The analyzed antibiotics are listed in Supplementary Material, Table S1 with their corresponding limit of quantification (LOQ). These antibiotics were chosen because they correspond to the ARGs targeted in this study and they are prescribed to humans in the Netherlands. The analytical procedures are based on previous research (Thomas *et al.* 2012; Hernández *et al.* 2018; Paulus *et al.* 2019) and changes are made to optimize the sample preparation and analysis for the analyzed antibiotics. Individual stock solutions of the reference standards, including mass labeled ones, were prepared at a concentration of 100 mg/L in acetonitrile with 5% dimethyl sulfoxide and 5% ultrapure water (18.2 MΩ/cm, ELGA LabWater, Lane End, UK). Individual stock solutions were stored at −20 °C. Work solutions containing all individual stocks were freshly prepared in ultrapure water (500 μg/L) each time a new set of samples was processed and analyzed. Samples for analysis by direct injection were prepared as follows: a 50-mL flask was rinsed three times with a small volume of the sample. Samples were subsequently transferred to the flasks and spiked with 100 μL internal standard. Samples were then filtered through a 0.20-μm pore-size filter and transferred to a 1.8-mL amber vial for analysis. For solid-phase extraction, 100 mL of each sample was transferred in a precleaned HDPE bottle. 100 μL of internal standard work solution was added to each sample, 0.1% EDTA (v/v) was added, and then adjusted to pH = 2.5 with hydrochloric acid. The samples were horizontally shaken for 5 min at 120 rpm. Samples were extracted with Strata-X SPE columns (6 mL, 200 mg, Phenomenex, USA). The columns were conditioned with two column volumes of methanol, followed by two column volumes of acidified ultrapure water (pH = 2.5). Samples were then gently loaded onto the cartridges. Subsequently, the columns were washed with two column volumes of acidified ultrapure water (pH = 2.5) and dried under vacuum for 1 h. Thereafter, the columns were eluted with 3 × 2 mL of MeOH. Eluates were collected in glass tubes and evaporated to dryness under a gentle stream of nitrogen at 40 °C. Eluates were then reconstituted in 500 μL of ultrapure water with 5% of MeOH and vortexed for 5 s. The extract was transferred in 1.8 mL vials with inserts for analysis.

2.3. Analyses

2.3.1. Microbiological analysis

Extracted DNA was used for qPCR analysis of total bacterial counts and ARGs. For the quantification of total bacterial counts, 16S rRNA gene counts were obtained using a SYBR Green qPCR assay. For quantification of specific ARGs (see Supplementary Material, Table S2), multiplex qPCR assays were performed. The analysis included standards and positive and a negative control to confirm multiplex qPCR quality in every assay. The standards were made up with five subsequent dilutions (2.5×10^4–2.5×10^0 gene copies/μL) and the multiplex qPCR assays were performed using a iQ™ Multiplex Powermix (Bio Rad, München, Germany) and qPCR reactions were performed using a CFX96™ Real-Time PCR Detection System (Bio Rad, München, Germany) as described by Paulus *et al.* (2019). Quantified genes were reported as copies/L by CFX Manager v.3.1.1517.0823 (Bio Rad, München, Germany). Data were reported if (a) the amplification efficiency was 85–105%, (b) $R^2 > 0.9$, (c) the efficiency of the internal control was $> 20\%$, and (d) the number of gene copies was higher than the LOQ. If the number of gene copies was between the limit of detection (LOD) and the LOQ, the data were qualitatively reported as 'indicative'. A $R^2 > 0.9$ and high efficiencies (from 81.7 to 104.8%) confirmed the linearity and the sensitivity of the qPCR assays. The absolute concentrations of the ARGs are shown in Supplementary Material, Table S3.

2.3.2. Chemical analysis

Analysis of selected antibiotics (Supplementary Material, Table S1) in the water samples for both direct injection and SPE was performed by liquid chromatography tandem mass spectrometer (LC-MS/MS). Chromatographic separation was achieved using a Nexera X2 LC-30 AD (Shimadzu, Tokyo, Japan) equipped with a Luna Omega Polar C18 (2.1 mm × 100 mm, 1.6 μm) column (Phenomenex, CA, USA). The injection volume was 100 μL and the column compartment was set to 30 °C. Mobile phases consisted of ultrapure water with 0.1% formic acid (A) and MeOH with 0.1% formic acid (B). A linear gradient from 95% A to 0% over 10 min was used. This was held for 3 min and then starting conditions were reset after 0.5 min and held for equilibration for another 2.5 min, for a total run time of 16 min. The flow rate was kept at 300 μL/min through the whole analysis for both direct injection and SPE. The mass analyser used was a Sciex Triple Quad 6500+ (SCIEX, MA, USA) equipped with an IonSpray source operated in positive and negative mode. Analyses were performed in multiple reaction monitoring mode (MRM) using an H-ESI interface. The spray voltage was set to 5,000 and −4,500 V for positive and negative ionization, respectively. Electrospray temperature (TEM) was set to 400 °C, the curtain gas (CUR) was set to 30 psi, collision gas (CAD) was 8 psi and the ion source gas pressures (GS1 and GS2) were set to 40 psi. Declustering potential (DP), collision energy (CE), collision cell (CXP), and resolution (Q1 and Q3) were optimized for each compound (Supplementary Material, Table S4). The dwell time varied between 3 and 50 ms, depending on the target antibiotic.

The validation was based on the guidelines developed by Peters *et al.* (2007) and the guidelines for bioanalytical method validation by the European Medicines Agency (EMA) (van Amsterdam *et al.* 2013). During the validation, performance parameters such as precision, LOD, LOQ, linearity, recovery, selectivity and carry-over were evaluated and met the criteria for the 25 antibiotics (Supplementary Material, Table S1b).

2.3.3. Data analysis

Influent and effluent qPCR data were normalized to the measured 16S rRNA gene copy number (genes/L), the flow (m^3/day) and the number of inhabitants connected to the WWTP (i.e., copies/16S gene/day/1,000 inhabitants). All antibiotic concentrations were also normalized to the number of citizens connected to the WWTP and the flow (i.e., mg/day/1,000 inhabitants). The latter was done to facilitate comparisons between the two locations which have a different number of inhabitants served.

A Shapiro–Wilk test (if $n \geq 3$) was used to test for normality of the data. Significant differences between location, time, and/or measurements were assessed by paired or unpaired *T*-tests with Bonferroni correction to counteract the problem of multiple comparisons. If the data did not meet the criteria for normality, a Wilcoxon signed-ranked test with Bonferroni correction or a Kruskal–Wallis one-way analysis of variance (ANOVA) was used instead.

Correlations between antibiotics and ARG mass loads were calculated using Pearson's Rank correlation coefficient (if the data were linear) or a Spearmen's correlation coefficient (for non-parametric data). Data were correlated if the p-value < 0.05 and if there were four or more data points available. Correlation was determined at $R^2 > 0.5$, with a moderate correlation if $0.5 > R^2 > 0.7$, and high correlation if $R^2 > 0.7$. All data were collected and analyzed with R version 3.6.1 (R, Vienna, Austria).

Redundancy analysis was conducted to find correlations between antibiotics and ARGs. If data were lower than the detection limit, ARG data were set to 0, only if the efficiency of the internal control was higher than 20%. If the efficiency was lower than 20%, the missing data were set to 'NA'. Antibiotic concentrations were set to half the LOQ if the target compound was detected in at least 70% of the samples at concentrations < LOQ.

3. RESULTS AND DISCUSSION

3.1. Antibiotic-resistant genes

Figure 1 and Figure 2 show the normalized ARG loads of 14 ARGs in the influent of WWTP 1 and 2 and Figure 3 and Figure 4 show the normalized ARG loads in the effluent. These ranged from 0.011 to 30,734 copies/16S/day/1,000 inhabitants for

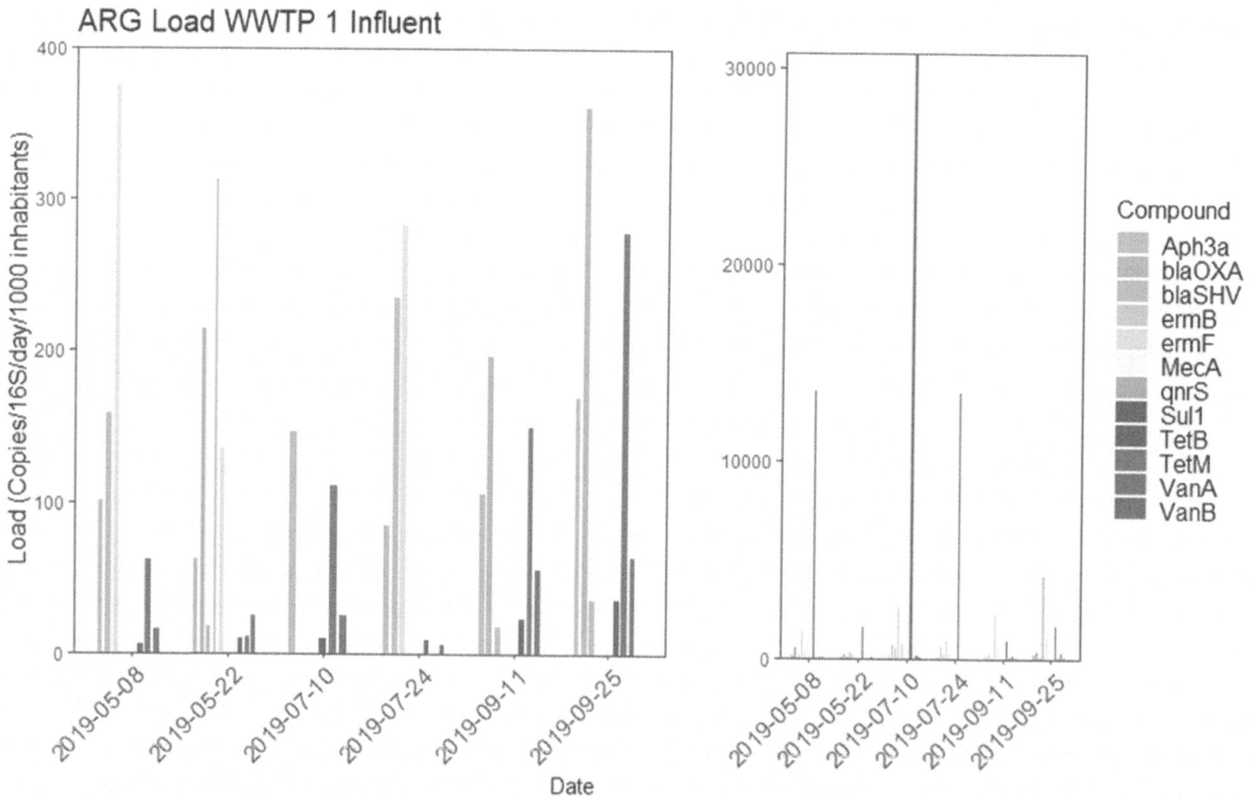

Figure 1 | ARG loads of the detected genes in influent wastewater of WWTP 1. In the left graph, the less prevalent ARGs are shown and on the right all the detected ARGs are presented.

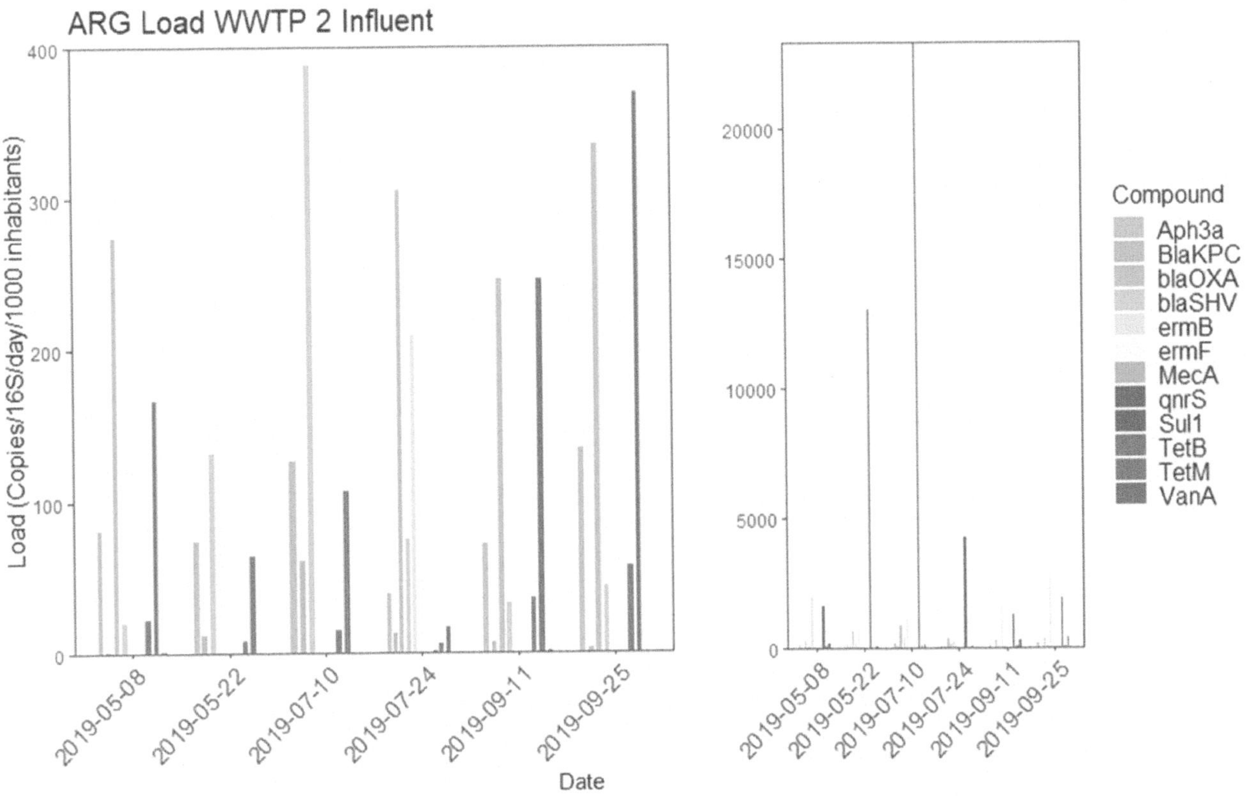

Figure 2 | ARG loads of the detected genes in influent wastewater of WWTP 2. In the left graph, the less prevalent ARGs are shown and on the right all the detected ARGs are presented.

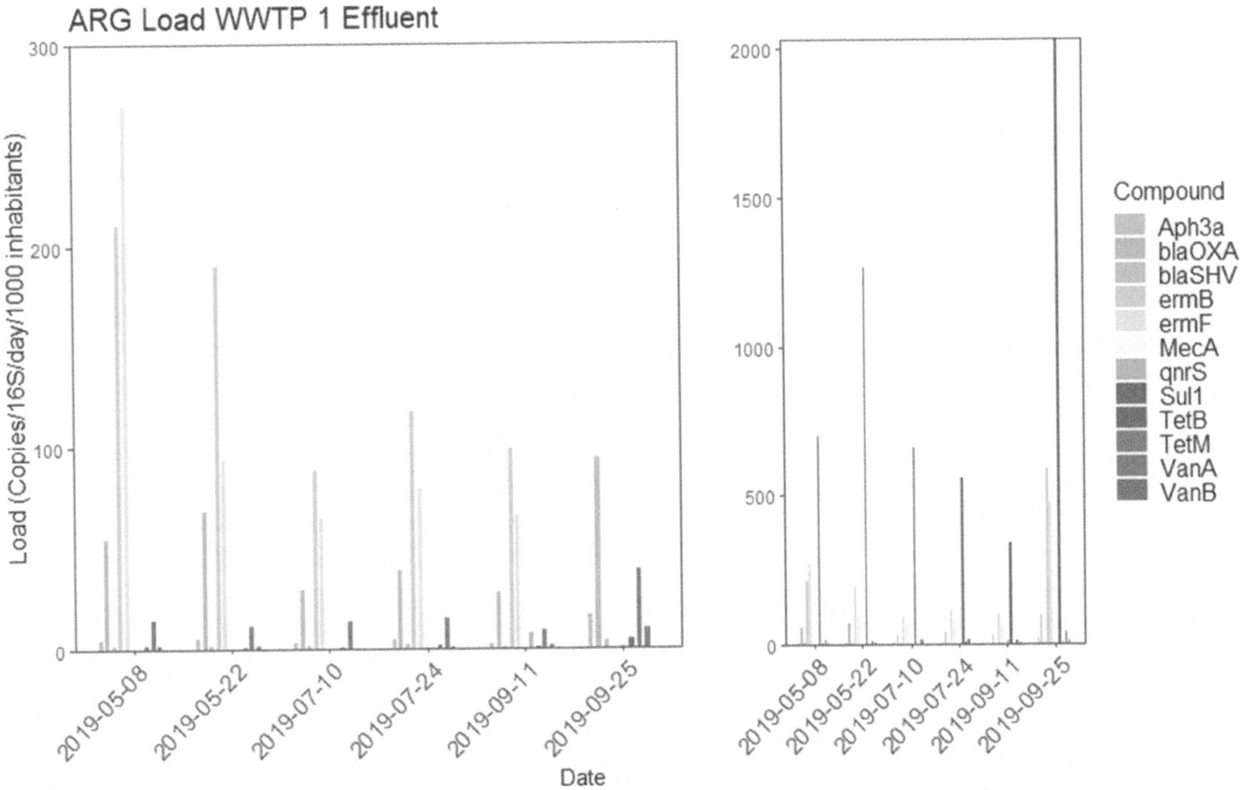

Figure 3 | ARG loads of the detected genes in effluent wastewater of WWTP 1. In the left graph, the less prevalent ARGs are shown and on the right all the detected ARGs are presented.

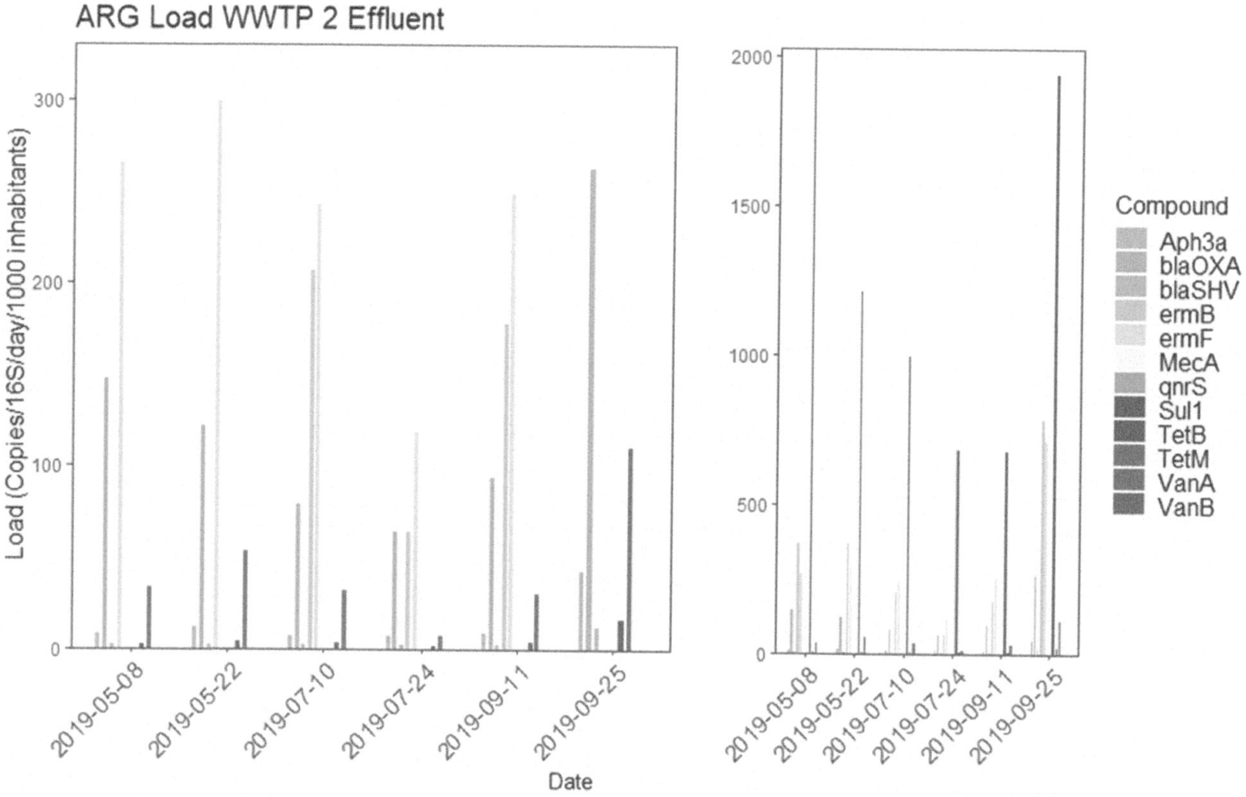

Figure 4 | ARG loads of the detected genes in effluent wastewater of WWTP 2. In the left graph, the less prevalent ARGs are shown and on the right all the detected ARGs are presented.

WWTP 1, with *sul1* as the most prevalent ARG. No significant differences were found neither between days nor months of sampling when considering the sum of all genes detected (one-way ANOVA or Kruskal–Wallis, *p*-value < 0.05). Higher mass loads were seen for most of the detected ARGs on the 25th September, a day with 5 mm of rainfall (KNMI n.d.). Overall, a substantial variability in mass loads between sampling days was observed for all considered ARGs as shown in Figure 1. For instance, *ermB* loads in influent and effluents varied up to one order of magnitude between July and September in WWTP 1. Substantial variations were also visible within the same month, with *blaOXA* for instance, decreasing by a factor of two within the month of May in the influent of WWTP 2. *Sul1* loads in the influent were 35 times higher on 10th July (also a day with 5 mm of rainfall) compared to the 11th September. Similarly, *ermF* loads in the influent were seven times higher on the 25th September compared to the 22nd May. For WWTP 2, normalized mass loads of ARGs ranged from 0.014 to 13,537 copies/16S/day/1,000 inhabitants, while *sul1* ranged between 0.044 and 23,280 copies/16S/day/1,000 inhabitants, making it again the most prevalent ARG (see Figure 2). As for WWTP 1, no significant differences between days nor months of sampling were found when considering the sum of all detected genes (one-way ANOVA or Kruskal–Wallis, *p*-value < 0.05). All detected genes showed higher mass loads on the 25th September, especially in the effluent. Similar to WWTP 1, a high variability between the measuring dates was clearly visible with highest loads in September, but for instance *sul1* mass loads in the influent were 18 times higher on the 10th July compared to the 11th September (i.e., 23,280 vs. 1,267 copies/16S/day/1,000 inhabitants). No significant differences were found between ARG mass loads in the influent nor effluent of WWTPs 1 and 2. With regards to ARGs concentrations, figures reported here were in the same range reported by Paulus *et al.* (2019), who measured ARGs in samples collected from the Netherlands. Additionally, a study from Finland and Estonia, reported similar *blaOXA* concentration in influents compared to here (mean concentration 1×10^{11} vs. 2.5×10^{11} genes/L) (Laht *et al.* 2014). *Sul1* and *tetM* concentrations were substantially higher compared to here, namely 3×10^{11} and 10^{12} genes/L compared to 4×10^{10} and 2.5×10^{9} genes/L. In a study from Michigan (USA), *sul1* concentrations in raw wastewater was approximately 10 times higher compared to findings from this study (Gao *et al.* 2012).

3.2. Antibiotics

In total, 11 antibiotics were found in the influent and effluent wastewater of the WWTP. The absolute concentrations (µg/L) of the detected antibiotics can be found in Supplementary Material, Table S5. The variability in measured mass loads antibiotics between sampling days can be seen in Figure 5 and Figure 6. For instance, ceftazidime concentration on 25th July in influent of WWTP 1 was 10 times higher than the 10th July (0.08–0.80 µg/L). Ceftazidime concentrations in effluent of WWTP 2 were almost three times as high on the 10th July compared to the 25th of July. Also, trimethoprim concentration in effluent of WWTP 2 was six times higher on the 8th May compared to the 25th July. No significant differences could be

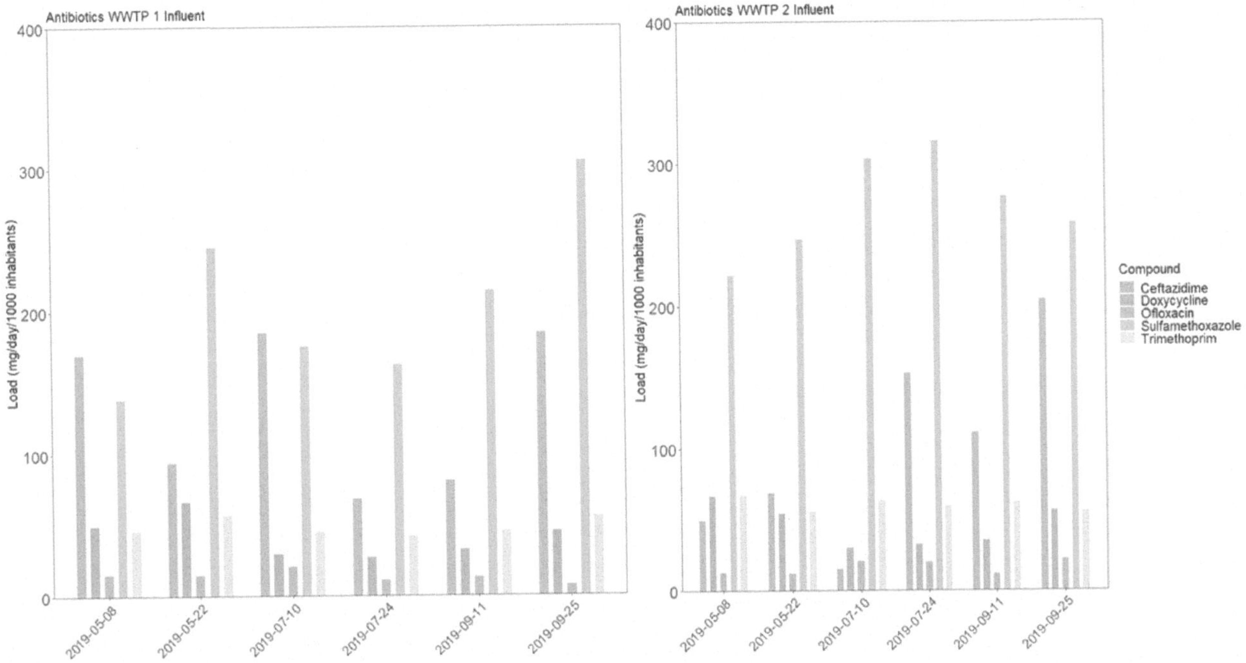

Figure 5 | Detected antibiotics in influent wastewater of WWTP 1 and WWTP 2.

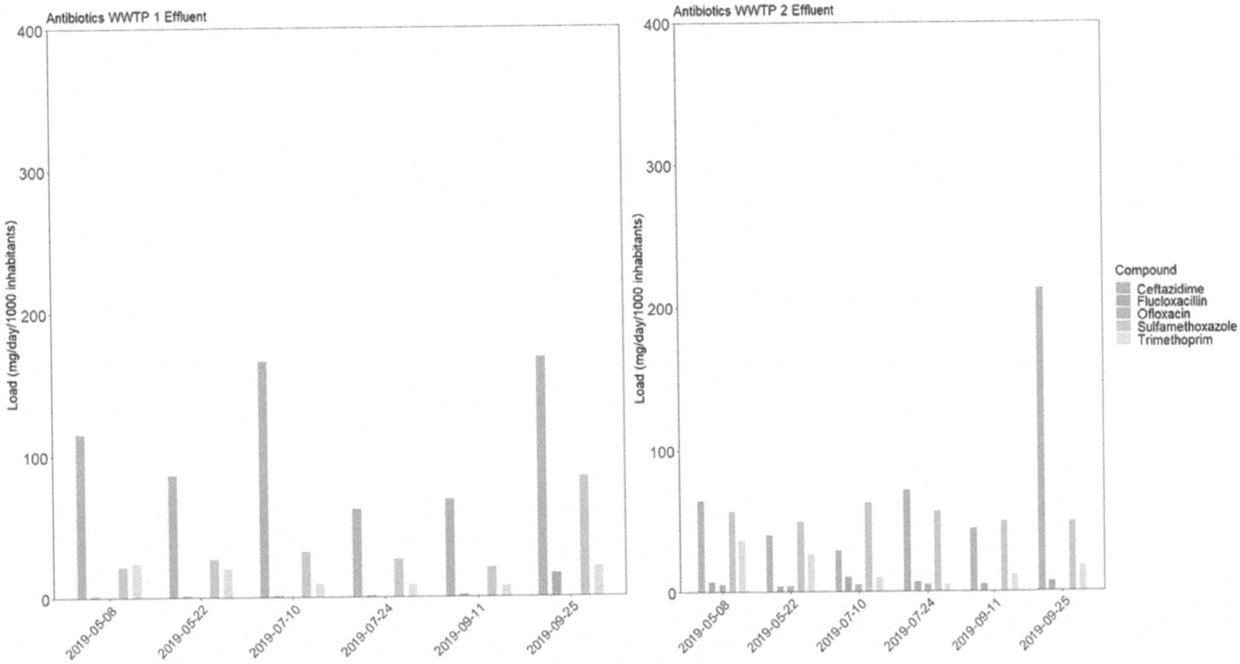

Figure 6 | Detected antibiotics in effluent wastewater of WWTP 1 and WWTP 2.

found between sampled months. Significant differences were found between influent and effluent for sulfamethoxazole in both WWTPs and for trimethoprim and doxycycline in WWTP 1 (*t*-test, *p*-value < 0.05). The concentration range of the detected antibiotics and their detection frequencies are listed in Table 1. Antibiotics measured in this context have also been reported in other studies. For instance, Rodriguez-Mozaz *et al.* (2015) reported maximum sulfamethoxazole concentrations of 0.38 µg/L in influents and 0.065 µg/L in effluents. As for the study by Gracia-Lor *et al.* (2014), these results are in the same range as those reported here. However, mass loads of ciprofloxacin measured by Castrignanò *et al.* (2020) in influents in Utrecht (i.e., 58.7 mg/1,000 people/day) were roughly three times lower compared to mass loads reported in this study (i.e., 176.5 mg/1,000 people/day). Similarly, ofloxacin mass loads reported here (i.e., 13.4 mg/1,000 people/day) were approximately three times lower compared to results reported by Castrignanò *et al.* (2020) (i.e., 4.3 mg/1,000 people/day). Influent concentrations for sulfamethoxazole reported here (i.e., 1.24 µg/L) were comparable to findings by Yang *et al.* (2005) (i.e. 1.09 µg/L), who analyzed samples collected from WWTPs in Fort Collins, Colorado, USA. The authors also reported lower effluent concentrations compared to findings from this study (i.e., 0.10 vs. 0.22 µg/L in this study).

3.3. Correlation between ARGs and antibiotics

Redundancy analysis was conducted to explore correlations between measured ARGs and antibiotics. When all samples were considered together (i.e., influents and effluents), significant correlations ($p > 0.50$) were found (Table 2). Significant correlations were also found between ARGs and antibiotics measured in effluents of both WWTPs as shown in Table 3. For instance, correlations were found between sulfomethoxazole and *sul1*, trimethoprim and *sul1* and between flucloxacillin

Table 1 | Mean values of detected antibiotics and detection frequency

	WWTP 1 [µg/L]		WWTP 2 [µg/L]	
Antibiotics	**Influent**	**Effluent**	**Influent**	**Effluent**
Ceftazidime	0.73 (100%)	0.59 (100%)	0.45 (100%)	0.32 (100%)
Trimethoprim	0.28 (100%)	0.08 (100%)	0.29 (100%)	0.09 (100%)
Sulfamethoxazole	1.16 (100%)	0.18 (100%)	1.33 (100%)	0.26 (100%)
Ofloxacin	0.08 (100%)	< LOQ	0.08 (83.3%)	0.02 (66.7%)
Ciprofloxacin	1.08 (33.3%)	< LOQ	1.47 (16.7%)	0.16 (83.3%)
Doxycycline	0.24 (100%)	0.01 (33.3%)	0.19 (100%)	0.02 (83.3%)
Piperacillin	< LOQ	0.01 (16.7%)	< LOQ	0.01 (33.3%)
Ceftriaxon	< LOQ	0.10–0.20 (83.3%)	< LOQ	0.20 (16.7%)
Flucloxacillin	< LOQ	0.04 (33.3%)	< LOQ	0.03 (100%)
Amoxicillin	< LOQ	0.02 (16.7%)	< LOQ	< LOQ
Sulfadiazine	0.06 (33.3%)	0.01 (16.7%)	0.07 (16.7%)	< LOQ

< LOQ, results were below the limit of quantification.

Table 2 | Significant correlation coefficients of all samples together

All	Intl1	qnrS	Sul1	TetB	blaSHV	MecA	blaOXA	Aph3a	VanA	VanB	TetM	ermB	ermF	blaKPC
Sulfamethoxazole	–	–	0.79	0.88	0.85	–	0.85	0.86	–	–	0.78	0.80	0.85	–
Ofloxacin	0.83	–	0.83	0.78	0.85	–	0.78	0.78	–	–	0.70	0.75	0.78	–
Ceftazidime	0	–	0	0	0	–	0	0	0.43	–	0	0	0	–
Trimethoprim	0.79	–	0.79	0.83	0.74	–	0.86	0.81	0	–	0.79	0.84	0.85	–
Ciprofloxacin	0	–	0	0	0	–	0	0	0	–	0	0	0	–
Doxycycline	0.76	–	0.79	0.84	0.75	–	0.86	0.84	–	–	0.78	0.88	0.85	–
Flucloxacillin	0.73	–	0.73	0.72	0.70	–	0.73	0.71	0.57	–	0.60	0.74	0.71	–
Ceftriaxon	–	–	–	–	–	–	–	–	–	–	–	–	–	–

Table 3 | Significant correlation coefficients of effluent samples together

Effluent	Intl1	qnrS	Sul1	TetB	blaSHV	MecA	blaOXA	Aph3a	VanA	VanB	TetM	ermB	ermF	blaKPC
Sulfamethoxazole	0.6	–	0.59	0.71	0.80	–	0.59	0.71	–	–	–	–	0.64	–
Ofloxacin	–	–	–	–	–	–	–	–	0.64	–	–	–	–	–
Ceftazidime	–	–	–	–	–	–	–	–	–	–	–	–	–	–
Trimethoprim	0.61	–	0.61	–	–	–	0.64	–	–	–	0.59	0.80	0.69	–
Ciprofloxacin	–	–		–	0.86	–	–	–	–	–	–	–	–	–
Doxycycline	–	–	0.64	–	0.63	–	0.70	–	–	–	0.69	0.72	0.58	–
Flucloxacillin	0.55	0.22	–	0.66	0.78	–	0.58	0.69	–	–	0.50	–	0.69	–
Ceftriaxon	–	–	–	–	–	–	–	–	–	–	–	–	–	–

and *blaSHV*. These correlations were found when considering all samples together (influents and effluents) as well as when considering only WWTP effluents. These findings might indicate that although WWTPs reduce total ARG and antibiotic concentrations, selective pressure for some of these ARGs might still be exerted in WWTPs. However, a larger sampling campaign needs to be carried out before these results can be confirmed. When considering the influents of both WWTPs, no significant correlations were found. This was also the case when influent and effluent samples of both WWTPs were investigated individually. The correlation found between sulfamethoxazole and total sulfonamide concentration with *sul1* was similar with findings from previous studies (Luo *et al.* 2010; Gao *et al.* 2012; Rodriguez-Mozaz *et al.* 2015; Xu *et al.* 2016). In all cases, higher antibiotic concentrations corresponded to higher ARG concentrations, which is in line with the hypothesis that exposure to antibiotics leads to selective pressure for ARGs (Li *et al.* 2012; Rodriguez-Mozaz *et al.* 2015).

Correlations between antibiotics and non-corresponding ARGs were also found (Tables 2 and 3). Although there is currently a lack of knowledge about the significance of such relationships, lateral gene transfer of plasmids containing multidrug resistance is plausible. In fact, it has been reported that *sul1* genes are probably carried on mobile genetic elements, explaining the frequently observed co-occurrence of *sul1* and the class 1 integron gene *Int1* (Luo *et al.* 2010; Xu *et al.* 2016; Subirats *et al.* 2018; Zhang *et al.* 2018). Class 1 integrons play an important role in antibiotic resistance dissemination in many multidrug resistant gram-negative bacteria (Strugeon *et al.* 2016). A positive correlation was also found between sulfamethoxazole and trimethoprim, which is not surprising as these antibiotics are often used in combination, both for human and veterinary purposes (Greeff & Mouton 2017). Results presented here show a strong correlation between *Int1* and *Sul1*, but also between *Int1* and other ARGs, such as *blaSHV*, *blaOXA* and *qnrS* (Spearman correlation > 0.8, $p < 0.05$). This suggests that these other ARGs might also be carried on mobile generic elements, which appears to be the case for most ARGs detected in WWTPs (Che *et al.* 2019). Furthermore, it has been postulated that human multidrug use aggravates co-selective pressure for antibiotic resistance (Xu *et al.* 2016). The occurrence of ARGs also correlates to heavy metals, implying that ARG proliferation is not only dependent on the presence of antibiotics, but also on the presence of heavy metals (Ohore *et al.* 2019).

Redundancy analysis was conducted to explore correlations between measured ARGs and antibiotics. When considering influents of both WWTPs and separately, no significant correlations were found.

3.4. Between-day variability

Particularly high mass loads were observed in concomitance of rain events. Potential dilution effects due to rain would be accounted for when considering mass loads, as concentrations are multiplied by wastewater flows. However, in this case, particularly high mass loads were still observed for some ARGs. This can be explained by the fact that rainfall increases the typical hydraulic load of the WWTP, which significantly reduces the efficiency of wastewater treatment removal of genes (Pallares-Vega *et al.* 2019). Previous studies have shown that rainfall can increase ARGs abundance in surface waters (Amos *et al.* 2014; Zhang *et al.* 2016; Di Cesare *et al.* 2017). In the specific case of antibiotic residues, previous studies have, however, shown that these hardly fluctuate on a day-to-day basis in wastewater influents (Coutu *et al.* 2013). Increased ARG loads in wastewater during rainfall could be due to various factors, such as flushing of biofilms from sewer walls due to sudden and dramatic increases in flows in sewers. In fact, microorganisms attach to surfaces to protect themselves from environmental hazards (Costerton *et al.* 1978) and studies that screened biofilms have shown that the latter are reservoirs of ARGs (Balcázar *et al.* 2015). For instance, in a study by Schwartz *et al.* (2003), *VanA* and *ampC* genes were detected in wastewater biofilms. Similarly, sediment resuspension during high flows in rivers might also contribute to the increase observed in ARG mass loads (He *et al.* 2018). A similar phenomenon could take place in sewers, as solids are resuspended due to higher flows (Bailey *et al.* 2020). Finally, surface runoffs (in combined sewers) during rainfall could also be involved and would thus suggest that urban runoff might contribute to ARG loads and spreading of AMR (Almakki *et al.* 2019). Regardless of the reason behind the increase in ARG mass loads observed during rain events, the substantial variability in mass loads observed between days, even when no rainfall was involved, has important implications for sampling schemes. In fact, as sewage surveillance (or WBE) is increasingly recognized as an important addition to existing methods used to monitor the burden of AMR (Hendriksen *et al.* 2019; Pärnänen *et al.* 2019) or to monitor the circulation of pathogens at the community level (Medema *et al.* 2020b), it is important to develop sampling schemes which can capture the overall variability of target markers (e.g., resistance genes or RNA). From the results obtained here, collection of samples on single days, even if 24-h composites are collected within the same month (thus precluding any seasonal effects), does not appear sufficient to capture the total variability of ARG loads. Hence, individual 24-h composite samples are not adequate to obtain a representative image of the AMR burden in a community. This becomes even more important when the goal is to compare different communities and assess differences in AMR burden. Incorrect conclusion might be drawn due to the variability in ARG loads at the influent of WWTPs. Although to a lesser extent, important variations were also observed in effluents suggesting, when monitoring AMR emissions to the environment, adequate sampling schemes need to be designed. Future studies should thus involve the collection of multiple (consecutive) samples rather than individual 24-h composites. Similarly, to what has been done for antibiotics, additional studies should be carried out to determine the short-term (e.g., day-to-day) fluctuations in ARG loads measured at the influent of WWTPs.

4. CONCLUSION

Among the measured antibiotics, trimethoprim, sulfamethoxazole and flucloxacillin showed a significant positive correlation with the corresponding resistance genes *sul1* and *blaSHV*. Significant and strong correlations were also found between the integron Int1 and other ARGs, as well as between antibiotics and non-corresponding ARGs. These correlations were observed in effluents only, not in influents of the tested WWTPs. Mass loads varied substantially over time, especially for ARGs, which varied up to a factor of 35 between July and September. Moreover, higher mass loads were observed in concomitance with rainfall. Biofilm release, resuspension of particulate matter and urban runoff are among the considered explanations for these observations. However, an important variability in mass loads was observed between days, in particular for ARGs and regardless of rainfall. This suggests that higher sampling frequencies (using 24-h composite samples) are needed to adequately capture the fluctuations throughout the year and obtain representative estimates of ARG mass loads. This has important implications for studies which use sewage surveillance to assess AMR burden, or monitor the circulation of pathogens, at the community level. Although being highly useful to complement conventional surveillance methods in assessing community-wide exposure to AMR, adequate sampling schemes need to be designed to ensure that conclusions are drawn from valid and representative data. Additionally, following current practice for monitoring trends in chemicals measured in wastewater, including antibiotics, we advocate for the use of mass loads to interpret levels of AMR measured in wastewater.

ACKNOWLEDGEMENTS

The authors thank the personnel of the WWTPs for providing the samples and relevant data which were used in this study. The authors also acknowledge the help of Meindert de Graaf during sample collection and preparation. This work was funded by the Joint Research Program of the Dutch and Belgian drinking water companies.

DATA AVAILABILITY STATEMENT

All relevant data are included in the paper or its Supplementary Information.

CONFLICT OF INTEREST

The authors declare there is no conflict.

REFERENCES

Aarestrup, F. M. & Woolhouse, M. E. J. 2020 Using sewage for surveillance of antimicrobial resistance. *Science* **367** (6478), 630–632. https://doi.org/10.1126/science.aba3432.

Almakki, A., Jumas-Bilak, E., Marchandin, H. & Licznar-Fajardo, P. 2019 Antibiotic resistance in urban runoff. *Science of The Total Environment* **667**, 64–76.

Amos, G. C. A., Zhang, L., Hawkey, P. M., Gaze, W. H. & Wellington, E. M. 2014 Functional metagenomic analysis reveals rivers are a reservoir for diverse antibiotic resistance genes. *Veterinary Microbiology* **171** (3–4), 441–447.

Asghar, H., Diop, O. M., Weldegebriel, G., Malik, F., Shetty, S., El Bassioni, L., Akande, A. O., Al Maamoun, E., Zaidi, S., Adeniji, A. J., Burns, C. C., Deshpande, J., Oberste, M. S. & Lowther, S. A. 2014 Environmental surveillance for polioviruses in the global polio eradication initiative. *Journal of Infectious Diseases* **210**, S294–S303.

Bailey, O., Zlatanovic, L., van der Hoek, J. P., Kapelan, Z., Blokker, M., Arnot, T. & Hofman, J. 2020 A stochastic model to predict flow, nutrient and temperature changes in a sewer under water conservation scenarios. *Water* **12** (4), 1187.

Balcázar, J. L., Subirats, J. & Borrego, C. M. 2015 The role of biofilms as environmental reservoirs of antibiotic resistance. *Frontiers in Microbiology* **6**, 1216.

Bouki, C., Venieri, D. & Diamadopoulos, E. 2013 Detection and fate of antibiotic resistant bacteria in wastewater treatment plants: a review. *Ecotoxicology and Environmental Safety* **91**, 1–9.

Castrignano, E., Yang, Z. E., Feil, E. J., Bade, R., Castiglioni, S., Causanilles, A., Gracia-Lor, E., Hernandez, F., Plosz, B. G., Ramin, P., Rousis, N. I., Ryu, Y., Thomas, K. V., de Voogt, P., Zuccato, E. & Kasprzyk-Hordern, B. 2020 Enantiomeric profiling of quinolones and quinolones resistance gene QnrS in European wastewaters. *Water Research* **175**, 115653. https://doi.org/10.1016/j.watres.2020.115653.

Che, Y., Xia, Y., Liu, L., Li, A. D., Yang, Y. & Zhang, T. 2019 Mobile antibiotic resistome in wastewater treatment plants revealed by Nanopore metagenomic sequencing. *Microbiome* **7** (1), 1–13.

Costerton, J. W., Geesey, G. G. & Cheng, K. J. 1978 How bacteria stick. *Scientific American* **238** (1), 86–95.

Coutu, S., Wyrsch, V., Wynn, H. K., Rossi, L. & Barry, D. A. 2013 Temporal dynamics of antibiotics in wastewater treatment plant influent. *Science of the Total Environment* **458–460**, 20–26.

Di Cesare, A., Eckert, E. M., Rogora, M. & Corno, G. 2017 Rainfall increases the abundance of antibiotic resistance genes within a riverine microbial community. *Environmental Pollution* **226**, 473–478.

Duarte, D. J., Oldenkamp, R. & Ragas, A. M. J. 2019 Modelling environmental antibiotic-resistance gene abundance: a meta-analysis. *Science of the Total Environment* **659**, 335–341. https://doi.org/10.1016/j.scitotenv.2018.12.233.

Gao, P., Munir, M. & Xagoraraki, I. 2012 Correlation of tetracycline and sulfonamide antibiotics with corresponding resistance genes and resistant bacteria in a conventional municipal wastewater treatment plant. *Science of the Total Environment* **421**, 173–183.

Gatica, J. & Cytryn, E. 2013 Impact of treated wastewater irrigation on antibiotic resistance in the soil microbiome. *Environmental Science and Pollution Research* **20** (6), 3529–3538.

Gracia-Lor, E., Ibáñez, M., Zamora, T., Sancho, J. V. & Hernández, F. 2014 Investigation of pharmaceutical metabolites in environmental waters by LC-MS/MS. *Environmental Science and Pollution Research* **21** (8), 5496–5510.

Gracia-Lor, E., Castiglioni, S., Bade, R., Been, F., Castrignanò, E., Covaci, A., González-Mariño, I., Hapeshi, E., Kasprzyk-Hordern, B., Kinyua, J., Lai, F. Y., Letzel, T., Lopardo, L., Meyer, M. R., O'Brien, J., Ramin, P., Rousis, N. I., Rydevik, A., Ryu, Y., Santos, M. M., Senta, I., Thomaidis, N. S., Veloutsou, S., Yang, Z., Zuccato, E. & Bijlsma, L. 2017 Measuring excretion biomarkers in wastewater as a new source of epidemiological information: current state and future perspectives. *Environment International* **99**, 131–150. https://doi.org/10.1016/j.envint.2016.12.016.

Greeff, S. C. d. & Mouton, J. W. 2017 *NethMap 2017: Consumption of Antimicrobial Agents and Antimicrobial Resistance among Medically Important Bacteria in the Netherlands/MARAN 2017: Monitoring of Antimicrobial Resistance and Antibiotic Usage in Animals in the Netherlands in 2016.*

He, Y., Nurul, S., Schmitt, H., Sutton, N. B., Murk, T. A. J., Blokland, M. H., Rijnaarts, H. H. M. & Langenhoff, A. A. M. 2018 Evaluation of attenuation of pharmaceuticals, toxic potency, and antibiotic resistance genes in constructed wetlands treating wastewater effluents. *Science of the Total Environment* **631**, 1572–1581.

Hendriksen, R. S., Munk, P., Njage, P., Van Bunnik, B., McNally, L., Lukjancenko, O., Roder, T., Nieuwenhuijse, D., Pedersen, S. K., Kjeldgaard, J., Kaas, R. S., Clausen, P. T. L. C., Vogt, J. K., Leekitcharoenphon, P., van der Schans, M. G. M., Zuidema, T., de RodaHusman, A. M., Rasmussen, S., Peterson, B., Amid, C., Cochrane, G., Sicheritz-Ponten, T., Schmitt, H., Alvarez, J. R. M, Aidara-Kane, A., Pamp, S. J., Lund, O., Hald, T., Woolhouse, M., Koopmans, M. P., Vigre, H., Petersen, T. N. & Aarestrup, F. M. 2019 Global monitoring of antimicrobial resistance based on metagenomics analyses of urban sewage. *Nature Communications* **10** (1), 1124.

Hernández, F., Castiglioni, S., Covaci, A., de Voogt, P., Emke, E., Kasprzyk-Hordern, B., Ort, C., Reid, M., Sancho, J. V., Thomas, K. V., van Nuijs, A. L. N., Zuccato, E. & Bijlsma, L. 2018 Mass spectrometric strategies for the investigation of biomarkers of illicit drug use in wastewater. *Mass Spectrometry Reviews* **37** (3), 258–280.

Huijbers, P. M. C., Flach, C.-F. & Joakim Larsson, D. G. 2019 A conceptual framework for the environmental surveillance of antibiotics and antibiotic resistance. *Environment International* **130**, 104880. https://doi.org/10.1016/j.envint.2019.05.074.

Hutinel, M., Huijbers, P. M. C., Fick, J., Åhrén, C., Larsson, D. G. J. & Flach, C.-F. 2019 Population-level surveillance of antibiotic resistance in Escherichia Coli through sewage analysis. *Eurosurveillance* **24** (37). https://doi.org/10.2807/1560-7917.ES.2019.24.37.1800497.

Koninklijk Nederlands Meteorologisch Instituut (KNMI) n.d *Archief maand/seizoen/jaaroverzichten*. Available from: https://www.knmi.nl/nederland-nu/klimatologie/maand-en-seizoensoverzichten/ (accessed 8 April 2022).

Laht, M., Karkman, A., Voolaid, V., Ritz, C., Tenson, T., Virta, M. & Kisand, V. 2014 Abundances of tetracycline, sulphonamide and beta-lactam antibiotic resistance genes in conventional wastewater treatment plants (WWTPs) with different waste load. *PloS one* **9** (8), e103705.

Li, J., Wang, T., Shao, B., Shen, J., Wang, S. & Wu, Y. 2012 Plasmid-mediated quinolone resistance genes and antibiotic residues in wastewater and soil adjacent to swine feedlots: potential transfer to agricultural lands. *Environmental Health Perspectives* **120** (8), 1144–1149.

Luo, Y., Mao, D., Rysz, M., Zhou, Q., Zhang, H., Xu, L. & Alvarez, J. J. 2010 Trends in antibiotic resistance genes occurrence in the Haihe River, China. *Environmental Science & Technology* **44** (19), 7220–7225.

Manaia, C. M., Rocha, J., Scaccia, N., Marano, R., Radu, E., Biancullo, F., Cerqueira, F., Fortunato, G., Iakovides, I. C., Zammit, I, Kampouris, I., Vaz-Moreira, I. & Nunes, O. C. 2018 Antibiotic resistance in wastewater treatment plants: tackling the black box. *Environment International* **115**, 312–324. https://doi.org/10.1016/j.envint.2018.03.044.

Medema, G., Been, F., Heijnen, L. & Petterson, S. 2020a Implementation of environmental surveillance for SARS-CoV-2 virus to support public health decisions: opportunities and challenges. *Current Opinion in Environmental Science & Health*. https://doi.org/10.1016/j.coesh.2020.09.006

Medema, G., Been, F., Heijnen, L. & Petterson, S. 2020b Implementation of environmental surveillance for SARS-CoV-2 virus to support public health decisions: opportunities and challenges. *Current Opinion in Environmental Science & Health.*

Muloi, D., Ward, M. J., Pedersen, A. B., Fèvre, E. M., Woolhouse, M. E. J. & van Bunnik, B. A. D. 2018 Are food animals responsible for transfer of antimicrobial-resistant escherichia coli or their resistance determinants to human populations? a systematic review. *Foodborne Pathogens and Disease* **15** (8), 467–474. https://doi.org/10.1089/fpd.2017.2411.

Ohore, O. E., Addo, F. G., Zhang, S., Han, N. & Anim-Larbi, K. 2019 Distribution and relationship between antimicrobial resistance genes and heavy metals in surface sediments of Taihu Lake, China. *Journal of Environmental Sciences* **77**, 323–335.

Pallares-Vega, R., Blaak, H., van der Plaats, R., de Roda Husman, A. M., Leal, L. H., van Loosdrecht, M. C., Weissbrodt, D. G. & Schmitt, H. 2019 Determinants of presence and removal of antibiotic resistance genes during WWTP treatment: a cross-sectional study. *Water Research* **161**, 319–328.

Pärnänen, K. M., Narciso-da-Rocha, C., Kneis, D., Berendonk, T. U., Cacace, D., Do, T. T., Elpers, C., Fatta-Kassinos, D., Henriques, I., Jaeger, T., Karkman, A., Martinez, J. L., Michael, S. G., Michael-Kordatou, I., O'Sullivan, K., Rodriquez-Mozaz, S., Schwartz, T., Sheng, H., Sorum, H., Stedtfeld, R. D., Tiedje, J. M., Giustina, S. V. D., Walsh, F., Vaz-Moreira, I., Virta, M. & Manaia, C. M. 2019 Antibiotic resistance in European wastewater treatment plants mirrors the pattern of clinical antibiotic resistance prevalence. *Science Advances* **5** (3), eaau9124.

Paulus, G. K., Hornstra, L. M., Alygizakis, N., Slobodnik, J., Thomaidis, N. & Medema, G. 2019 The impact of on-site hospital wastewater treatment on the downstream communal wastewater system in terms of antibiotics and antibiotic resistance genes. *International Journal of Hygiene and Environmental Health* **222** (4), 635–644.

Peters, F. T., Drummer, O. H. & Musshoff, F. 2007 Validation of new methods. *Forensic Science International* **165** (2–3), 216–224.

Rizzo, L., Manaia, C., Merlin, C., Schwartz, T., Dagot, C., Ploy, M. C., Michael, I. & Fatta-Kassinos, D. 2013 Urban wastewater treatment plants as hotspots for antibiotic resistant bacteria and genes spread into the environment: a review. *Science of the Total Environment* **447**, 345–360.

Rodriguez-Mozaz, S., Chamorro, S., Marti, E., Huerta, B., Gros, M., Sànchez-Melsió, A., Borrego, C. M., Barcelo, D. & Balcázar, J. L. 2015 Occurrence of antibiotics and antibiotic resistance genes in hospital and urban wastewaters and their impact on the receiving river. *Water Research* **69**, 234–242.

Schlüter, A., Szczepanowski, R., Pühler, A. & Top, E. M. 2007 Genomics of IncP-1 antibiotic resistance plasmids isolated from wastewater treatment plants provides evidence for a widely accessible drug resistance gene pool. *FEMS Microbiology Reviews* **31** (4), 449–477.

Schwartz, T., Kohnen, W., Jansen, B. & Obst, U. 2003 Detection of antibiotic-resistant bacteria and their resistance genes in wastewater, surface water, and drinking water biofilms. *FEMS Microbiology Ecology* **43** (3), 325–335.

Strugeon, E., Tilloy, V., Ploy, M. C. & Da Re, S. 2016 The stringent response promotes antibiotic resistance dissemination by regulating integron integrase expression in biofilms. *MBio* **7** (4), e00868–16.

Subirats, J., Timoner, X., Sànchez-Melsió, A., Balcázar, J. L., Acuña, V., Sabater, S. & Borrego, C. M. 2018 Emerging contaminants and nutrients synergistically affect the spread of class 1 integron-integrase (intiI) and sulI genes within stable streambed bacterial communities. *Water Research* **138**, 77–85.

Thomas, K. V., Bijlsma, L., Castiglioni, S., Covaci, A., Emke, E., Grabic, R., Hernandez, F., Karolak, S., Kasprzyk-Hordern, B., Lindberg, R. H., Lopez de Alda, M., Meierjohann, A., Ort, C., Pico, Y., Quintana, J. B., Reid, M., Riekermann, J., Terzic, S., van Nuijs, A. L. N. & de Voogt, P. 2012 Comparing illicit drug use in 19 European cities through sewage analysis. *Science of the Total Environment* **432**, 432–439.

van Amsterdam, P., Companjen, A., Brudny-Kloeppel, M., Golob, M., Luedtke, S. & Timmerman, P. 2013 The European Bioanalysis Forum community's evaluation, interpretation and implementation of the European Medicines Agency guideline on Bioanalytical Method Validation. *Bioanalysis* **5** (6), 645–659.

Vaz-Moreira, I., Nunes, O. C. & Manaia, C. M. 2014 Bacterial diversity and antibiotic resistance in water habitats: searching the links with the human microbiome. *FEMS Microbiology Reviews* **38** (4), 761–778.

Wang, J., Chu, L., Wojnárovits, L. & Takács, E. 2020 Occurrence and fate of antibiotics, antibiotic resistant genes (ARGs) and antibiotic resistant bacteria (ARB) in municipal wastewater treatment plant: an overview. *Science of The Total Environment* **744**, 140997. https://doi.org/10.1016/j.scitotenv.2020.140997.

Wullings, B., Wubbels, G., Veenendaal, H. & Kool, D. V. D. 2007 Snelle, kwantitatieve detectie van Legionella pneumophilia met Q-PCR. *H 2 O* **40** (5), 39.

Xu, Y., Guo, C., Luo, Y., Lv, J., Zhang, Y., Lin, H., Wang, L. & Xu, J. 2016 Occurrence and distribution of antibiotics, antibiotic resistance genes in the urban rivers in Beijing, China. *Environmental Pollution* **213**, 833–840.

Yang, S., Cha, J. & Carlson, K. 2005 Simultaneous extraction and analysis of 11 tetracycline and sulfonamide antibiotics in influent and effluent domestic wastewater by solid-phase extraction and liquid chromatography-electrospray ionization tandem mass spectrometry. *Journal of Chromatography A* **1097** (1–2), 40–53.

Zhang, S., Pang, S., Wang, P., Wang, C., Han, N., Liu, B., Han, B., Li, Y. & Anim-Larbi, K. 2016 Antibiotic concentration and antibiotic-resistant bacteria in two shallow urban lakes after stormwater event. *Environmental Science and Pollution Research* **23** (10), 9984–9992.

Zhang, A. N., Li, L. G., Ma, L., Gillings, M. R., Tiedje, J. M. & Zhang, T. 2018 Conserved phylogenetic distribution and limited antibiotic resistance of class 1 integrons revealed by assessing the bacterial genome and plasmid collection. *Microbiome* **6** (1), 1–14.

First received 14 January 2022; accepted in revised form 20 June 2022. Available online 4 July 2022

doi: 10.2166/wh.2022.094

Modeling infection from SARS-CoV-2 wastewater concentrations: promise, limitations, and future directions

Jeffrey Soller[a], Wiley Jennings[b], Mary Schoen[a], Alexandria Boehm[c], Krista Wigginton[d], Raul Gonzalez[e], Katherine E. Graham[c], Graham McBride[f], Amy Kirby[b] and Mia Mattioli[b,*]

[a] Soller Environmental, LLC, 3022 King St, Berkeley, CA 94703, USA
[b] Waterborne Disease Prevention Branch, Division of Foodborne, Waterborne, and Environmental Diseases, National Center for Emerging and Zoonotic Infectious Diseases, Centers for Disease Control and Prevention, Atlanta, Georgia, USA
[c] Stanford University Department of Civil and Environmental Engineering, Stanford, California, USA
[d] Department of Civil and Environmental Engineering, University of Michigan, Ann Arbor 48109, Michigan, USA
[e] Hampton Roads Sanitation District, 1434 Air Rail Avenue, Virginia Beach, VA 23455, USA
[f] National Institute of Water & Atmospheric Research Ltd (NIWA), Hillcrest, Hamilton, New Zealand
*Corresponding author. E-mail: mmattioli@cdc.gov

ABSTRACT

Estimating total infection levels, including unreported and asymptomatic infections, is important for understanding community disease transmission. Wastewater can provide a pooled community sample to estimate total infections that is independent of case reporting biases toward individuals with moderate to severe symptoms and by test-seeking behavior and access. We derive three mechanistic models for estimating community infection levels from wastewater measurements based on a description of the processes that generate SARS-CoV-2 RNA signals in wastewater and accounting for the fecal strength of wastewater through endogenous microbial markers, daily flow, and per-capita wastewater generation estimates. The models are illustrated through two case studies of wastewater data collected during 2020–2021 in Virginia Beach, VA, and Santa Clara County, CA. Median simulated infection levels generally were higher than reported cases, but at times, were lower, suggesting a discrepancy between the reported cases and wastewater data, or inaccurate modeling results. Daily simulated infection estimates showed large ranges, in part due to dependence on highly variable clinical viral fecal shedding data. Overall, the wastewater-based mechanistic models are useful for normalization of wastewater measurements and for understanding wastewater-based surveillance data for public health decision-making but are currently limited by lack of robust SARS-CoV-2 fecal shedding data.

Key words: COVID-19, infectious disease surveillance, mechanistic modeling, SARS-CoV-2, wastewater

HIGHLIGHTS

- Reported COVID-19 cases do not capture total infections which is important for understanding community disease transmission.
- We present three wastewater mechanistic simulation models to estimate community infection levels and demonstrate them through two case studies.
- Wastewater-based mechanistic models presented are useful for public health decisions but currently limited by limited SARS-CoV-2 fecal shedding data.

GRAPHICAL ABSTRACT

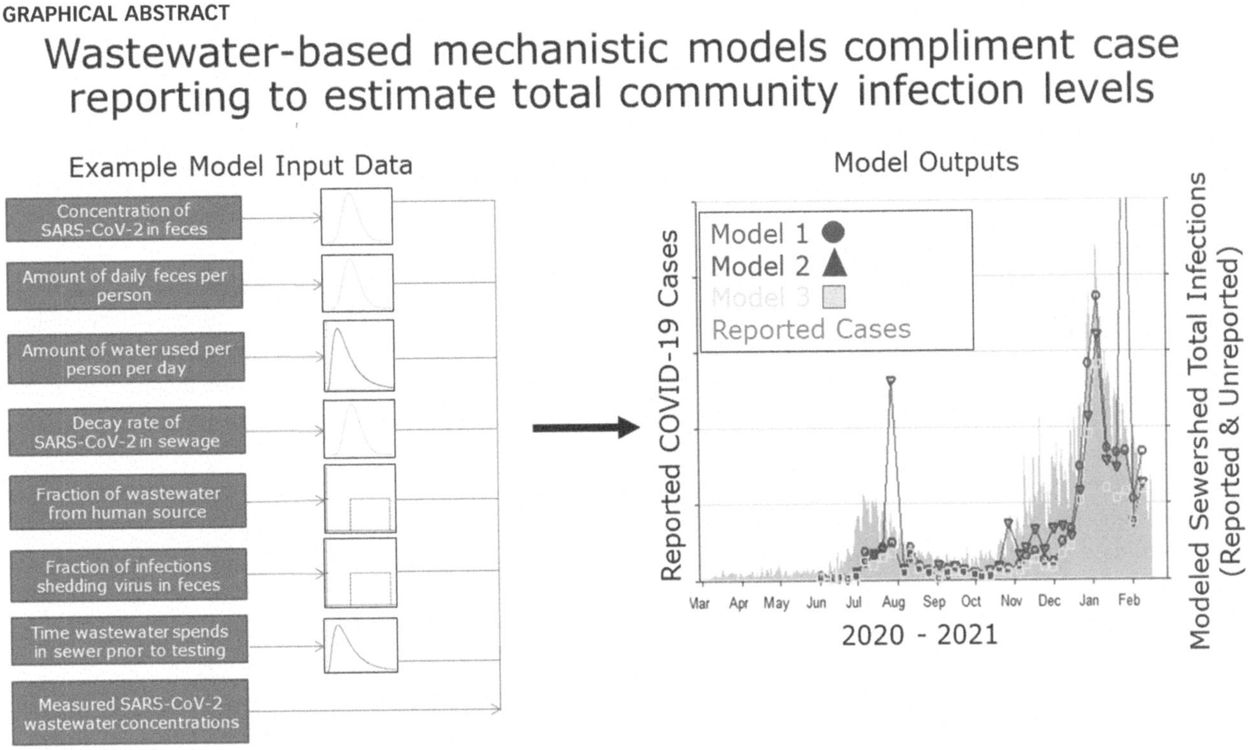

INTRODUCTION

SARS-CoV-2, the virus that causes COVID-19 infections, primarily causes respiratory illness (CDC 2021a). However, the RNA from this virus is also present in feces of infected symptomatic, pre-symptomatic, post-symptomatic, and asymptomatic individuals (Cheung *et al.* 2020; Foladori *et al.* 2020; WHO 2020a; Wolfel *et al.* 2020; Zhang *et al.* 2020; Zheng *et al.* 2020). Environmental surveillance through the testing of wastewater for evidence of pathogens has a long history of use in public health, particularly for poliovirus and more recently antimicrobial resistance (Asghar *et al.* 2014; WHO 2020b, 2020c). SARS-CoV-2 RNA has been reported in untreated wastewater and settled solids (e.g., sludge) in a number of countries (WHO 2020a), and wastewater SARS-CoV-2 RNA monitoring data have proven useful as an indicator of community illness in conjunction with traditional case reporting and surveillance methods (Medema *et al.* 2020; Peccia *et al.* 2020; D'Aoust *et al.* 2021; Fernandez-Cassi *et al.* 2021; Saguti *et al.* 2021; Weidhaas *et al.* 2021; Hewitt *et al.* 2022). Because of this, wastewater surveillance systems are being implemented for the COVID-19 response to provide data on overall infection trends and variant tracking within specific populations as a complement to clinical- and individual-based surveillance data for public health decision-making (Bivins *et al.* 2020; Medema *et al.* 2020; Peccia *et al.* 2020; Betancourt *et al.* 2021; Graham *et al.* 2021; Zhu *et al.* 2021; Kirby *et al.* 2022).

In addition to community infection trends, wastewater-based surveillance data have been proposed as a tool for estimating absolute community-level COVID-19 infections (Ahmed *et al.* 2020; Bivins *et al.* 2020; Hart & Halden 2020; Gerrity *et al.* 2021; Wurtzer *et al.* 2021). Estimating absolute infection levels (reported and unreported) is important for understanding disease transmission and designing effective mitigation strategies but has proven difficult to achieve using traditional surveillance indicators due to the large and variable relative burden of unreported cases (Bivins *et al.* 2020; Medema *et al.* 2020). To date, methods for estimating wastewater-based SARS-CoV-2 infections have included mechanistic-, statistical-, and epidemiological-based numerical modeling approaches (Baud *et al.* 2020; Ceylan 2020; Medema *et al.* 2020; Paul *et al.* 2020; Turk *et al.* 2020; Gerrity *et al.* 2021; Huisman *et al.* 2022; Kaplan *et al.* 2022; Weidhaas *et al.* 2021). For a mechanistic example, Ahmed *et al.* (2020) proposed an approach for COVID-19 wastewater surveillance in Australia based on the concentration of SARS-CoV-2 RNA in wastewater, the volume wastewater generated daily in a catchment, the number of SARS-CoV-2 RNA copies shed in stool by an infected individual each day, and the concentration of SARS-CoV-2 RNA in feces, each of which is represented by a combination of point estimate values and statistical distribution estimates in a Monte Carlo-based numerical simulation. Kaplan *et al.* (2022) proposed an epidemiologically- and statistically-calibrated scaling model to estimate incidence in a community from sludge RNA, a transmission dynamics model that aligns lagged epidemic indicators, and a site-specific scaling factor.

Mechanistic wastewater models are based on the processes that generate SARS-CoV-2 RNA signals in wastewater without additional statistical calibration against reported cases. Although wastewater-based disease research is rapidly advancing, the previously proposed mechanistic models fail to account for salient attributes of the wastewater processes, such as target decay within the system, and to account for variability in available data to estimate human contribution to the wastewater stream. In addition, the proposed statistical and epidemiological wastewater-infection-based models are either site-specific, using statistical correction factors dependent on reported cases, or are dependent on other clinical surveillance data; both limit the ability of wastewater to provide estimates independent of the variability in these clinical surveillance data (Li *et al.* 2020; Wu *et al.* 2020; Fernandez-Cassi *et al.* 2021). Moreover, epidemiologic compartmental models anchored to hospitalization or death rates are limited in their ability to provide near real-time infection estimates due to delayed nature of these indicators (Kaplan *et al.* 2021).

In this study, we derive three mechanistic models for estimating community infection levels from wastewater measurements that are agnostic to location. Each model uses a distinct approach to account for the human fecal content of wastewater, including endogenous microbial markers, wastewater flow measurements, and population-level estimates of per-capita wastewater generation. A numerical method is used to simulate the models, which are illustrated through two case studies of wastewater data and COVID-19 case data collected over several months in 2020–2021 (Virginia Beach, VA, and Santa Clara County, CA). We also propose a mathematical approach for connecting wastewater-based infection estimates to clinical surveillance data for public health interpretation. Finally, we describe the uncertainty associated with each modeling approach and the critical data needed to effectively use wastewater pathogen measurements to estimate community infections for public health decision-making.

METHODS

Our model formulations were developed to estimate the fraction of infected individuals in a community that could be shedding SARS-CoV-2 RNA either through respiratory or fecal secretions, or both, regardless of symptoms (referred to hereafter as 'infected'). Some of these individuals may not be considered 'infected' from a clinical perspective because the fecal shedding of viral RNA does not necessarily correspond with respiratory symptoms or shedding (Wolfel *et al.* 2020; Zheng *et al.* 2020). We use a stochastic, Monte Carlo numerical simulation-based mechanistic modeling approach (Soller & Eisenberg 2008) encompassing the following three related formulations: (1) a flow generation formulation based on per-capita domestic potable water use adjusted for wastewater generation from non-domestic sources; (2) a fecal strength formulation that is a modified flow generation formulation to include an endogenous human fecal control as a proxy for dilution and loss of human feces and associated pathogens during the sewer transport and testing processes; and (3) a flow receipt formulation based on collection site daily flow measurements and wastewater utility provided sewershed population estimates. The inputs of the models are microbial wastewater influent concentrations (i.e., SARS-CoV-2 RNA and endogenous control when applicable), where influent is defined as the untreated wastewater entering a treatment plant containing both solids and liquids.

The three formulations are summarized below, and detailed mathematical specifications are provided in Supplementary Material (SI) Section A. The model formulations, as applicable, incorporate the following: variability in each model parameter using data from the peer reviewed scientific literature; differences between disease presentation that exhibit gastrointestinal (GI) symptoms and those that do not (i.e., account for differences in total fecal matter shed for those with and without GI symptoms); per-capita water use; the decay of viral RNA during transport within the wastewater collection system, and finally, the fraction of wastewater that results from human fecal waste (versus commercial, industrial, run-off, etc.) (refer to Supplementary Material, Table A1 for a comparison of the three formulations). All three modeling approaches are fundamentally dependent on the levels of SARS-CoV-2 RNA in the feces of infected individuals and are related as shown mathematically in Supplementary Material, Section A.

Flow generation formulation

The flow generation formulation is based on a mechanistic conceptualization of daily SARS-CoV-2 RNA input into the wastewater stream resulting from the fraction of SARS-CoV-2-infected individuals fecally shedding the virus (F_{shed}), both with (F_{diar},

D_{diar}, V_{diar}) and without GI symptoms (M_{nodiar} D_{nodiar}), over the course of an individual's daily water use (V_{ww}). The formulation then accounts for time within the sewer system (T_{sewer}) and resulting decay over that period (k_{virus}) to estimate the total daily SARS-CoV-2 RNA concentration at the wastewater treatment plant influent (or settled solids) collection point (D_{ww}) (Equation (1)). Settled solids concentrations are converted to estimated wastewater influent concentrations via the use of a viral-specific solid-to-liquid partitioning coefficient (K_d) following methods described in Kim *et al.* (2022) and Graham *et al.* (2021). Of note, T_{sewer} does not include time spent in the sedimentation basin for settled solids measurements, which in some systems could be longer than time spent in the sewer system.

$$D_{WW} = \frac{[D_{diar}V_{diar}F_{diar} + D_{nodiar}M_{nodiar}(1 - F_{diar})]e^{(-k_{virus}T_{sewer})}F_{shed}}{V_{WW}}F_{active}F_{cont} \tag{1}$$

where F_{shed} is the fraction of infected individuals that shed viral RNA in feces (unitless); F_{diar} is the fraction of infected individuals with diarrhea (unitless); D_{diar} is the SARS-CoV-2 RNA concentration in feces among infected individuals with diarrhea (viral copies/mL); V_{diar} is the volume of feces per person per day among infected individuals with diarrhea (mL/day); D_{nodiar} is the SARS-CoV-2 RNA concentration in feces among infected individuals without diarrhea (viral copies/g); M_{nodiar} is the mass of feces per person per day among infected individuals without diarrhea (g/day); V_{ww} is the wastewater generated per person per day (L/person day); k_{virus} is the pseudo-first-order decay coefficient for SARS-CoV-2 RNA in sewage (1/day); T_{sewer} is the time wastewater spends in the sewer prior to reaching the wastewater treatment plant (days); F_{cont} is the volume fraction of wastewater coming from a source that could potentially contain SARS-CoV-2 RNA (unitless).

By rearranging the equation, the fraction of the population with an active infection (F_{active}) can be estimated for a specified median level of SARS-CoV-2 in the wastewater influent (Equation (2)).

$$F_{active} = \frac{D_{WW}V_{WW}}{[D_{diar}V_{diar}F_{diar} + D_{nodiar}M_{nodiar}(1 - F_{diar})]e^{(-k_{virus}T_{sewer})}F_{shed}F_{cont}} \tag{2}$$

The defining attribute of this formulation is a variable that defines the fraction of daily municipal wastewater that could be contributing SARS-CoV-2 RNA into the waste stream (F_{cont}). F_{cont} represents a variable, but time-invariant, estimate of the 'human fecal strength' (FS) of the wastewater. The salient assumption for this formulation is that it is feasible (e.g., for a wastewater treatment plant manager at the site monitored) to estimate with reasonable accuracy.

Fecal strength formulation

The flow generation formulation is limited when F_{cont} cannot be estimated accurately or when temporal variation of fecal strength is important. Such temporal variation could occur because of increased flows due to wet weather or variable wastewater composition due to dynamic water usage and human behavioral patterns (e.g., governmental restricted water usage or communities with large population influxes such as tourist locations). To address this, we reconfigured the flow generation model to define FS-based on an endogenous microbial control – a wastewater constituent that is present in relatively stable, high, and measurable concentrations in human excreta – rather than an expert judgement point estimate. We used Pepper Mild Mottle Virus (PMMoV) as the example human fecal-specific endogenous control (Zhang *et al.* 2006; Hamza *et al.* 2011).

This alternative approach for parameterization of fecal strength is the ratio of the concentration of endogenous control measured in wastewater, adjusted for decay in the sewer, to the expected wastewater endogenous control concentration if only human wastewater sources were contributing (e.g., sources of wastewater that could potentially contain SARS-CoV-2 RNA). A theoretical fecal strength of 1 would mean that human-associated wastewater is undiluted by other sources (e.g., industrial or run-off). The notation for the variables in the fecal strength formulation and the flow receipt formulation follow those presented above for the flow generation formulation where D refers to the concentration in feces and M refers to the mass of feces, with subscripts providing details about the specific parameters. FS is defined as follows, noting that this quotient is a unitless fraction:

$$FS = \frac{D_{WW-endog}}{e^{(-k_{endog}T_{sewer})}(D_{stool-endog} \times M_{stool}/V_{WW})} \tag{3}$$

where k_{endog} is the pseudo-first-order decay coefficient for endogenous control in sewage (1/day); $D_{stool-endog}$ is the endogenous control concentration in stool (gc/g); and M_{stool} is the stool mass generated per person per day (g/person day).

To demonstrate the mechanistic relatedness of the flow generation and fecal strength formulations, *FS* can be substituted for F_{cont} in Equation (2) of the fecal strength formulation as follows:

$$F_{active-SARS}=\frac{D_{WW-SARS}V_{WW}}{[D_{diar}V_{diar}F_{diar}+D_{nodiar}M_{nodiar}(1-F_{diar})]e^{(-k_{virus}T_{sewer})}F_{shed-SARS}((D_{WW-endog})/e^{(-k_{endog}T_{sewer})}(D_{stool-endog}\times M_{stool}/V_{WW}))} \tag{4a}$$

Comparison of Equations (2) and (4a) indicates that two additional pieces of information are needed for this formulation compared to the flow generation formulation: (1) the concentration of the endogenous control in stool (D_{stool_endog}), which does not distinguish concentrations between diarrheal and non-diarrheal stool and (2) the corresponding decay coefficient (k_{endog}).

Assuming that the concentration in stool is ~1.0 g/mL (Penn *et al.* 2018) and that the decay rate constants of the endogenous control and SARS-CoV-2 RNA are negligible or similar (which is most plausible for an RNA viral control such as PMMoV), the fecal strength formulation simplifies as follows (details provided in Supplementary Material, Section A):

$$F_{active-SARS}=\frac{D_{stool-endog}D_{WW-SARS}}{D_{stool-SARS}D_{WW-endog}F_{shed-SARS}}. \tag{4b}$$

This fecal strength formulation (Equation (4a) and (4b)) has both advantages and disadvantages compared to the flow generation formulation (Equation (2)). The advantages are that the endogenous control (e.g., PMMoV RNA) measurements can theoretically serve as a normalizing control in that they can reflect the strength of human feces in the wastewater, the viral losses and dilution that occur as stool is transported from excreta through the sewage collection system to the sampling location, and the viral losses that occur during laboratory processing and testing. The primary disadvantage of this formulation is that it relies on measurements of the endogenous control concentrations in human feces, of which data are limited, and that endogenous control measurements must be made in each sample. This disadvantage is further complicated by the fact that wastewater processing and testing methods are not standardized, meaning that absolute concentrations generally cannot be compared across laboratories (without first conducting inter-lab comparisons).

Flow receipt formulation

Like the flow generation formulation, the flow receipt formulation relies on flow rate measurements to estimate the number of active SARS-CoV-2 infections within a sewershed (F_{active}) (Equation (5d)). However, in this formulation, the flow estimates are measured at the sampling site, often at the influent to a wastewater treatment plant (V_{WWTP}) over the course of one day, which enables flow to vary from sample-to-sample (or day to day). To put flow on a per-capita basis, this formulation also requires an explicit sewershed population estimate (N) and estimate of the number of infected individuals in a population (I_{active}). This formulation is derived from a mass balance equation on the number of RNA copies in wastewater:

$$D_{WW}V_{WWTP}=D_{stool}V_{stool}e^{-k_{virus}T_{sewer}}F_{shed}I_{active} \tag{5a}$$

where

$$I_{active}=F_{active}N \tag{5b}$$

Thus,

$$\frac{D_{WW}V_{WWTP}}{N}=D_{stool}V_{stool}\,e^{-k_{virus}T_{sewer}}F_{shed}F_{active} \tag{5c}$$

or

$$F_{active} = \frac{D_{WW} V_{WWTP}}{D_{stool} V_{stool} e^{-k_{virus} T_{sewer}} F_{shed} N} \tag{5d}$$

Similar to the derivations provided above, this flow receipt formulation is equivalent to the fecal strength formulation under specific, realistic conditions (mathematical details are provided in the Supplementary Material, refer to formulas A3.6–A3.11). This formulation also assumes that the SARS-CoV-2 RNA wastewater measurement multiplied by the average daily flow rate reasonably represents the average SARS-CoV-2 RNA load over one day. A primary advantage of this formulation is that, for samples collected at treatment plants, flow measurements and population estimates are commonly available. Therefore, the flow receipt formulation reduces the requisite number of laboratory measurements compared to the other two formulations. Moreover, flow measurements and population estimates may have lower relative error and/or variability than microbial endogenous control measurements, which vary on a log-scale basis. The disadvantages of the flow receipt formulation are that it does not reflect SARS-CoV-2 RNA losses that occur during laboratory processing or unpredictable SARS-CoV-2 RNA losses in the sewage system that are attributable to processes other than the modeled gene target decay. This formulation also requires a static wastewater contributing population estimate, total daily flow measurements, and an aggregated average value for D_{stool} (used to account for variability in influent over the course of a day into a treatment facility).

Wastewater modeling case studies

The characteristics of the wastewater case studies are summarized in Table 1, including information on geographic area and population represented, sampling and testing methodologies, and clinical case data. For both case studies, viral wastewater concentration data (SARS-CoV-2 and PMMoV RNA) were measured by reverse transcriptase quantitative polymerase chain reaction (RT-qPCR) using digital droplet PCR technology (ddPCR). Note that the SARS-CoV-2 RNA and PMMoV RNA data were collected from primary settled solids rather than wastewater influent for the San Jose, CA case study (SJ). For the purposes of numerical modeling, the observed settled solids concentrations of SARS-CoV-2 RNA and PMMoV RNA were converted to estimated wastewater influent concentrations using a solid-to-liquid partitioning coefficient (K_d) following methods described in Kim *et al.* (2022) and Graham *et al.* (2021). Use of a fixed K_d assumes the viral RNA remains at a consistent level of equilibrium between water and solids through the sample collection period.

Model parameterization and implementation

To parameterize the model for SARS-CoV-2, we used SARS-CoV-2 data from the scientific literature, other surrogate pathogen data when not available for SARS-CoV-2, and professional judgement as needed – details are provided in Supplementary Material, Section B. The model variables, corresponding statistical distributions, and distribution parameters used in the model formulations are shown in Table 2.

For each formulation, we ran simulations using Eqns 2, 4a and 5 (for the flow generation, fecal strength, and flow receipt formulations, respectively) using case study SARS-CoV-2 influent concentration data to estimate F_{active}. For the flow generation and fecal strength formulations, the measured daily average SARS-CoV-2 RNA concentration (via 24-h composite sampling) was converted to an estimated median SARS-CoV-2 RNA concentration. This conversion is necessary so that subsequent calculations are not biased by the measured SARS-CoV-2 RNA concentrations, which are inherently mean values and right skewed rather than the median values that are required for the subsequent calculations. For each observation, this conversion was accomplished by running an initial set of simulations to solve for an F_{active} value in which the simulated mean concentration equaled the observed mean. The median value of the simulated distribution was then used in the subsequent calculations as indicated above. For the flow receipt formulation, we computed an average value of D_{stool} through simulation for use in Eqn 5c for consistency with the conceptual model of flow receipt at a treatment plant rather than variation in D_{stool} from individuals. During each iteration of the numerical simulation, values for the parameters were drawn from their respective statistical distributions to estimate F_{active}, and each Monte Carlo simulation comprised of 100,000 iterations for each case study and model formulation.

Table 1 | Overview of case studies including information on wastewater service area, sampling, testing, and clinical case data

Characteristic	Virginia Beach, VA	San Jose, CA
Geographic area	Atlantic Treatment Plant, Hampton Roads region, VA, USA	San Jose wastewater treatment plant, Santa Clara County, CA, USA
Sewershed population	343,016	1,458,000
Wastewater sampling dates	6/2/20 to 2/9/21	3/18/20 to 3/31/21
Sampling frequency	1 per week	Weekly from 3/18/20 to 7/14/20, then 5 × /week
Wastewater matrix	Primary influent	Primary settled solids
Number wastewater samples	37	195
Sample collection	24-h flow-weighted composite	24-h composite solids sample
Sample concentration	Electronegative membrane filtration with $MgCl_2$ addition and acidification followed by extraction via NucliSENS easyMag. Solids were not removed or allowed to settle prior to filtration.	Dewatering by centrifugation at 24,000 g at 4 °C for 30 min and decanting supernatant. Extraction of 1.8 – 2.8 g of sample using the RNeasy PowerSoil total RNA kit with further purification using Zymo OneStep PCR inhibitor removal columns.
SARS-CoV-2 RNA RT-PCR target	N1, N2[a]	N
Detection technology	1-step RT-ddPCR	1-step RT-ddPCR
Case study reference	Gonzalez *et al.* (2020, 2021)	Graham *et al.* (2021); Kim *et al.* (2022); Wolfe *et al.* (2021)
Case data geographic area	The sewershed area determined from the treatment plant service area polygon and address-level case data	The sewershed service area of the treatment plant comprises approximately 75% of the cumulative cases for the entire county
Case data dates	Earliest available data hierarchy: (1) illness onset date; (2) specimen collection date of earliest associated lab; (3) date of diagnosis; (4) earliest date received by local/county health department, date received by state, date of report, investigations start date, confirmation date, investigation create date	1/27/2020 to 11/4/2021 reported by Santa Clara County Health District.

PCR, polymerase chain reaction; ddPCR, digital droplet PCR.

[a]In 2021, this changed to using a single N target, a slight modification of the N2 target published by Gonzalez *et al.* (2020, 2021).

Model performance evaluations

We conducted a series of in-depth analyses to determine the extent to which selected parameter value choices in the three model formulations affected the output (see Supplementary Material, Section D for further detail). The parameters evaluated include the following: (1) the concentration of SARS-CoV-2 RNA in feces of infected individuals, (2) the duration of SARS-CoV-2 RNA shedding in feces, and (3) the partitioning coefficients for SARS-CoV-2 and PMMoV (for the San Jose, CA case study where primary sludge was sampled). Not all model parameters were evaluated in this way and other potentially important considerations related to sensitivity include, but are not limited to (1) the gene target used to quantify SARS-CoV-2 RNA in sewage; (2) uncertainty or variability of sewage concentration measurements including the approach to handling SARS-CoV-2 RNA values reported below quantifiable limits; and (3) relative contributions of non-fecal (mucosal) SARS-CoV-2 RNA shedding to the wastewater stream. Consideration of those additional factors could serve as a guide for future study.

RESULTS

Daily confirmed cases, cumulative confirmed cases, observed wastewater SARS-CoV-2 RNA and PMMoV RNA concentrations from Virginia Beach, VA (VB) are presented in Figure 1(a). These case data represent the sewershed area determined from the Atlantic wastewater treatment plant service area polygons and address-level cases as reported in

Table 2 | Model variables, distributions, and distribution parameters

Model parameter	Description	Distribution	Distribution parameters
D_{diar}	SARS-CoV-2 density in feces among infected individuals with diarrhea (log10 virus copies/mL)	Normal	(5.1, 0.76)
V_{diar}	volume of feces per person per day among infected individuals with diarrhea (mL/day)	Normal	(1000, 100)
F_{diar}	Fraction of infections with diarrhea (unitless)	Uniform	(0.2, 0.3)
D_{nodiar}	SARS-CoV-2 density in feces among infected individuals without diarrhea (log10 virus copies/g)	Normal	(4.067, 1.591)
M_{nodiar}	Mass of feces per person per day for infected individuals without diarrhea (g/day)	Lognormal	(4.84, 0.4)
V_{ww}	Wastewater generated per person per day (L/day)	Lognormal	(5.397, 0.1595)
T_{sewer}	time wastewater spends in the sewer prior to reaching the wastewater treatment plant (h)	Lognormal	(1.2, 0.85)
F_{shed}	Fraction of infections resulting in shedding of viral RNA in feces (unitless)	Uniform	(0.6, 0.8)
F_{cont}	Volume fraction of wastewater coming from a source that could potentially be contributing to the observed reported influent SARS-CoV-2 concentration (unitless)	Uniform	(0.10, 0.85)
k_{virus}	Pseudo-first-order decay coefficient estimate for SARS-CoV-2 RNA target (day^{-1})	Point Estimate	0.29
k_{PMMoV}	Pseudo-first-order decay coefficient estimate for PMMoV target (day^{-1})	Point Estimate	0.29
$k\_d_{SARS\_CoV2}$	SARS-CoV-2 Partitioning coefficient for conversion of settled solids density to wastewater influent density (unitless)	Point Estimate – Median	900
		Point Estimate (lower, upper)	(280, 10,000)
k_d_{PMMoV}	PMMoV Partitioning coefficient for conversion of settled solids density to wastewater influent density (unitless)	Point Estimate – Median	2000
		Point Estimate (lower, upper)	(1000, 30,000)
D_{WW_endog}	PMMoV density in feces (log10 virus copies/g)	Uniform from min-median;	(5.58, 8.28)
		Uniform from median-max	(8.28, 9.99)

Gonzalez *et al.* (2020, 2021). The daily confirmed cases, cumulative confirmed cases, and computed influent SARS-CoV-2 RNA and PMMoV RNA data for the San Jose, CA (SJ) case study are presented below in Figure 1(b). The SJ case data represent the City of San Jose treatment plant service area within Santa Clara County, which comprises approximately 75% of the population and 75% of the cumulative cases for the county.

Median F_{active} values from the case study simulations are presented in Figure 2 alongside the reported daily case numbers (sewershed level for Virginia Beach and county level for San Jose). Note that the ratio of F_{active} to the confirmed daily cases varies between the two case studies as shown on Figure 2(a) and 2(b). In general, the median F_{active} values from all three formulations followed the reported case data trends well. However, daily simulated infection estimates exhibited large ranges (two orders of magnitude), in part due to the strong model dependence on highly variable viral fecal shedding data (Supplementary Material, Figure D1). Given the assumed uncertainty in the limited fecal SARS-CoV-2 RNA shedding data (further detail in Supplementary Material, Section C), the precision and accuracy of the estimated median F_{active} values is unknown within the range of simulated values.

To evaluate the simulated F_{active} results within the context of the reported number of cases at each study location, we relied on published data indicating that infected individuals shed detectable SARS-CoV-2 RNA in their feces for a median of 21 days (Weiss *et al.* 2020; Zhou *et al.* 2020). Because the model formulations treat fecal shedding as a time-invariant parameter, we computed the cumulative number of confirmed cases over 21 days and plotted those against F_{active} multiplied by the wastewater utility estimated sewershed contributing population (Figure 3). The slopes of the lines shown represent the average

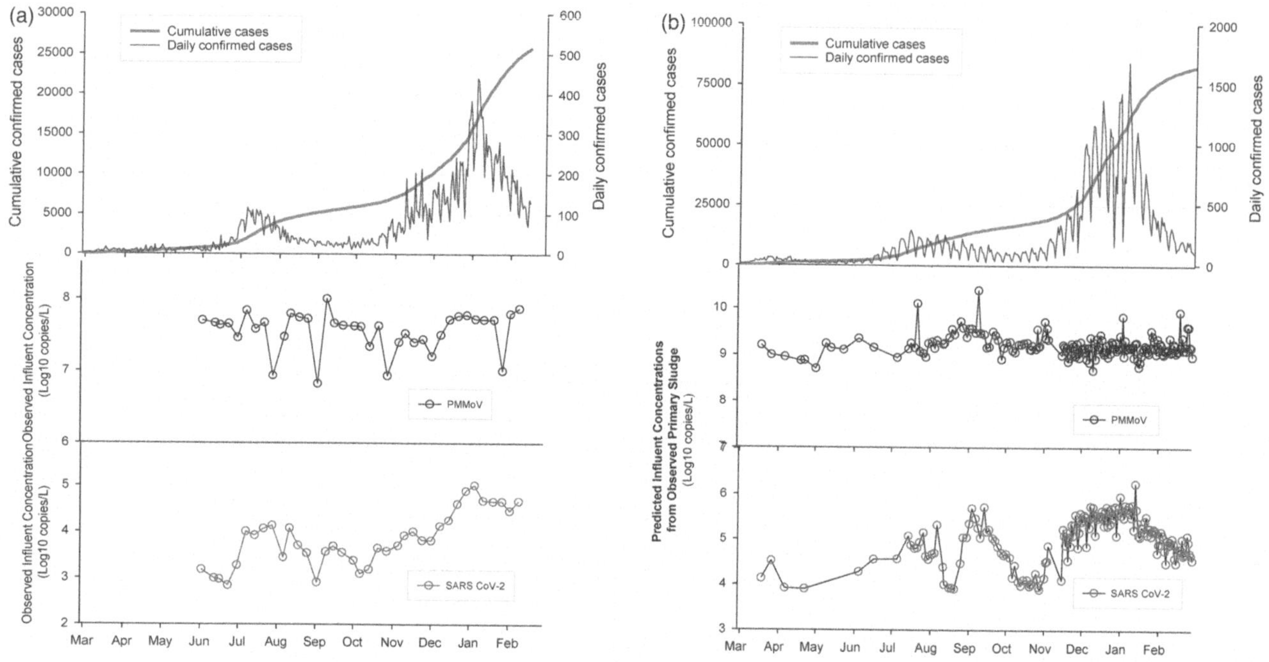

Figure 1 | March 2020–February 2021 daily confirmed cases, cumulative confirmed cases, SARS-CoV-2 RNA and PMMoV RNA concentrations for (a) Virginia Beach, VA and (b) San Jose, CA.

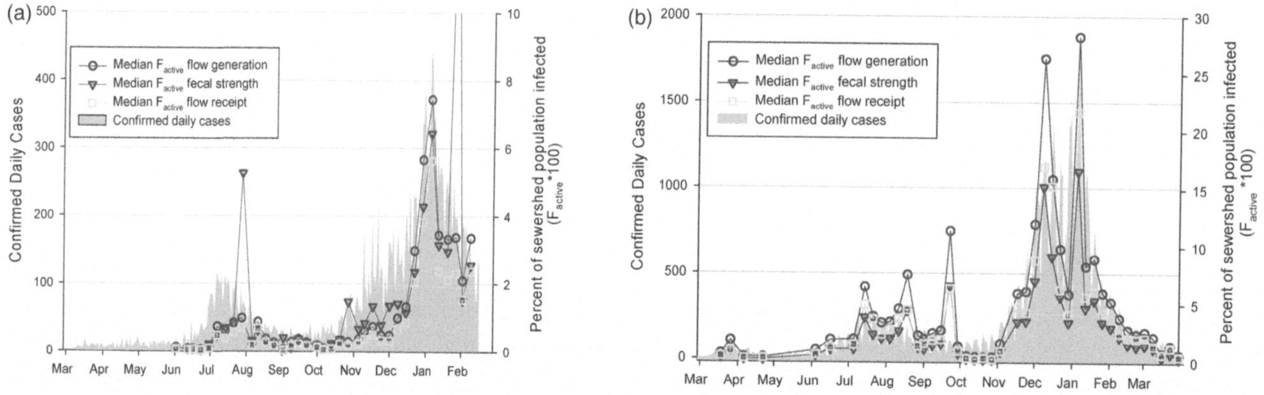

Figure 2 | Estimated percent of population actively infected each day monitored for the three model formulations compared to reported cases for (a) Virginia Beach, VA and (b) San Jose, CA beginning March 2020. Note that the ratio of the left and right *y*-scales differs between plots (a) and (b).

'ascertainment ratio (AR)' (i.e., ratio of predicted to reported cases) over the period evaluated. As shown in Figure 3, at times the median simulated infection levels were lower than reported case burden. This observation suggests that there could be a discrepancy between the dates of reported cases and wastewater data (refer to Discussion), temporally varying nature of the true AR, and/or inaccuracies in the modeling results.

Following Wolfe *et al.* (2021), we plotted the observed daily cases as a function of the influent concentrations of SARS-CoV-2 RNA for both case studies, and influent concentrations of SARS-CoV-2 RNA normalized by observed PMMoV RNA concentrations to evaluate and compare across case studies the association between wastewater concentrations of SARS-CoV-2 RNA and reported cases (Figure 4). As shown, the datasets from the two case studies are more comparable when normalized by PMMoV highlighting a potentially important implementation consideration.

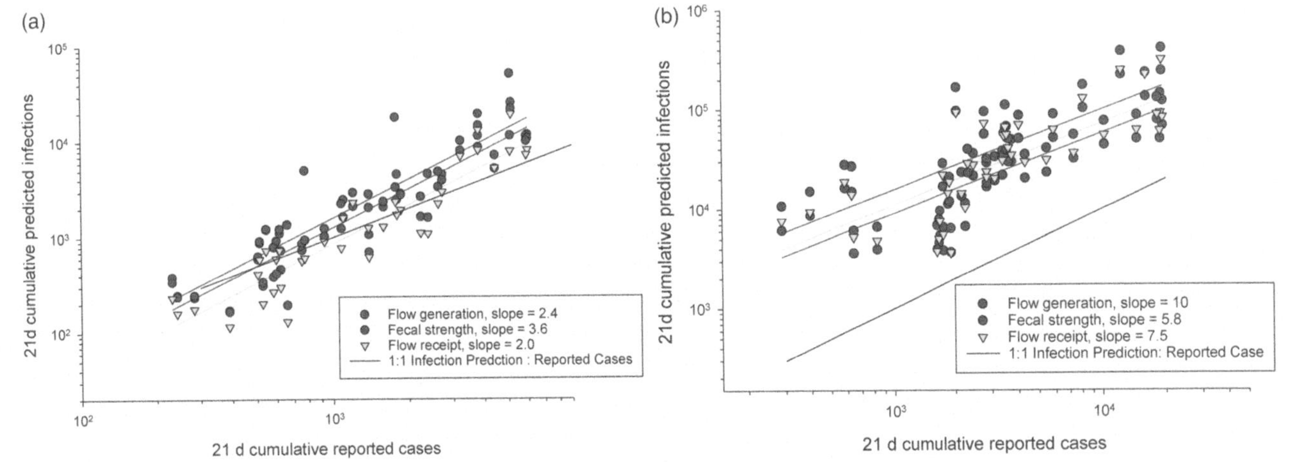

Figure 3 | 21-day reported cases versus predicted cases for the median outcome value of three model formulation simulations (a) Virginia Beach, VA and (b) San Jose, CA beginning March 2020. Reported slopes, representing ARs over surveyed period, are derived from linear regressions of linear scaled reported and estimated cases using the three model formulations.

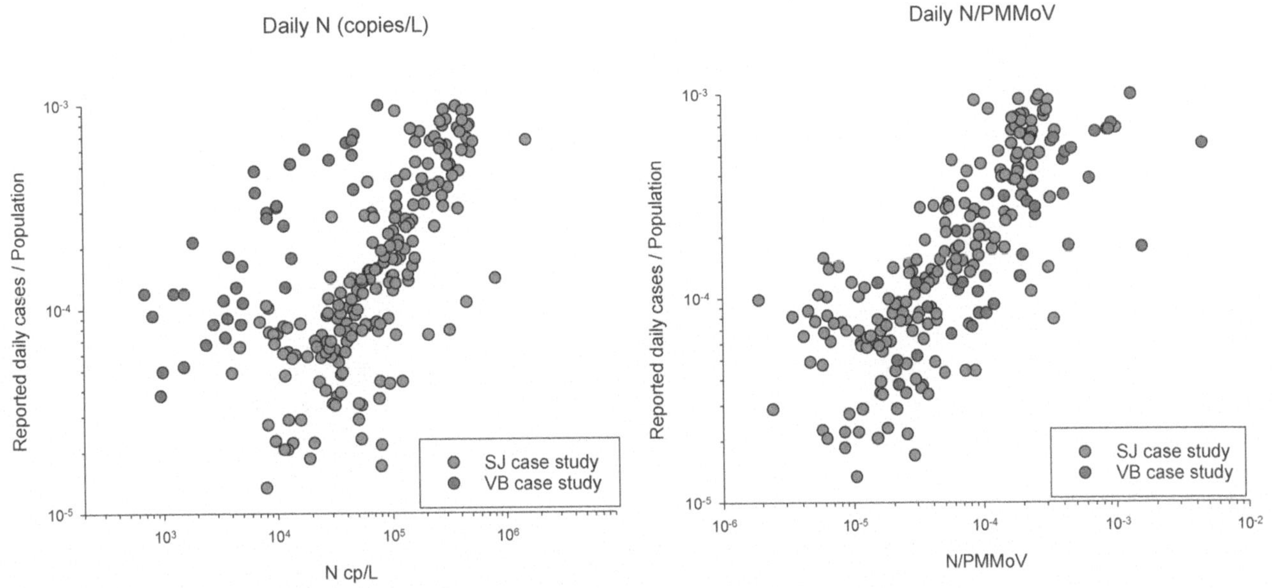

Figure 4 | Reported daily cases versus observed influent concentrations of SARS-CoV-2 (left) and SARS-CoV-2/PMMoV (right) for Virginia Beach, VA and San Jose, CA case studies.

Model performance evaluations

Detailed results from the sensitivity evaluations are presented in Supplementary Material, Sections C and D. Section C includes the fecal shedding distribution evaluation, and Section D includes evaluation of the uncertainty in the simulation results (D1), variability/uncertainty in computed influent concentrations for the San Jose case study based on the range of partitioning coefficients (D2), fecal shedding duration (D3), and model formulation sensitivity from inter-site variability (D4). The sensitivity evaluations indicate that: (1) there is substantial variability/uncertainty in the model output from individual simulations (Supplementary Material, Figure D1). These results are likely driven primarily by the highly uncertain fecal shedding data (Supplementary Material, Section C); (2) computed influent concentrations are strongly influenced by the selected values for the partitioning coefficients for conditions using sludge data as input (Supplementary Material, Figure D2); and (3) the general trends observed within any given site were not dependent on the specific fecal shedding duration used (Supplementary Material, Figure D3). Further details are presented in Supplementary Material, Sections C and D.

DISCUSSION

This study describes a modeling framework for estimating community-level COVID-19 infection based on wastewater monitoring data that are independent of individual and clinical surveillance data. However, the case studies presented highlight the limited available data to parameterize key model inputs, resulting in high outcome variability and presumed uncertainty, in turn, limiting model application to support public health decision-making for the COVID-19 response. The mechanistic modeling framework described encompasses three complementary formulations, each of which requires a unique set of inputs for estimating human contribution to wastewater (referred to as fecal strength) and size of population represented by a wastewater sample. A principal benefit of the mechanistic modeling approach is that it provides the conversion of wastewater (or sludge) levels into estimated total cases for direct comparison and interpretation against clinical surveillance data for public health decision-making. For example, this wastewater modeling approach can aid in understanding the variable effective reproductive number through the epidemic by estimating a total community infection burden that includes unreported and sub-clinical infections (Huisman *et al.* 2022). An additional benefit of the mechanistic modeling approach is that it can be updated to aid in understanding relative differences in community infection as the underlying drivers of transmission change temporally, such as changing variants and increasing vaccination levels. It can also be useful to provide insights about relative infection levels during time periods where community-level testing is changing (increasing or decreasing) or reporting of testing results is delayed or aggregated (such as holidays or some weekends).

Available wastewater-related data, and public health decision-making needs will determine which of these three wastewater-based infection model formulations would be most suitable for a community. For example, if it is reasonable to estimate the fraction of wastewater that derives from industrial and/or other sources that are unlikely to contain viral contributions and to assume those estimates are stable over the modeled period; or if endogenous control data are not available or feasible to collect, the flow generation formulation may be most applicable. Conversely, if a community is generating paired SARS-CoV-2 RNA and endogenous control wastewater surveillance data, the fecal strength formulation may be desirable over the flow generation formulation to provide enhanced potential for inter-site comparability. Moreover, Figure 4 suggests that, when applying the model across lab methods, sample types, and locations, the FS model maybe more applicable because the fecal normalization incorporated into that model likely makes the ratio of SARS-CoV-2 to reported cases more consistent across sites and sample types.

Differences between wastewater characteristics in communities may also influence which formulation is most appropriate. For example, in communities with substantial agricultural contributions to the community wastewater, the use of PMMoV RNA or another endogenous control may be more complicated as PMMoV may not be human fecal-specific in those settings (Zhang *et al.* 2006), and thus, make the flow generation or flow receipt formulation more appropriate. Another consideration is the population served by a treatment facility and whether the population is relatively static (i.e., are there large changes over short periods of time or does a large fraction of the population live in one sewershed and work in another). The FS formulation may be more representative of the true population contributing feces to wastewater, but it is also likely more vulnerable to outliers, as seen in Figure 2, because the formulation is dependent on two microbial measurements that can easily vary across orders of magnitude. Interpretation of the modeling results is complicated in cases where the population is more dynamic, and in such cases, careful consideration of these important factors is warranted.

Average modeled AR values ranged from 2 to 10 across the two case studies for the period of March 2020 to March 2021. While this range appears plausible given the reported levels of asymptomatic COVID-19 infections and estimates of unreported cases (Ma *et al.* 2021; Sah *et al.* 2021), the range of model results and sensitivity evaluations limit confidence in the model output for any specific data point. In particular, fecal shedding data for SARS-CoV-2 RNA and PMMoV RNA are presently limited, constraining the utility of the model. Moreover, multiple wastewater testing method evaluation studies have shown that absolute SARS-CoV-2 RNA wastewater measurements are not necessarily comparable across laboratory methods or wastewater sample types (Deng *et al.* 2021; Pecson *et al.* 2021; Foster *et al.* 2022; Kim *et al.* 2022). Therefore, it is unknown whether the differences in AR values between the two case studies reported here (mean AR values of 2–4 in VB compared to 6–10 in SJ) represent true differences in community infection levels captured by case reporting over the modeled period or are a function of differences in virus recovery, testing methods, and/or sample types (i.e., influent versus sludge). However, one interesting perspective from this modeling approach is the evaluation of the temporally varying nature of AR, which can vary due to many factors including community mitigation behavior (e.g., mask wearing, physical distancing), testing availability, circulating variants, and vaccination status. Few other analytical approaches have the ability to evaluate this characteristic on a time frame relevant for decision-making (Fernandez-Cassi *et al.* 2021; Huisman *et al.* 2022).

Although we conducted an extensive analysis of available data related to fecal shedding of SARS-CoV-2 RNA, there are limitations and uncertainties associated with those data. To mitigate potential bias associated with high levels of shedding, we selected best fitting distributions with an emphasis on the upper tails (refer to Supplementary Material, Section C for evaluation of this assumption on model estimates). However, future work is critically needed to characterize fecal shedding of SARS-CoV-2 RNA and human-specific microbial fecal targets, such as PMMoV RNA, to enhance the utility of this modeling framework. The model sensitivity evaluations showed that the mechanistic model was most sensitive to the fecal shedding parameter, which due to lack of available data was treated stochastically rather than temporally varying over the estimated shedding period. Temporal characterization of SARS-CoV-2 RNA fecal shedding over the full duration of infection, from exposure to recovery could allow for a time dependent model and in turn may reduce the uncertainty in community SARS-CoV-2 infection level estimates from wastewater. Similarly, inter-laboratory method evaluations that identify critical methodological impacts on measurement comparability would allow for model output comparability across sites. Finally, while shedding contributions via other bodily fluid types (e.g., mucous, urine, etc.) was found to be negligible compared to fecal contributions (Supplementary Material, Appendix B), these other shedding routes should be considered as variants emerge with potentially different clinical presentations.

Until more well characterized shedding data are available, site-specific relative ascertainment ratios (RAR) (relative to specific times of known infection and transmission profiles as described in the Supplementary Material, Section A for methods and Supplementary Material, Figures D3 and D4 for examples) are a present use for the described SARS-CoV-2 infection modeling framework given the uncertainties in the available input data. Relative AR values could provide insight into coverage of testing resources in relation to trends in total community cases relative to reported cases (Supplementary Material, Section D). This may become particularly useful to continue monitoring SARS-CoV-2 infection levels as mass screening of asymptomatic patients is no longer conducted due to elevated costs and increased vaccination coverage. Another possible approach that could help to minimize bias associated with the proposed mechanistic model uncertainty is the use of multiple formulations. This general approach has proven to be useful in other unrelated applications that need to account for substantial uncertainty in the modeling (i.e., long-range weather forecasting).

The described mechanistic modeling framework can be used for future wastewater surveillance targets, such as those diseases currently endemic in the United States, like norovirus, or other pathogens where sufficient high-quality input data are available. One potential complicating factor is the presence of pathogens that are also shed from non-human sources (i.e., parasitic protozoans such as *Cryptosporidium spp.*) as those other sources to the waste stream are not accounted for in the present framework. Antibiotic resistance genes (ARGs) are another promising target for wastewater surveillance, but in terms of modeling infection burden, ARGs have the added complication of being carried by multiple health relevant microorganisms. However, the current framework could be used for emerging pathogens whose characteristics are similar to currently well characterized pathogens – new SARS variants may fall into this category in the near term. In this case, the framework can be useful for complementing traditional epidemiological approaches (i.e., case counts, etc.) for broadly understanding pathogen spread within a community while the AR is unknown or changing.

Because of the public health utility of wastewater surveillance, the Centers for Disease Control and Prevention in the United States (CDC) launched the National Wastewater Surveillance System (NWSS) (CDC 2021b). NWSS provides novel national wastewater-based disease surveillance infrastructure to support public health action during and after the COVID-19 pandemic by providing coordination, laboratory and epidemiology capacity, and technical guidance for wastewater surveillance programs implemented by state, tribal, local, and territorial health departments (Kirby *et al.* 2021). NWSS runs a centralized data system that standardizes wastewater data submission, analysis, visualization, and sharing between jurisdictions and with the public. If the uncertainty in wastewater-infection modeling is sufficiently reduced, NWSS is positioned to rapidly integrate this analytic capacity into the data dashboard for public health use, which may become more relevant as CDC adds new targets, including emerging pathogens and antimicrobial resistance genes (CDC 2021b).

The interpretation of wastewater SARS-CoV-2 RNA data will continue to be multi-faceted and complicated as vaccination and variant emergence evolve. Understanding the extent to which these emerging factors will affect SARS-CoV-2 RNA fecal shedding will be critical to understand the utility of this modeling framework. Nevertheless, interpretation of wastewater SARS-CoV-2 RNA data will continue to be important to complement traditional surveillance data and the modeling framework presented can be useful for researchers and public health decision-makers.

ACKNOWLEDGEMENTS

We thank the participating wastewater utilities, Hampton Roads Sanitation District and San José-Santa Clara Regional Wastewater Facility, for the use of their data and dedication to protecting public health during the COVID-19 response. We also thank Santa Clara County and Virginia Beach Health Department for gathering and publicly sharing their data used in this study, as well as their dedication to the COVID-19 response. Finally, we acknowledge Daniel Gerrity, Nasa Armstrong, and Zach Marsh for their intellectual feedback and discussion regarding model development. Use of trade names and commercial sources is for identification only and does not imply endorsement by the Centers for Disease Control and Prevention (CDC). The findings and conclusions in this report are those of the authors and do not necessarily represent those of the CDC.

DATA AVAILABILITY STATEMENT

All relevant data are included in the paper or its Supplementary Information.

CONFLICT OF INTEREST

The authors declare there is no conflict.

REFERENCES

Ahmed, W., Angel, N., Edson, J., Bibby, K., Bivins, A., O'Brien, J. W., Choi, P. M., Kitajima, M., Simpson, S. L., Li, J., Tscharke, B., Verhagen, R., Smith, W. J. M., Zaugg, J., Dierens, L., Hugenholtz, P., Thomas, K. V. & Mueller, J. F. 2020 First confirmed detection of sars-cov-2 in. untreated wastewater in australia: A proof of concept for the wastewater surveillance of covid-19 in.the community. *The Science of the Total Environment* **728**, 138764.

Asghar, H., Diop, O. M., Weldegebriel, G., Malik, F., Shetty, S., El Bassioni, L., Akande, A. O., Al Maamoun, E., Zaidi, S., Adeniji, A. J., Burns, C. C., Deshpande, J., Oberste, M. S. & Lowther, S. A. 2014 Environmental surveillance for polioviruses in the global polio eradication initiative. *J Infect Dis* **210** Suppl 1, S294–303.

Baud, D., Qi, X., Nielsen-Saines, K., Musso, D., Pomar, L. & Favre, G. 2020 Real estimates of mortality following covid-19 infection. *The Lancet Infectious diseases* **20**(1), 773.

Betancourt, W. Q., Schmitz, B. W., Innes, G. K., Prasek, S. M., Pogreba Brown, K. M., Stark, E. R., Foster, A. R., Sprissler, R. S., Harris, D. T., Sherchan, S. P., Gerba, C. P. & Pepper, IL. 2021 Covid-19 containment on a college campus via wastewater-based epidemiology, targeted clinical testing and an intervention. *Sci Total Environ* **779**, 146408.

Bivins, A., North, D., Ahmad, A., Ahmed, W., Alm, E., Been, F., Bhattacharya, P., Bijlsma, L., Boehm, A. B., Brown, J., Buttiglieri, G., Calabro, V., Carducci, A., Castiglioni, S., Cetecioglu Gurol, Z., Chakraborty, S., Costa, F., Curcio, S., de Los Reyes 3rd, F. L., 3rd, Vela J, Delgado, Farkas, K., Fernandez-Casi, X., Gerba, C., Gerrity, D., Girones, R., Gonzalez, R., Haramoto, E., Harris, A., Holden, P. A., Islam, M. T., Jones, D. L., Kasprzyk-Hordern, B., Kitajima, M., Kotlarz, N., Kumar, M., Kuroda, K., La Rosa, G., Malpei, F., Mautus, M., McLellan, S. L., Medema, G., Meschke, J. S., Mueller, J., Newton, R. J., Nilsson, D., Noble, R. T., van Nuijs, A., Peccia, J., Perkins, T. A., Pickering, A. J., Rose, J., Sanchez, G., Smith, A., Stadler, L., Stauber, C., Thomas, K., van der Voorn, T., Wigginton, K., Zhu, K. & Bibby, K. 2020 Wastewater-based epidemiology: Global collaborative to maximize contributions in the fight against covid-19. *Environ Sci Technol* **54**, 7754–7757.

CDC 2021a Covid-19, https://www.Cdc.Gov/coronavirus/2019-ncov/index.Html. Available: https://www.cdc.gov/coronavirus/2019-ncov/cases-updates/geographic-seroprevalence-surveys.html [accessed May 24, 2021 2021].

CDC 2021b National wastewater surveillance system (nwss), https://www.Cdc.Gov/coronavirus/2019-ncov/cases-updates/wastewater-surveillance.Html. [accessed May 24, 2021 2021].

Ceylan, Z. 2020 Estimation of covid-19 prevalence in italy, spain, and france. *Science of The Total Environment* **729**, 138817.

Cheung, K. S., Hung, I. F., Chan, P. P., Lung, K. C., Tso, E., Liu, R., Ng, Y. Y., Chu, M. Y., Chung, T. W., Tam, A. R., Yip, C. C., Leung, K. H., Yim-Fong Fung, A., Zhang, R. R., Lin, Y., Cheng, H. M., Zhang, A. J., To, K. K., Chan, K. H., Yuen, K. Y. & Leung, W. K. 2020 Gastrointestinal manifestations of sars-cov-2 infection and virus load in fecal samples from the hong kong cohort and systematic review and meta-analysis. *Gastroenterology* **159**(1), 81–95.

D'Aoust, P. M., Graber, T. E., Mercier, E., Montpetit, D., Alexandrov, I., Neault, N., Baig, A. T., Mayne, J., Zhang, X., Alain, T., Servos, M. R., Srikanthan, N., MacKenzie, M., Figeys, D., Manuel, D., Juni, P., MacKenzie, A. E. & Delatolla, R. 2021 Catching a resurgence: Increase in sars-cov-2 viral rna identified in wastewater 48 h before covid-19 clinical tests and 96 h before hospitalizations. *Sci Total Environ* **770**, 145319.

Deng, K., Uhlig, S., Ip, H. S., Lea Killian, M., Goodman, L. B., Nemser, S., Ulaszek, J., Pickens, S., Newkirk, R., Kmet, M., Frost, K., Hettwer, K., Colson, B., Nichani, K., Schlierf, A., Tkachenko, A., Reddy, R. & Reimschuessel, R. 2021 Interlaboratory comparison of sars-cov2 molecular detection assays in use by u.S. Veterinary diagnostic laboratories. Journal of veterinary diagnostic investigation: official publication of the American Association of Veterinary Laboratory Diagnosticians, Inc **33**, 1039–1051.

Fernandez-Cassi, X., Scheidegger, A., Bänziger, C., Cariti, F., Tuñas Corzon, A., Ganesanandamoorthy, P., Lemaitre, J. C., Ort, C., Julian, T. R. & Kohn, T. 2021 Wastewater monitoring outperforms case numbers as a tool to track covid-19 incidence dynamics when test positivity rates are high. *Water Research* **200**, 117252.

Foladori, P., Cutrupi, F., Segata, N., Manara, S., Pinto, F., Malpei, F., Bruni, L. & La Rosa, G. 2020 Sars-cov-2 from faeces to wastewater treatment: What do we know? A review. *Sci Total Environ* **743**, 140444.

Foster, C. S. P., Stelzer-Braid, S., Deveson, I. W., Bull, R. A., Yeang, M., Au, J.-P, Ruiz Silva, M., van Hal, S. J., Rockett, R. J., Sintchenko, V., Kim, K. W. & Rawlinson, W. D. 2022 Assessment of inter-laboratory differences in sars-cov-2 consensus genome assemblies between public health laboratories in australia. *Viruses* **14**, 185.

Gerrity, D., Papp, K., Stoker, M., Sims, A. & Frehner, W. 2021 Early-pandemic wastewater surveillance of sars-cov-2 in.southern nevada: Methodology, occurrence, and incidence/prevalence considerations. *Water research X* **10**, 100086.

Gonzalez, R., Curtis, K., Bivins, A., Bibby, K., Weir, M. H., Yetka, K., Thompson, H., Keeling, D., Mitchell, J. & Gonzalez, D. 2020 Covid-19 surveillance in southeastern virginia using wastewater-based epidemiology. *Water Res* **186**, 116296.

Gonzalez, R., Larson, A., Thompson, H., Carter, E. & Cassi, X. F. 2021 Redesigning sars-cov-2 clinical rt-qpcr assays for wastewater rt-ddpcr. medRxiv:DOI 10.1101/2021.1103.1102.21252754.

Graham, K. E., Loeb, S. K., Wolfe, M. K., Catoe, D., Sinnott-Armstrong, N., Kim, S., Yamahara, K. M., Sassoubre, L. M., Mendoza Grijalva, L. M., Roldan-Hernandez, L., Langenfeld, K., Wigginton, K. R. & Boehm, A. B. 2021 Sars-cov-2 rna in wastewater settled solids is associated with covid-19 cases in a large urban sewershed. *Environ Sci Technol* **55**, 488–498.

Hamza, I. A., Jurzik, L., Uberla, K. & Wilhelm, M. 2011 Evaluation of pepper mild mottle virus, human picobirnavirus and torque teno virus as indicators of fecal contamination in river water. *Water Res* **45**, 1358–1368.

Hart, O. E. & Halden, R. U. 2020 Computational analysis of sars-cov-2/covid-19 surveillance by wastewater-based epidemiology locally and globally: Feasibility, economy, opportunities and challenges. *Sci Total Environ* **730**, 138875.

Hewitt, J., Trowsdale, S., Armstrong, B. A., Chapman, J. R., Carter, K. M., Croucher, D. M., Trent, C. R., Sim, R. E. & Gilpin, B. J. 2022 Sensitivity of wastewater-based epidemiology for detection of sars-cov-2 rna in a low prevalence setting. *Water Res* **211**, 118032.

Huisman, J., Scire, J., Caduff, L., Fernandez-Cassi, X., Ganesanandamoorthy, P., Kull, A., Scheidegger, A., Stachler, E., Boehm, A., Hughes, B., Knudson, A., Topol, A., Wigginton, K., Wolfe, M., Kohn, T., Ort, C., Stadler, T. & Julian, T. 2022 Wastewater-based estimation of the effective reproductive number of sars-cov-2. *Environ Health Perspect* **130**, 057011.

Kaplan, E. H., Wang, D., Wang, M., Malik, A. A., Zulli, A. & Peccia, J. 2021 Aligning sars-cov-2 indicators via an epidemic model: Application to hospital admissions and rna detection in sewage sludge. *Health care management science* **24**, 320–329.

Kaplan, E. H., Zulli, A., Sanchez, M. & Peccia, J. 2022 Scaling sars-cov-2 wastewater concentrations to population estimates of infection. *Scientific Reports* **12**, 3487.

Kim, S., Kennedy, L. C., Wolfe, M. K., Criddle, C. S., Duong, D. H., Topol, A., White, B. J., Kantor, R. S., Nelson, K. L., Steele, J. A., Langlois, K., Griffith, J. F., Zimmer-Faust, A. G., McLellan, S. L., Schussman, M. K., Ammerman, M., Wigginton, K. R., Bakker, K. M. & Boehm, A. B. 2022 Sars-cov-2 rna is enriched by orders of magnitude in primary settled solids relative to liquid wastewater at publicly owned treatment works. *Environmental Science: Water Research & Technology* **8**, 757–770.

Kirby, A. E., Walters, M. S., Jennings, W. C., Fugitt, R., LaCross, N., Mattioli, M., Marsh, Z. A., Roberts, V. A., Mercante, J. W., Yoder, J. & Hill, V. R. 2021 Using wastewater surveillance data to support the covid-19 response - united states, 2020-2021. *MMWR Morb Mortal Wkly Rep* **70**, 1242–1244.

Kirby, A. E., Welsh, R. M., Marsh, Z. A., Yu, A. T., Vugia, D. J., Boehm, A. B., Wolfe, M. K., White, B. J., Matzinger, S. R., Wheeler, A., Bankers, L., Andresen, K., Salatas, C. New York City Department of Environmental P, Gregory, D. A., Johnson, M. C., Trujillo, M., Kannoly, S., Smyth, D. S., Dennehy, J. J., Sapoval, N., Ensor, K., Treangen, T., Stadler, L. B. & Hopkins, L. 2022 Notes from the field: Early evidence of the sars-cov-2 b.1.1.529 (omicron) variant in community wastewater - united states, november-december 2021. *MMWR Morb Mortal Wkly Rep* **71**, 103–105.

Li, R., Pei, S., Chen, B., Song, Y., Zhang, T., Yang, W. & Shaman, J. 2020 Substantial undocumented infection facilitates the rapid dissemination of novel coronavirus (sars-cov-2). *Science* **368**, 489–493.

Ma, Q., Liu, J., Liu, Q., Kang, L., Liu, R., Jing, W., Wu, Y. & Liu, M. 2021 Global percentage of asymptomatic sars-cov-2 infections among the tested population and individuals with confirmed covid-19 diagnosis: A systematic review and meta-analysis. *JAMA Network Open* **4**, e2137257–e2137257.

Medema, G., Heijnen, L., Elsinga, G., Italiaander, R. & Brouwer, A. 2020 Presence of sars-coronavirus-2 rna in sewage and correlation with reported covid-19 prevalence in the early stage of the epidemic in the netherlands. *Environmental Science & Technology Letters* **7**, 511–516.

Paul, R., Arif, A. A., Adeyemi, O., Ghosh, S. & Han, D. 2020 Progression of covid-19 from urban to rural areas in the united states: A spatiotemporal analysis of prevalence rates. *The Journal of Rural Health* **36**, 591–601.

Peccia, J., Zulli, A., Brackney, D. E., Grubaugh, N. D., Kaplan, E. H., Casanovas-Massana, A., Ko, A. I., Malik, A. A., Wang, D., Wang, M., Warren, J. L., Weinberger, D. M., Arnold, W. & Omer, S. B. 2020 Measurement of sars-cov-2 rna in wastewater tracks community infection dynamics. *Nature biotechnology* **38**, 1164–1167.

Pecson, B. M., Darby, E., Haas, C. N., Amha, Y. M., Bartolo, M., Danielson, R., Dearborn, Y., Di Giovanni, G., Ferguson, C., Fevig, S., Gaddis, E., Gray, D., Lukasik, G., Mull, B., Olivas, L., Olivieri, A., Qu, Y. & Consortium, SA-C-I. 2021 Reproducibility and sensitivity of 36 methods to quantify the sars-cov-2 genetic signal in raw wastewater: Findings from an interlaboratory methods evaluation in the u.S. *Environmental Science: Water Research & Technology* **7**, 504–520.

Penn, R., Ward, B. J., Strande, L. & Maurer, M. 2018 Review of synthetic human faeces and faecal sludge for sanitation and wastewater research. *Water Res* **132**, 222–240.

Saguti, F., Magnil, E., Enache, L., Churqui, M. P., Johansson, A., Lumley, D., Davidsson, F., Dotevall, L., Mattsson, A., Trybala, E., Lagging, M., Lindh, M., Gisslen, M., Brezicka, T., Nystrom, K. & Norder, H. 2021 Surveillance of wastewater revealed peaks of sars-cov-2 preceding those of hospitalized patients with covid-19. *Water Res* **189**, 116620.

Sah, P., Fitzpatrick, M. C., Zimmer, C. F., Abdollahi, E., Juden-Kelly, L., Moghadas, S. M., Singer, B. H. & Galvani, A. P. 2021 Asymptomatic sars-cov-2 infection: A systematic review and meta-analysis. *Proceedings of the National Academy of Sciences* **118**, e2109229118.

Soller, J. A. & Eisenberg, JNS. 2008 An evaluation of parsimony for microbial risk assessment models. *Environmetrics* **19**, 61–78.

Turk, P. J., Chou, S-H, Kowalkowski, M. A., Palmer, P. P., Priem, J. S., Spencer, M. D., Taylor, Y. J. & McWilliams, A. D. 2020 Modeling covid-19 latent prevalence to assess a public health intervention at a state and regional scale: Retrospective cohort study. *JMIR Public Health Surveill* **6**, e19353.

Weidhaas, J., Aanderud, Z. T., Roper, D. K., VanDerslice, J., Gaddis, E. B., Ostermiller, J., Hoffman, K., Jamal, R., Heck, P., Zhang, Y., Torgersen, K., Laan, J. V. & LaCross, N. 2021 Correlation of sars-cov-2 rna in wastewater with covid-19 disease burden in sewersheds. *Sci Total Environ* **775**, 145790.

Weiss, A., Jellingso, M. & Sommer, M. O. A. 2020 Spatial and temporal dynamics of sars-cov-2 in.covid-19 patients: A systematic review and meta-analysis. *EBioMedicine* **58**, 102916.

WHO 2020a Antimicrobial resistance surveillance system (glass). Available from: Https://www.Who.Int/glass/en/. Accessed 19 august 2020.

WHO 2020b Guidelines for environmental surveillance of poliovirus circulation. Available from https://apps.Who.Int/iris/handle/10665/67854?Locale=en&mode=full. Accessed 19 august 2020.

WHO 2020c Status of environmental surveillance for sars-cov-2 virus, scientific brief, available from: Https://apps.Who.Int/iris/bitstream/handle/10665/333670/who-2019-ncov-sci_brief-environmentalsampling-2020.1-eng.Pdf. Accessed 19 august 2020.

Wolfe, M. K., Topol, A., Knudson, A., Simpson, A., White, B., Vugia, D. J., Yu, A. T., Li, L., Balliet, M., Stoddard, P., Han, G. S., Wigginton, K. R. & Boehm, AB. 2021 High-frequency, high-throughput quantification of sars-cov-2 rna in wastewater settled solids at eight publicly owned treatment works in northern california shows strong association with covid-19 incidence. *mSystems* **6**, e0082921.

Wolfel, R., Corman, V. M., Guggemos, W., Seilmaier, M., Zange, S., Muller, M. A., Niemeyer, D., Jones, T. C., Vollmar, P., Rothe, C., Hoelscher, M., Bleicker, T., Brunink, S., Schneider, J., Ehmann, R., Zwirglmaier, K., Drosten, C. & Wendtner, C. 2020 Virological assessment of hospitalized patients with covid-2019. *Nature* **581**, 465–469.

Wu, S. L., Mertens, A. N., Crider, Y. S., Nguyen, A., Pokpongkiat, N. N., Djajadi, S., Seth, A., Hsiang, M. S., Colford Jr., J. M., Jr., Reingold, A., Arnold, B. F., Hubbard, A. & Benjamin-Chung, J. 2020 Substantial underestimation of sars-cov-2 infection in the united states. *Nature communications* **11**, 4507.

Wurtzer, S., Waldman, P., Ferrier-Rembert, A., Frenois-Veyrat, G., Mouchel, J. M., Boni, M., Maday, Y., consortium O., Marechal, V. & Moulin, L. 2021 Several forms of sars-cov-2 rna can be detected in wastewaters: Implication for wastewater-based epidemiology and risk assessment. *Water Res* **198**, 117183.

Zhang, T., Breitbart, M., Lee, W. H., Run, J. Q., Wei, C. L., Soh, S. W., Hibberd, M. L., Liu, E. T., Rohwer, F. & Ruan, Y. 2006 Rna viral community in human feces: Prevalence of plant pathogenic viruses. *PLoS biology* **4**, e3.

Zhang, W., Du, R-H, Li, B., Zheng, X-S, Yang, X-L, Hu, B., Wang, Y-Y, Xiao, G-F, Yan, B., Shi, Z-L & Zhou, P. 2020 Molecular and serological investigation of 2019-ncov infected patients: Implication of multiple shedding routes. *Emerging Microbes & Infections* **9**, 386–389.

Zheng, S., Fan, J., Yu, F., Feng, B., Lou, B., Zou, Q., Xie, G., Lin, S., Wang, R., Yang, X., Chen, W., Wang, Q., Zhang, D., Liu, Y., Gong, R., Ma, Z., Lu, S., Xiao, Y., Gu, Y., Zhang, J., Yao, H., Xu, K., Lu, X., Wei, G., Zhou, J., Fang, Q., Cai, H., Qiu, Y., Sheng, J., Chen, Y. & Liang, T. 2020 Viral load dynamics and disease severity in patients infected with sars-cov-2 in.zhejiang province, china, january-march 2020: Retrospective cohort study. *BMJ* **369**, m1443.

Zhou, F., Yu, T., Du, R., Fan, G., Liu, Y., Liu, Z., Xiang, J., Wang, Y., Song, B., Gu, X., Guan, L., Wei, Y., Li, H., Wu, X., Xu, J., Tu, S., Zhang, Y., Chen, H. & Cao, B. 2020 Clinical course and risk factors for mortality of adult inpatients with covid-19 in.wuhan, china: A retrospective cohort study. *The Lancet* **395**, 1054–1062.

Zhu, Y., Oishi, W., Maruo, C., Saito, M., Chen, R., Kitajima, M. & Sano, D. 2021 Early warning of covid-19 via wastewater-based epidemiology: Potential and bottlenecks. *Sci Total Environ* **767**, 145124.

First received 15 April 2022; accepted in revised form 29 June 2022. Available online 11 July 2022

doi: 10.2166/wh.2022.026

Research needs for optimising wastewater-based epidemiology monitoring for public health protection

Katie Robins [a,b], Anne F. C. Leonard [a,c,*], Kata Farkas[d], David W. Graham [b], David L. Jones[d,e], Barbara Kasprzyk-Hordern [f], Joshua T. Bunce[a,b], Jasmine M. S. Grimsley[a], Matthew J. Wade [a,b], Andrew M. Zealand[a] and Shannon McIntyre-Nolan[a,g]

[a] Environmental Monitoring for Health Protection, UK Health Security Agency, Nobel House, London SW1P 3HX, UK
[b] School of Engineering, Newcastle University, Cassie Building, Newcastle-upon-Tyne NE1 7RU, UK
[c] University of Exeter Medical School, European Centre for Environment and Human Health, University of Exeter, Cornwall TR10 9FE, UK
[d] School of Natural Sciences, Bangor University, Bangor, Gwynedd LL57 2UW, UK
[e] SoilsWest, Centre for Sustainable Farming Systems, Food Futures Institute, Murdoch University, Murdoch, WA 6105, Australia
[f] Department of Chemistry, University of Bath, 1 South, Bath BA2 7AY, UK
[g] Her Majesty's Prison and Probation Service, Ministry of Justice, London, SW1H 9AJ, UK
*Corresponding author. E-mail: anne.leonard@exeter.ac.uk

KR, 0000-0003-1419-8057; AFCL, 0000-0002-8231-602X; DWG, 0000-0002-9753-496X; BK-H, 0000-0002-6809-2875; MJW, 0000-0001-9824-7121

ABSTRACT

Wastewater-based epidemiology (WBE) is an unobtrusive method used to observe patterns in illicit drug use, poliovirus, and severe acute respiratory syndrome coronavirus-2 (SARS-CoV-2). The pandemic and need for surveillance measures have led to the rapid acceleration of WBE research and development globally. With the infrastructure available to monitor SARS-CoV-2 from wastewater in 58 countries globally, there is potential to expand targets and applications for public health protection, such as other viral pathogens, antimicrobial resistance (AMR), pharmaceutical consumption, or exposure to chemical pollutants. Some applications have been explored in academic research but are not used to inform public health decision-making. We reflect on the current knowledge of WBE for these applications and identify barriers and opportunities for expanding beyond SARS-CoV-2. This paper critically reviews the applications of WBE for public health and identifies the important research gaps for WBE to be a useful tool in public health. It considers possible uses for pathogenic viruses, AMR, and chemicals. It summarises the current evidence on the following: (1) the presence of markers in stool and urine; (2) environmental factors influencing persistence of markers in wastewater; (3) methods for sample collection and storage; (4) prospective methods for detection and quantification; (5) reducing uncertainties; and (6) further considerations for public health use.

Key words: antimicrobial resistance, chemicals, metabolites, pathogens, public health wastewater-based epidemiology

HIGHLIGHTS

- Wastewater-based epidemiology (WBE) has been successfully used to track SARS-CoV-2.
- WBE can monitor populations which are not captured through routine clinical surveillance.
- WBE has the potential for monitoring many additional indicators of public health.
- Research is required to validate the use of WBE in public health management.
- Considerations include integration of WBE with existing public health systems and ethics.

GRAPHICAL ABSTRACT

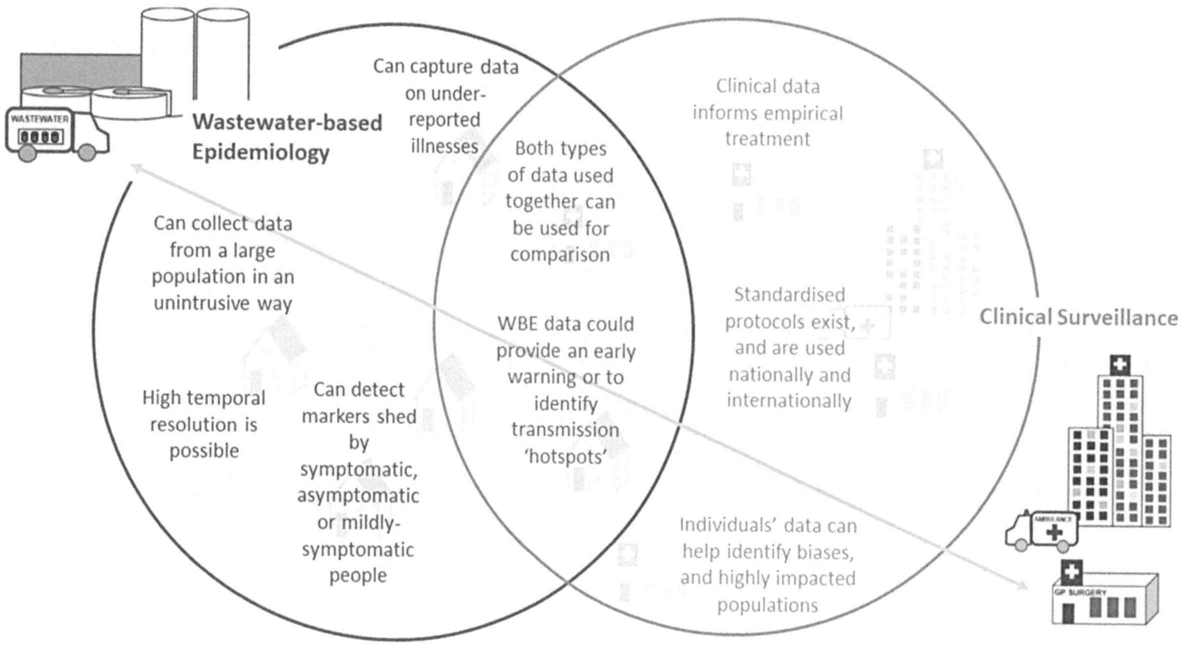

INTRODUCTION

The COVID-19 pandemic has had a major impact on public health, well-being, and economic systems globally (Kolahchi *et al.* 2021). The causative agent, severe acute respiratory syndrome coronavirus-2 (SARS-CoV-2) has spread rapidly since 2019. Clinical presentation has ranged from asymptomatic cases to hospitalisation or death (Weiss *et al.* 2020; Zheng *et al.* 2020; Park *et al.* 2021). One method for monitoring levels of SARS-CoV-2 that is inclusive, prone to fewer population biases, and less invasive than testing individuals is surveillance of wastewater in sewer systems, where the level of infection in the catchment community is associated with viral copies in sewage (Polo *et al.* 2020; Sims & Kasprzyk-Hordern 2020). This approach, termed wastewater-based epidemiology (WBE), has been used to complement existing public health data and clinical monitoring procedures (Naughton *et al.* 2021), with most data coming from high-income countries, for example, in the United Kingdom (Wade *et al.* 2020), the Netherlands (Medema *et al.* 2020), and the United States (Kirby *et al.* 2021). There are WBE studies in low and middle-income countries (LMICs), such as Tunisia (Jmii *et al.* 2021), Morocco (Amahmid *et al.* 2021), Brazil (Barbosa *et al.* 2022), where challenges for conducting this work are greater and complex (Pandey *et al.* 2021). The response to the pandemic has led to rapid acceleration of WBE research and development, and monitoring is now being performed at national, regional, and local scales (University of California 2022). Additionally, WBE has provided strategically important information through the development of new bioinformatics pipelines for variant detection and advanced algorithms for data interpretation, all of which have helped reduce uncertainty in the data (Karthikeyan *et al.* 2021; Habtewold *et al.* 2022). WBE has also provided early warning of local disease outbreaks, thus facilitating public health actions, such as focusing subsequent deployment of face masks, mobile testing, and vaccination facilities (e.g., surge testing; Hillary *et al.* 2021) and increasing bed capacity at hospitals (Brown *et al.* 2021; Crits-Christoph *et al.* 2021; Herold *et al.* 2021; Swift *et al.* 2021).

WBE is not a new concept, having been used to monitor illicit drug consumption as early as 2005 (Zuccato *et al.* 2005; Biello 2007) and the global distribution of poliovirus over the last decade (Hovi *et al.* 2012). Aside from these applications and monitoring of SARS-CoV-2, WBE has been used on a more *ad hoc* basis, with the implementation of a wide range of temporally and spatially distinct use cases, such as consumption of the antiviral oseltamivir (Tamiflu) during the 2009 H1N1 pandemic (Singer *et al.* 2013), the presence of influenza A and H1N1 in wastewater in the Netherlands (Heijnen &

Medema 2011), consumption of illicit and licit drugs (including antidepressants) during the 2010–2014 Greek economic crisis (Thomaidis *et al.* 2016), and the global distribution of antimicrobial resistance (AMR) (Hendriksen *et al.* 2019).

WBE has demonstrated potential in monitoring different human health and behaviour markers to provide valuable information for public health activities. However, this remains a relatively novel epidemiological tool, and to facilitate the expansion of WBE to other health targets and integration into public health surveillance activities research gaps need to be identified, so further study can be directed in these areas.

The aim of this paper is to review the applications of WBE for public health surveillance, focusing on the following three broad areas: (1) pathogenic viruses; (2) AMR; and (3) chemicals. It describes the additional value WBE contributes to public health surveillance, evaluates the evidence base supporting these applications, and identifies research gaps that need to be addressed to permit large-scale adoption of WBE as a central pillar in public health surveillance and decision-making.

PATHOGENIC VIRUSES

Added value of WBE monitoring

Viruses that cause diseases in humans significantly impact life quality and expectancy, put strain on healthcare settings, and negatively impact economies (Shang *et al.* 2021). Many novel pathogens to emerge in the past 100 years have been zoonotic in origin, including SARS-CoV-1 and Middle East respiratory syndrome coronavirus (MERS-CoV); human immunodeficiency virus (HIV); novel influenza viruses H1N1, H3, N2, and H2N2; Zika virus; and Ebola virus (Sharp & Hahn 2011; Morse *et al.* 2012; Recht *et al.* 2020). Furthermore, the most recent pandemic virus, SARS-CoV-2, is likely to have originated in a zoonotic reservoir (although the zoonotic origin has not been conclusively identified) (Holmes *et al.* 2021). The conditions for the emergence of novel viruses with pandemic potential remain (Woolhouse *et al.* 2008; Plowright *et al.* 2021) and the development of robust plans and resources to monitor the progress of pathogen spread and inform public health interventions is essential to adequately prepare for subsequent pandemics. Endemic pathogens, such as influenza virus, non-influenza RNA respiratory viruses (e.g., respiratory syncytial virus (RSV)), and enteric viruses (e.g., norovirus, rotavirus, etc.) also have a significant health burden, particularly among children, the elderly, and the immunocompromised (Modjarrad *et al.* 2016; Paget *et al.* 2019). Enhanced monitoring for pathogenic viruses imposing high burdens of disease by testing wastewater could help to reduce the number of respiratory infections and sequelae (Jartti *et al.* 2006; Ruotsalainen *et al.* 2013).

Currently, clinical surveillance for pathogenic viruses is limited to those who seek treatment, which excludes those who are asymptomatic, mildly symptomatic, or who do not present for treatment for other reasons (e.g., lack of access to healthcare or mistrust in healthcare systems; Tsimtsiou *et al.* 2021). There have been attempts to use data collection methods that would include information on populations not seeking clinical treatment, as well as people who do. This has been done using surveys of symptoms, such as the FluSurvey in Europe (FluSurvey 2017), or Flu Near You in the United States and Canada (FluSurvey 2017; Baltrusaitis *et al.* 2017), and analysis of digital search engine data on influenza-like illness by FluModel (i-sense-flu) (Lampos *et al.* 2015; Wagner *et al.* 2018). However, the results of surveys and search engine data analyses are also affected by biases such as a greater proportion of participants from more affluent and educated backgrounds than the population baseline tend to take part in voluntary surveys (Baltrusaitis *et al.* 2017; FluSurvey 2017). Models using digital search engine data to track search trends are biased towards populations with regular internet usage, and for whom surveillance data exists to train the model.

Clinical diagnostics, such as PCR-based methods or rapid antigen testing of individuals' samples, allow correct identification of an infective agent to inform empirical treatment of symptomatic patients, which may reduce the consumption of antimicrobials and allow the infected patients to isolate and limit the spread of the disease (Chen *et al.* 2011). However, the lack of community-level surveillance means infected patients are not identified until they reach clinical settings, at which point the disease may have spread significantly. WBE offers the opportunity to provide community-level data on pathogen presence to supplement and/or enhance existing clinical surveillance methods and public health activities (Figure 1), namely through (1) detecting the spread of viruses in individuals who have not interacted with healthcare systems; (2) identifying strategic locations for implementation of disease interventions and preventative measures; and (3) tracking changes in the epidemiology of the target that warrant further attention (Tang *et al.* 2017).

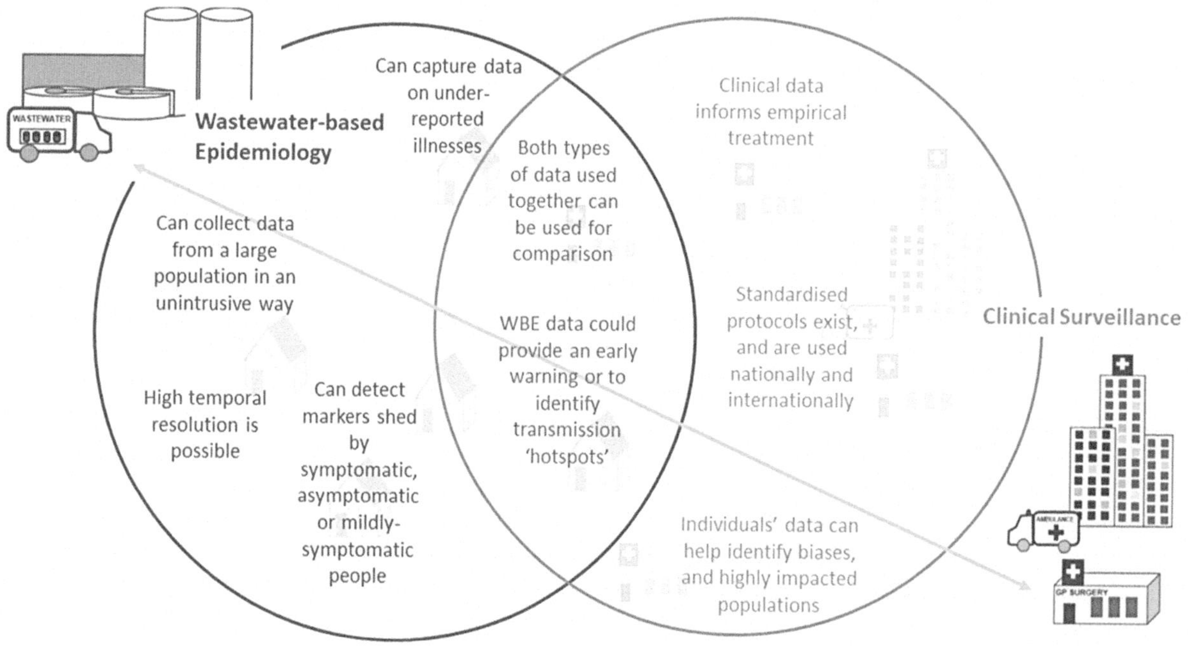

Figure 1 | Diagram showing the advantages of WBE and clinical surveillance techniques for pathogenic virus detection, and the potential benefits of integrating WBE with clinical surveillance data for public health.

Presence of pathogens in human stool and urine

For WBE to be useful for viral pathogen surveillance, markers of specific viral infection must be present in faeces, urine, or vomit (Jones *et al.* 2020). The shedding patterns of viral particles are likely to be affected by several factors, including virus-specific factors (e.g., infection site, incubation period) and host-specific factors (e.g., symptom severity, susceptibility to infection, and age).

The evidence base for faecal shedding of enteric viruses is well established (Supplementary Material, S1), (Allard *et al.* 1992; Papapetropoulou & Vantarakis 1998; Chitambar *et al.* 2001; Maunula *et al.* 2009; Lion *et al.* 2010; Jeulin *et al.* 2011; Ganesh & Lin 2013), since the transmission of such viruses is understood to be via the faecal–oral route. However, quantitative estimates of faecal shedding of respiratory viruses are less well explored (Supplementary Material, S1) since their primary transmission route is through respiratory aerosol droplets, rather than water or food contamination.

Interest in WBE for the pandemic respiratory virus, SARS-CoV-2, has resulted in numerous studies producing quantitative and temporal information related to faecal shedding of SARS-CoV-2 RNA (Cheung *et al.* 2020; Jones *et al.* 2020; Hoffmann & Alsing 2021). Evidence suggests SARS-CoV-2 shedding is greatest early in the course of the infection cycle and then decreases and continues for several days after symptoms and respiratory shedding have ceased (Gupta *et al.* 2020b; Van Doorn *et al.* 2020; Xu *et al.* 2020; Cavany *et al.* 2022; Hoffmann & Alsing 2021; Zhang *et al.* 2021b). For most non-enteric viruses, shedding studies are limited to a binary positive/negative result and are conducted when gastrointestinal symptoms are also experienced and may only be detected in a minority of confirmed cases (von Listow 2006; Minodier *et al.* 2015). Furthermore, the faecal and urinary shedding dynamics of respiratory viruses may not align with shedding at other sites. For example, prolonged shedding of H7N9 Avian Influenza has been observed in faecal and urine samples after respiratory samples became negative and viral detection was more common in the urine of severely ill patients (Hu *et al.* 2013a; Zhu *et al.* 2015).

For most viral pathogens, including SARS-CoV-2, our knowledge of shedding rates in faeces and urine have been obtained from hospitalised individuals, thus biasing the results towards cases presenting with more severe symptoms or comorbidities. Additionally, there is scant understanding of how shedding varies with the age of the patient, comorbidities, vaccination status, or re-infection. Data are therefore not representative of the general population, and future research needs to address this bias by ensuring diversity in symptom phenotypes and immunological states is represented in study cohorts. Limited available evidence (e.g., from norovirus, Sabrià *et al.* 2016) underscores the importance of understanding the differences in shedding of different viruses in faeces and urine from both symptomatic cases and asymptomatic carriers (Jones *et al.* 2020), who will both contribute waste to sewer systems. Further data are also needed on shedding dynamics, as existing evidence rarely captures the pre-symptomatic phase when shedding rates may be high (Hoffmann & Alsing 2021). The ability to detect outbreaks before cases present in healthcare settings is extremely valuable for public health planning and infection control. Information is also needed on the persistence of faecal shedding following the infectious period to improve the ability to detect active cases (Santos *et al.* 2020). Understanding how immunity (whether attained through vaccination or previous infection) impacts shedding is also necessary, as this could potentially influence the interpretation of epidemiological estimates through different phases of an outbreak.

In general, urine samples provide lower sensitivity for viral detection, particularly in enteric pathogens such as Enterovirus and Parechovirus (>90% in faeces vs. <60% in urine) (Crom et al. 2013). However, variability exists and for some pathogens shedding in other excretions may be the predominant source: the measles virus (measles morbillivirus) has been found in the urine of ~88% of infected patients in one study (van Binnendijk *et al.* 2003), and viral markers may be present in body fluids which also enter sewer systems (Crank *et al.* 2022), such as sputum (e.g., SARS-CoV-2 (Li *et al.* 2022)), saliva (e.g., human herpes virus (Fox *et al.* 1990), Epstein–Barr virus (Niederman *et al.* 1976)), semen (e.g., HIV-1 (Kariuki *et al.* 2020), hepatitis C virus (Cavalheiro *et al.* 2008), and Zika virus (Atkinson *et al.* 2016)). This is particularly relevant when developing the sampling framework to account for differences where samples may be more heterogenous (e.g., building-level surveillance; Crank *et al.* 2022).

Impacts of wastewater environments on viral detection

The stability of pathogen markers in the environment and in wastewater are highly variable (Bofill-Mas *et al.* 2006). In general, DNA targets are more stable than RNA targets, although physiochemical characteristics of the environment influence the persistence of both (Ward & Ashley 1980; Bofill-Mas *et al.* 2006; Muirhead *et al.* 2020). Enveloped viruses (viruses with a lipid envelope that include influenza viruses, coronaviruses, and parainfluenza viruses) are usually less stable in the environment than non-enveloped viruses (without a lipid envelope, including enteroviruses, noroviruses, and adenoviruses). Noroviruses and adenoviruses can survive up to months in the environment (Kotwal & Cannon 2014) and have frequently been detected in wastewater and surface waters (Lago *et al.* 2003; La Rosa *et al.* 2010; Rajko-Nenow *et al.* 2013; Guillois *et al.* 2016; Farkas *et al.* 2018; Falman *et al.* 2019; Mancini *et al.* 2019; McCall *et al.* 2021). However, very limited information currently exists on the persistence of the genetic material of enveloped viruses in the environment, especially in wastewater. It is only recently that Ebola (Zaire ebolavirus) was found to persist in water for 8 days (Bibby *et al.* 2015), whereas it was previously assumed that enveloped viruses were quickly deactivated in the environment (Wigginton *et al.* 2015; Ye *et al.* 2016).

Due to the requirement for biosafety level (BSL) 3 laboratories to perform infectivity studies, surrogate viruses such as Phi6 bacteriophage have been used to assess the potential for survivability (Aquino de Carvalho *et al.* 2017). The persistence of Phi6 bacteriophage under different temperatures, biological activity, and liquid media was found to vary significantly, with temperature being particularly important (Aquino de Carvalho *et al.* 2017). Even though temperature greatly affects viral survival, most WBE studies do not take temperature or seasonality into account and risk mis-quantifying results at different times of the year, particularly in temperate regions (Hart & Halden 2020). The presence of stormwater in combined sewers can also result in dilution which can limit the accurate quantification of microbial communities (Wade *et al.* 2022). Even in separated systems, fluctuating groundwater infiltration has the potential to impact observed results, in a way that is not immediately apparent (Choi *et al.* 2018).

Matrix composition has repeatedly been found to influence virus detection efficiency and survival in wastewater. For example, organic matter has a great impact on molecular detection as they inactivate the enzymes used for DNA amplification (Matheson *et al.* 2010; Hata *et al.* 2015). Ammonia can inactivate viruses through genome degradation and viruses could be aggregated or adsorbed into liquid media (Bibby *et al.* 2015; Aquino de Carvalho *et al.* 2017). A study found that aggregation of viruses in the solid fraction of the wastewater was higher for enveloped viruses than non-enveloped viruses at 26 and 6% of virus adsorbed, respectively (Ye *et al.* 2016). Therefore, sampling only the liquid fraction may exclude a large proportion of the viral target.

Biotic factors may also impact detection of viral targets in wastewater. Wastewater is a complex and dynamic mix of microbial communities and competition, predation and selective pressures are not well understood (Choi *et al.* 2018). In some locations, urban wastewater may also contain animal viruses (e.g., influenza A; Heijnen & Medema 2011), which can confound the signal due to difficulties in separating human and animal contributions without sequencing-based approaches.

It is evident that mechanisms of virus marker inactivation in wastewater environments for different viruses of interest require further investigation. The observed differences between faecal and sewage samples indicate environmental conditions that alter microbial communities and further research is required to be able to interpret viral signals in WBE accurately.

Sampling methodology and storage requirements for viral pathogens

Wastewater sampling for viral pathogens may use flow-based interval approaches, but is typically achieved through grab or composite sampling. Grab samples are collected at a single point in time, often at a time of day that is estimated to be 'peak flow' to maximise the target load from human sources (Polo *et al.* 2020). However, this can vary greatly depending on the location of sampling sites as network travel times of wastewater can typically range between 2 and 12 h (Castiglioni *et al.* 2013) and peak shower usage may dilute the signal (Ahmed *et al.* 2021). Some markers can have greater daily variability than others (Ahmed *et al.* 2021). Due to the temporal variability of wastewater concentrations, this is often not representative of the entire population input over 24 h, therefore a composite sample may be used. Composite sampling is a means to try and overcome this, typically via the use of autosamplers, which sample at set intervals (ranging from minutes to hours) and can be used to develop a time series or mixed to make a composite sub-sample. Grab samples appear to be suitable for enteric viruses in wastewater effluent (i.e., treated wastewater) (Sidhu *et al.* 2017; Farkas *et al.* 2018), however, variation in wastewater influent means daily composite samples are the best approach for enteric viruses (~4 samples per day; Farkas *et al.* 2018) and SARS-CoV-2 detection in influent wastewater (24 samples per day; Ahmed *et al.* 2021). Further temporal studies to determine enteric pathogen variation over 24 h are recommended (Farkas *et al.* 2018) and whether these patterns are consistent across various wastewater systems. There is a lack of research comparing sampling techniques for other respiratory viruses. The use of autosamplers requires specific infrastructure and is associated with a higher operational workload and cost, as well as the need for on-site sample refrigeration. These facilities are not always available, particularly in resource-poor areas.

Passive samplers are absorbent devices that can accumulate markers of interest when submerged in wastewater and may offer a good alternative to composite sampling. They do not require specialised equipment to be deployed (Habtwold *et al.* 2022; Schang *et al.* 2021), and can be used for sampling in the sewer network where autosampler use is impractical (Corchis-Scott *et al.* 2021) and are considerably less costly (Petrie *et al.* 2016). Passive samplers have been successfully used to detect viruses, such as hepatitis A and E, norovirus (Vincent-Hubert *et al.* 2017), poliovirus (Sattar & Westwood 1977; Cassemiro *et al.* 2016), SARS-CoV-2 (Schang *et al.* 2021). However, they may saturate quickly, and daily exposure times may be too long to obtain representative results. Furthermore, as passive samples are submerged and accumulate biological material over time, often to the point of saturation, viral concentration data obtained from passive samplers cannot be easily normalised for factors such as wastewater flow. Thus, the results are considered semi-quantitative. Overall, insight into wastewater dynamics and temporal variation of the marker of interest is needed for different pathogens to determine the most accurate and cost-effective sampling method. This could vary significantly between different sites depending on the pathogen of interest, the availability of infrastructure to permit composite sampling, and whether quantitative data are required for public health use.

There is currently limited standardised guidance on wastewater sample storage conditions and duration prior to molecular marker testing for viruses. It has been suggested that samples be kept refrigerated and analysed as soon as possible after collection, although temperatures and duration of storage vary, depending on the type of viral analyte (e.g., RNA/DNA, enveloped vs. non-enveloped viruses). For example, the US Centers for Disease Control and Prevention (CDC) recommends immediate sample refrigeration at 4 °C, and processing with 24 h (CDC 2022), whereas the US Environment Protection Agency (USEPA) recommends that samples be stored and chilled for up to 48–72 h prior to viable bacteriophage testing (USEPA 2001). For enterovirus viability and norovirus RNA testing in drinking water, the USEPA suggests storage for up to 24 h prior to analysis (Fout *et al.* 2010). Standardised methods need to be developed and agreed upon to allow comparability between studies, times and locations, and improve quality assurance.

Laboratory analysis methodologies for viral pathogens

Different methods for optimising RNA extraction, concentration, and quantification of SARS-CoV-2 have been explored. Concentration methods for viral pathogens in wastewater, primarily developed for non-enveloped enteric viruses, have since been applied for SARS-CoV-2 surveillance (Ahmed *et al.* 2021). These include precipitation with salt or polyethene glycol (PEG) (Farkas *et al.* 2021; Kevill *et al.* 2022), electrostatically charged membrane filtration (Ahmed *et al.* 2021), ultrafiltration (Izquierdo-Lara *et al.* 2021), adsorption-extraction (AE) with aluminium chloride, silica (Randazzo *et al.* 2020) or magnesium chloride ($MgCl_2$) (Ahmed *et al.* 2021), or using an automatic concentrating pipette (CP) instrument (Ahmed *et al.* 2021; McMinn *et al.* 2021). Concentration methods have been found to be one of the greatest sources of variation in the quantification of SARS-CoV-2 in the English wastewater monitoring programme, based on the recovery of Phi6 bacteriophage, which was used as a process control (Wade *et al.* 2022).

To improve the efficiency of wastewater sampling, testing samples for the presence of multiple viral markers would necessitate optimising viral recovery for multiple pathogens. However, a study by Ahmed *et al.* (2021) evaluated two concentration techniques (AE supplemented with $MgCl_2$ and CP) against concentration of SARS-CoV-2 and enteric pathogens. They found the CP method achieved the greatest overall recovery of SARS-CoV-2, however recovery did vary and was not consistent across all viral targets. Another study found viral recovery was reduced when CP was applied to high turbidity samples (Kevill *et al.* 2022). Further research should investigate concentration methods for different pathogens in wastewater and recovery efficiencies. This is key to optimising multi-pathogen recovery methods, minimising equipment requirements, and streamlining lab throughput. Further method development could additionally focus on optimisation in terms of concentration methods to enable both RNA and DNA extraction, where the latter could also be analysed using culture-independent methods for AMR surveillance, for example.

The gold standard for the targeted detection of viral pathogens in wastewater is quantitative polymerase chain reaction (qPCR). Although ubiquitous in much of WBE programmes, the method is sensitive to biological and chemical inhibitors, which are often present in wastewater extracts (Volkmann *et al.* 2007). Commercially available reaction mixes provide different levels of resistance to inhibitors, which may result in false-negative results and generate inter-laboratory variation (Farkas *et al.* 2017). The validation and implementation of strict quality control measures for each qPCR assay are crucial to achieving reliable results (Bustin *et al.* 2009). For example, one meta-analysis found that only 26% of studies published on SARS-CoV-2 detection in wastewater reported qPCR quality control details and where they were reported did not meet the minimum criteria (Bivins *et al.* 2021), questioning the reliability of the results.

Emerging targeted technologies such as digital droplet PCR (ddPCR), isothermal amplification (e.g., LAMP) assays and biosensors have also been developed; however, some of these are often less sensitive than qPCR (Haramoto *et al.* 2018; Farkas *et al.* 2020b). To date, besides qPCR, only ddPCR has been routinely used for WBE (Jahne *et al.* 2020; Al-Duroobi *et al.* 2021; Bivins & Bibby 2021), as this technique may be more sensitive than qPCR (Zhao *et al.* 2016), particularly for low viral loads or in matrices with more variable contamination (Taylor *et al.* 2017; Falzone *et al.* 2020).

Sequencing-based approaches have been trialled and implemented in WBE to generate data describing viral communities, including variant-level identification of circulating viruses (Fernandez-Cassi *et al.* 2018; Adriaenssens *et al.* 2018, 2021; Guajardo-Leiva *et al.* 2020) and identification of novel viruses and their variants (Rios *et al.* 2021). However, existing methods are time-consuming and often biased, tending to over-represent abundant or large DNA genomes and miss rare, shorter genomes (Garner *et al.* 2021). Amplicon sequencing increases sensitivity and accuracy for known viruses and has been effective in wastewater research to describe certain viral groups, including astroviruses (Hata *et al.* 2018), adenoviruses (Ogorzaly *et al.* 2015; Iaconelli *et al.* 2017), saproviruses (Mancini *et al.* 2019), enteroviruses (Tao *et al.* 2020), and noroviruses (Fumian *et al.* 2019). A similar approach has been used for SARS-CoV-2 by amplifying its whole genome or certain genes using the Illumina ARTIC pipeline (Avgeris *et al.* 2021; Dharmadhikari *et al.* 2021; Hillary *et al.* 2021).

Ultimately, the methodology for detection and quantification of viral pathogens will depend on the requirements of the data for public health. Research on the application of these technologies to wastewater samples for various pathogens is necessary to demonstrate their detection and quantification capability. Additionally, if it is necessary to detect several pathogens in one sample, development of high-throughput quantification tools for multiple targets would be necessary.

Comparison of WBE and public health data for viral pathogens

Clinical data, such as microbiological diagnostic data, may be used to assess how well markers in sewers correlate temporally and spatially with specific health outcomes in a community of interest. Comparison of WBE data should encompass two broad aspects: (1) that the markers reliably measure the intended target of surveillance (i.e., sensitivity and specificity) and (2) that data are collected at a temporal and spatial resolution that enable public health decision-making.

For SARS-CoV-2 in the UK, WBE data could be compared with public health data, including hospital admissions (Kaplan *et al.* 2020; Peccia *et al.* 2020; Galani *et al.* 2022), case data from the national and regional surveillance programmes (Wade *et al.* 2022), the Coronavirus Infection Survey (CIS) (Pouwels *et al.* 2020), and Real-time Assessment of Community Transmission studies (REACT) (Elliott *et al.* 2021). Comparisons between datasets also reveals potential population behaviours driving transmission, including social response to the pandemic, temporally resolved variations in population size (e.g., commuting, weekend breaks), sampling biases in wastewater, or biases, gaps, or reporting lags in clinical testing (Galani *et al.* 2022).

While the extensive clinical testing available throughout the pandemic meant comparable data was readily available (noting limitations within these data (Elliott *et al.* 2021)), similar comparisons might not be as feasible for pathogens like norovirus or RSV, where monitoring is dependent on data from healthcare settings. Therefore, to ensure the data collected through WBE is an accurate reflection of infection rates, future research must ensure these comparisons are designed into studies (e.g., by collecting samples or data from healthcare institutions, the wider community, and open data sources, such as social media).

Summary and suggestions for further research

Asymptomatic carriage of pathogenic viruses by members of the community is currently under-estimated but can play an important role in disease transmission. WBE has the potential to provide additional information on viral transmission dynamics that are currently missed by existing surveillance programmes, which are biased towards symptomatic and severely ill patients. However, knowledge gaps remain that must be addressed before WBE can be useful as a stand-alone surveillance tool to inform public health practice.

- Further quantitative studies on the shedding of viral pathogens (particularly for enveloped viruses such as influenza) are needed. Studies should ideally monitor patients throughout the entire course of the infection, with a range of variants, and recruit diverse cohorts to ensure generalisability of results. Shedding should be evaluated in faeces and urine. Impacts of immunity against viruses (achieved through vaccination or infection) on viral shedding dynamics need to be elucidated.
- Further studies on the persistence of various viruses in wastewater are necessary (particularly enveloped viruses), including evaluation of the impact of factors such as temperature, matrix composition, and biotic processes on viral detection. Researching this within the context of WBE may inform data interpretation for public health.
- Different sampling procedures and analytical techniques should be explored for pathogens of interest to optimise methods. The extent to which viral pathogen detection in wastewater samples can pre-empt or match clinical and community-level data must be evaluated to ensure that WBE data are an accurate reflection of clinical outcomes of interest.

ANTIMICROBIAL RESISTANCE

Added value of WBE for AMR

AMR, where microbes develop resistance against antimicrobial drugs, poses a threat to public health that is only likely to worsen. Antimicrobials are becoming less effective over time with continued overuse (Prestinaci *et al.* 2015) and fewer are being brought to market. Economic impacts are expected due to reduced productivity, resulting from lost days-of-work and longer hospital admissions (WEF 2021).

Compared to infectious viral pathogens described in the previous section, monitoring AMR is complicated given there is no one singular infectious agent responsible, nor single reservoirs for resistant organisms. There are numerous pathogenic antibiotic-resistant bacteria (ARBs) of concern, and antibiotic-resistant genes (ARGs), which can also be released in faecal matter and transferred between even distantly related bacteria by horizontal gene transfer (HGT). Beyond bacteria, resistance to antimicrobial treatment arises in other microorganisms such as azole antifungal resistance in pathogenic fungi, like *Candida albicans* (Whaley *et al.* 2017), antiviral resistance, including resistance to antivirals for hepatitis B (Locarnini 2004) and influenza virus (Lampejo 2020; Van Poelvoorde *et al.* 2020) and resistance among parasites, like Plasmodium spp.

AMR is an issue that is now recognised as extending beyond healthcare settings, with resistance determinants emerging in environmental settings (Larsen *et al.* 2022; Poirel *et al.* 2002) and selected for by antimicrobial use in agriculture systems (Woolhouse *et al.* 2015; Cheng *et al.* 2019; Pollock *et al.* 2020), the release of resistance in sewage into waterways and general environmental pollution (Harris *et al.* 2014; Agga *et al.* 2015; Murray *et al.* 2018; Hiller *et al.* 2019; Mughini-Gras *et al.* 2019). Therefore, surveillance and management of AMR are often considered through a One Health lens including environmental and animal monitoring (Mendelson & Matsoso 2015; WHO 2021a).

With current monitoring focussing on healthcare settings and a small group of key pathogens (e.g., English Surveillance for Antimicrobial use and Resistance (ESPAUR), (UKHSA 2021); European Antimicrobial resistance surveillance (EARS-Net), and WHO Global AMR Surveillance System (GLASS), (WHO 2021a)), the true scale and potential burden of AMR may be under-estimated (Burnham *et al.* 2019). ARGs, which are abundant in stool and wastewater samples, are the main way of indirectly monitoring AMR prevalence at large scales. The most comprehensive resistance monitoring study quantified ARGs in wastewater metagenomes in over 70 countries and found major differences in relative AMR abundances and classes in community sewage, reflecting large regional differences in 'local guts' (Hendriksen *et al.* 2019). Furthermore, a recent review by Pruden *et al.* (2021) described a range of applications of WBE for AMR monitoring including the following: (1) identification of hotspots to guide policy action on interventions; (2) inform medical and veterinary practitioners of the best antibiotics to use to avoid therapeutic failure; (3) inform risk assessments for discharge of antibiotics/ARB/ARGs into the wider environment; and (4) provide data for forecasting models and key drivers. Overall, the use cases for AMR surveillance using wastewater demonstrate the potential to add significantly to public health monitoring and therefore warrant further exploration.

Presence of AMR in human stool and urine

AMR is represented in stool by a diverse array of ARGs and mobile genetic elements (MGEs), primarily associated with bacterial hosts. Resistance carriage in humans varies substantially internationally due to different environmental exposures, diet, and antimicrobial use (Hu *et al.* 2013b; Pal *et al.* 2016; Feng *et al.* 2018). The abundance and diversity of ARGs also varies between body sites/tissues, with one study using metagenomic sequencing on oral and stool samples in different countries and finding a higher abundance, but lower diversity, of ARGs present in oral samples compared to stool samples (Carr *et al.* 2020). Studying stool samples may provide information not captured from data such as antibiotic prescription rates, which do not always correlate with ARGs present in faeces (Collignon *et al.* 2018; Hendriksen *et al.* 2019; Carr *et al.* 2020).

Quantification of ARGs present in stool using metagenomics and qPCR has been well studied (Hu *et al.* 2013b; Rose *et al.* 2017; Feng *et al.* 2018) and in urine mostly studied in the limited context of urinary tract infections (Li *et al.* 2020; Kaye *et al.* 2021). However, the clinical relevance of ARG detection in metagenomic data is poorly understood since the presence of ARGs in faecal samples does not necessarily result in AMR infections, even in high-risk patients (Tacconelli *et al.* 2009; Wryes *et al.* 2021). Furthermore, some genes (e.g., tetracycline genes) perform multiple functions unrelated to AMR, including signal tracking and hydrophobic protein transport (Martínez 2008; Dantas & Sommer 2012; Feng *et al.* 2018). There is also the question as to whether ARGs present in the gut and stool are an accurate indication of resistance at other parts of the body (Piddock 2016; Carr *et al.* 2020), the duration of colonisation, and risk factors extending shedding in faeces (Leonard *et al.* 2018). Further research in this area is also required for ARGs in urine samples, since there is evidence to suggest enhanced resistance following kidney transplantation is reflected in urine (Rani *et al.* 2020). Overall, the link between ARG detection in stool and urine to treatment-resistant infections among various at-risk groups needs further exploration.

Impact of wastewater on AMR detection

Like viral markers, the stability and abundance of AMR indicators in wastewater are complicated by many abiotic and biotic factors that influence microbial populations and resistomes (collection of AMR genes within the microbiome). As with viral pathogens, animal hosts of AMR can contribute ARGs to wastewater, making it difficult to precisely identify which AMR genes in a sample originated from humans (McCann *et al.* 2019). Biofilms present in sewers may also serve as a source of ARGs in wastewater independent of human waste inputs (Medina *et al.* 2020), and dispersion of ARG-bearing microbes may be triggered by the conditions in sewer environments (Kaplan 2010). In addition, bacteria and ARGs are capable of multiplying in the environment, and the presence of antibiotic residues, heavy metals, and biocides in wastewater can influence the selection of certain bacteria and their genes, and therefore their abundance, outside the human body (Baquero *et al.* 2008; Knapp *et al.* 2008; Martinez 2009; Gullberg *et al.* 2011; Murray *et al.* 2018).

There are many additional factors in wastewater that have an impact on the microbial community and resistome dynamics, including local residence times, geographic area, antibiotic prescribing, temperature, salinity, dissolved oxygen, micro-ecological factors such as predation, and the presence of organic matter (Caucci *et al.* 2016; Sun *et al.* 2016; Jiao *et al.* 2018;

Lamba *et al.* 2018; Jong *et al.* 2020; Pallares-Vega *et al.* 2021). Additionally, further research on different wastewater treatment plant types and maintenance and impacts of combined sewer overflows (CSOs) on AMR detection is critical in places with aging or poorly maintained sewer systems.

Due to the higher antibiotic usage in hospital settings, previous studies have investigated the impact of hospital wastewater mixing with community wastewater on resistomes, with mixed results (Islam *et al.* 2017; Lamba *et al.* 2017; Buelow *et al.* 2018, 2020; Quintela-Baluja *et al.* 2019, 2021; Li *et al.* 2021). Future work should focus on evaluating the factors that affect the influence of sewer conditions and inputs to the sewer systems that impact measures of AMR. Such research will aid interpretations of AMR data derived from wastewater samples.

Sampling methodology and storage requirements for AMR

As with viral markers, methods for the collection and storage of samples must be evaluated to ensure robust data are extracted from samples. A systematic review by Chau *et al.* (2021) compared composite and grab samples to evaluate performance according to association with clinical AMR data and found greater agreement with composite samples. The authors observed that several studies did not report their sampling methodologies, limiting comparability of studies and highlighting the urgent need for standardisation of approaches (Hassoun-Kheir *et al.* 2020). There are also few studies on the use of passive samplers for AMR surveillance, which may provide a more cost-effective and practical approach to sampling than composite samplers. Further research should seek to determine the most accurate and cost-effective sampling methodologies to inform standard practice, to allow data to be compared regionally and internationally. This has commenced through the EMBRACE-WATERS (rEporting antiMicroBial ResistAnCE in WATERS) initiative, which is aiming to develop reporting standards for AMR research in wastewater and other aqueous environments (Hassoun-Kheir *et al.* 2021).

Poulsen *et al.* (2021) compared storage conditions on the microbiome and resistome of sewage samples and pig faeces. They observed that sample storage conditions had an influence on microbial community and resistance genes, where freezing wastewater changed the microbiome in comparison to processing the samples immediately. AMR abundance in frozen sewage samples was also overall significantly higher than samples stored at 22 or 5 °C (Poulsen *et al.* 2021). Therefore, it was recommended that sewage samples are processed the same day, frozen immediately, or stored in the fridge and processed during the day or on the following day. However, the opposite finding was observed for pig faeces, indicating that stool samples, or perhaps even near-source samples have different storage requirements (Poulsen *et al.* 2021). Further research on the storage requirements of near-source samples would inform standard practice and standard protocols.

Access to storage facilities and appropriate alternatives need to be considered, to ensure sampling is consistent and samples are comparable.

Laboratory analysis methodologies for AMR

AMR detection can be conducted through culturing and gene-based methods (or a combination) (Figure 2). Culturing faecal indicators or pathogens, such as extended-spectrum beta-lactamases (ESBL) *Escherichia coli* is recommended by the WHO Tricycle Protocol (WHO 2021a, 2021b). Quantifying ESBL-producing Enterobacteriaceae (Marano *et al.* 2020) and carbapenem-resistant bacteria (Blaak *et al.* 2021) is useful as they are important but relatively rare phenotypes and can be effective indicators of clinically relevant bacteria displaying resistance to essential and last-resort antibiotics (WHO 2017). Antimicrobial susceptibility test methods can also indicate levels of resistance and have been well studied and standardised (Jorgensen & Turnidge 2015). Culturing is cost-effective and technically easy, and therefore more accessible for LMICs (Pruden *et al.* 2021). However, it provides comparatively narrow information about the phenotypic characteristics of target organisms. Therefore, combining genotypic methods, such as qPCR, extends monitoring to gene-specific targets to accompany phenotypic data. Targeting MGEs which may be high-risk also can be valuable (Zhang *et al.* 2021a), or have been found to closely reflect anthropogenic resistance, using the Intl1 (Gillings *et al.* 2015) or aintl1 (Quintela-Baluja *et al.* 2021) assays. Genotypic methods like qPCR can provide accurate quantitative data on AMR abundance which can be normalised to cell abundance through co-quantification of genes like 16S rRNA. However, to be meaningful in a health protection context, such analyses should be corroborated a priori by parallel culturing methods to ensure they mirror health consequences.

The use of multi-array high-throughput qPCR arrays has shown promise because they can be designed to capture many different indicator genes at once (Wang *et al.* 2014). These have already been used successfully for wastewater applications (Quintela-Baluja *et al.* 2019).

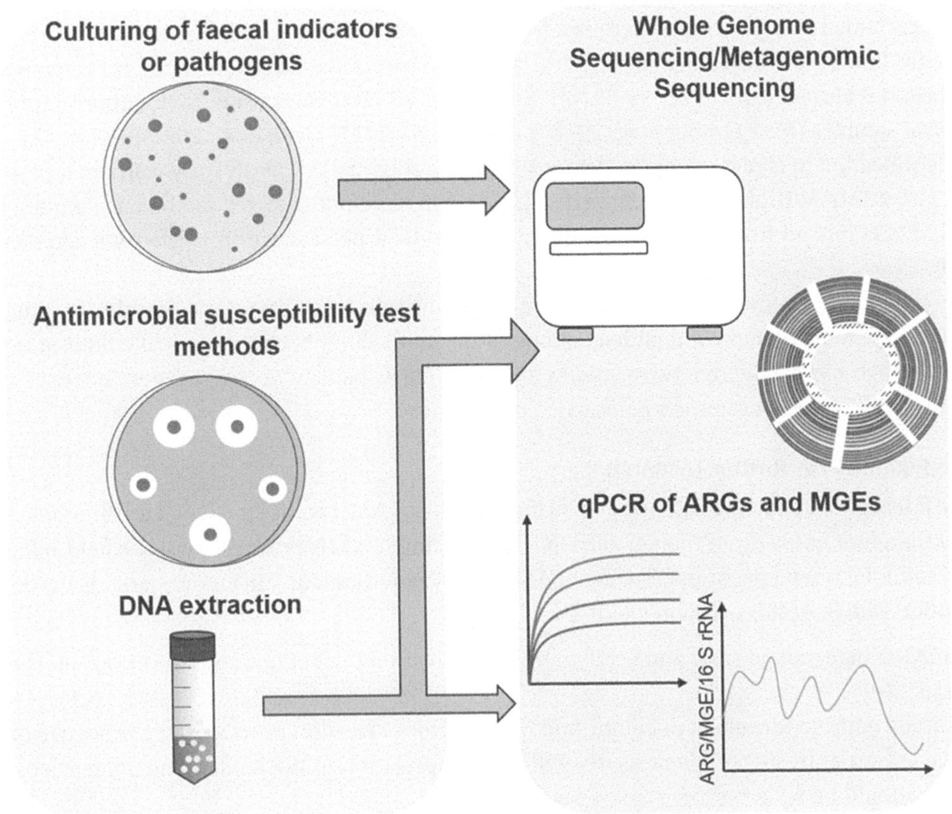

Figure 2 | Summary of laboratory techniques for analysis of AMR that could be used for WBE.

In recent years, metagenomic sequencing has proved to be a powerful approach for the detection of AMR. Metagenomic sequencing has the potential to reveal the diversity of ARGs, from natural ocean and soils to samples from the guts of animals, humans, and sewage (Hendriksen *et al.* 2019; Zeng *et al.* 2019). Factors included in metagenomic pipelines such as the Meta-Compare's relative resistome risk (Martínez *et al.* 2015), can assign a value to the mobility of ARGs or the potential to occur in pathogens which could be useful for determining clinical risk using metagenomic data. Sequencing-based approaches can also be used with and without a culture phase: targeted metagenomics permits the quantification of thousands of ARGs in culturable and potentially pathogenic bacteria (Leonard *et al.* 2018). As with viruses, detection of novel and rare genes will be limited.

Multiple pipelines for detection and enumeration of AMR exist (Gupta *et al.* 2020a), but a standardised method is needed for WBE if opportunities to compare data internationally are to be realised. With the high cost of sequencing-based approaches, this approach may not be accessible to some regions, therefore centralised sample processing and data analysis resources and funding to LMICs is needed (Aarestrup & Woolhouse 2020). Development of combined sequencing with multiplex qPCR and culturing methods would make monitoring more accessible across different socioeconomic settings.

Comparison of WBE and public health data for AMR

There is some evidence that WBE AMR data are correlated with data on regional clinical infections. Hutinel *et al.* (2019) demonstrated strong and significant correlations between *E. coli* isolated from wastewater samples and patient samples from Gothenburg, Sweden. Significant correlations were also found between samples taken from 10 wastewater treatment plants in Europe and clinical samples from EARS-Net, where resistance of *E. coli* to four antibiotics was analysed (Huijbers *et al.* 2020). Genotypic data gathered via qPCR arrays have also been found to broadly reflect clinical resistance in European sites (Pärnänen *et al.* 2019). Chau *et al.* (2021) identified a strong association with phenotypic and genotypic data from WBE to data gathered from clinical settings, indicating WBE provided insight into the resistance profiles of a community.

Comparisons between clinical and wastewater data have also shown WBE can be used to inform colonisation of specific pathogens, such as carbapenem-producing Enterobacteriaceae (CPE). CPEs are a major threat to healthcare settings and screening is currently limited to those who are most at risk (i.e., who have travelled, came into contact with a positive case, or visited a hospital abroad (Brown & Fry 2020)). Studies have found wastewater CPE cultures are reflective of what is observed in clinical samples (Meir-Gruber *et al.* 2016; Ludden *et al.* 2017; Flach *et al.* 2021). Isolates have been identified in WBE prior to identification in clinical samples, demonstrating its potential use as an early warning tool to prevent hospital outbreaks (Flach *et al.* 2021). WBE has also identified strains which were not detected in clinical samples, interpreted to be due to a carrier that was missed from clinical screening, or a result of persistence in wastewater after previous exposure (White *et al.* 2016).

While there are correlations between WBE and clinical resistance datasets, there is currently too much uncertainty to enable public health response. An improved understanding of the connection between clinical colonisation and wastewater could be achieved through more comprehensive studies including participants who are deemed low risk, and further investigation into the impact of wastewater on the persistence of resistance.

Summary and suggestions for further research

Surveillance of AMR using wastewater is complicated by the ubiquity of AMR across the One Health triad. Detection of AMR in wastewater may therefore not be from human carriers experiencing or at risk of treatment-resistant infection, but may be there as a result of zoonotic reservoirs, or persistence of AMR in the environment. Further research is needed before this tool can be used to support public health decision-making.

- The link between ARG detection in stool and urine to treatment-resistant infections among various at-risk groups warrants further exploration.
- Further studies should consider the effect of abiotic and biotic factors, selection processes, the influence of seasonality, and different types of wastewater treatment plants as this will aid interpretation of WBE data. The influence of wastewater from healthcare settings should be reviewed on a case-by-case.
- Further research should determine the best sampling and storage methodologies with consideration to resource constraints to inform standard practice and facilitate data comparisons.
- A standardised methodology for wastewater sample laboratory analysis is needed to allow international collaboration and comparisons. This approach needs to be accessible for LMICs.
- Improved understanding of the link between WBE and clinical resistance could be achieved through a comprehensive study incorporating participant and wastewater sampling and further investigation into the impact of wastewater on the persistence of resistance.

CHEMICALS

Added value of WBE for chemical surveillance

Chemical analysis of wastewater provides a broad range of information that can be used to understand multiple parameters influencing public health. Potential insights include but are not limited to (1) illicit drug use (Thomas *et al.* 2012; Ort *et al.* 2014; González-Mariño *et al.* 2020), (2) alcohol and tobacco consumption (Baz-Lomba *et al.* 2016), (3) consumption of over-the-counter medicines (Harman *et al.* 2011; Fattore *et al.* 2016; Melchor-Martínez *et al.* 2021; Rice *et al.* 2020), (4) consumption of antimicrobials (Castrignanò *et al.* 2020; Elder *et al.* 2021), (5) exposure to chemicals such as pesticides (Rousis *et al.* 2020), toxicants from everyday products including parabens, UV filters, bisphenols (BPAs), and phthalates (Lopardo *et al.* 2019a; Senta *et al.* 2020), and (6) biomarkers for certain non-communicable diseases, for example, cocaine, F2-isoprostanes, Prostaglandin E2 (Thomas & Reid 2011; O'Brien *et al.* 2019).

WBE has long been recognised as an approach for monitoring illicit drug use. Illicit drug use data are especially challenging in terms of surveillance, due to the behaviour being illegal, and highly stigmatised (Huizer *et al.* 2021). For this purpose, it fills significant gaps in other monitoring techniques such as surveys, drug-related hospital admissions, and arrest data, which do not capture the extent of consumption and are not temporally and spatially sensitive (González-Mariño *et al.* 2020; Huizer *et al.* 2021). This approach has been used widely, including an international study in 2011, where five illicit drugs (cocaine, cannabis, amphetamine, methamphetamine, and methylenedioxymethamphetamine (MDMA)) were analysed in the wastewater of initially 19 European cities, increasing to a total of 73 cities globally by 2017, including locations in LMICs

(Thomas *et al.* 2012; Gonzalez *et al.* 2020). Ongoing monitoring in 80 European cities is supported by the European Monitoring Centre for Drugs and Drug Addiction (EMCDDA) which hosts an interactive website (EMCDDA 2021). Antibiotic consumption has also been investigated within the context of ARG prevalence in wastewater, showing strong correlations between antibiotic levels, ARGs, and population size (Castrignanò *et al.* 2020; Elder *et al.* 2021).

Data gap exist in understanding consumption of pharmaceuticals being prescribed and purchased online or in pharmacies. Patients often do not follow dosage or disposal recommendations, stockpiling leftover doses, or self-prescribing (Escolà Casas *et al.* 2021). One benefit of WBE for pharmaceutical use is the ability to differentiate between metabolised vs. discarded medication levels. The example of carbamazepine (Figure 3) shows the importance of measuring metabolic products compared to drugs themselves as direct disposal of unused drugs might result in overestimation of intake. Furthermore, information on public health trends tend to rely on individual data collection and biological samples or surveys, such as NHS England's health survey (Rice & Kasprzyk-Hordern 2019). These data are often not available in real-time due to the need to collate data before assessing and formulating appropriate responses.

Chemical analysis of wastewater could therefore provide spatially and temporally comprehensive epidemiological information to aid public health and healthcare practitioners. Longitudinal analysis would allow regional and temporal comparisons to be made and help to understand principal mechanisms driving public health trends. WBE could also provide information in data-poor regions, where antimicrobials, prescription drugs, and exposure to pollutants may be less regulated.

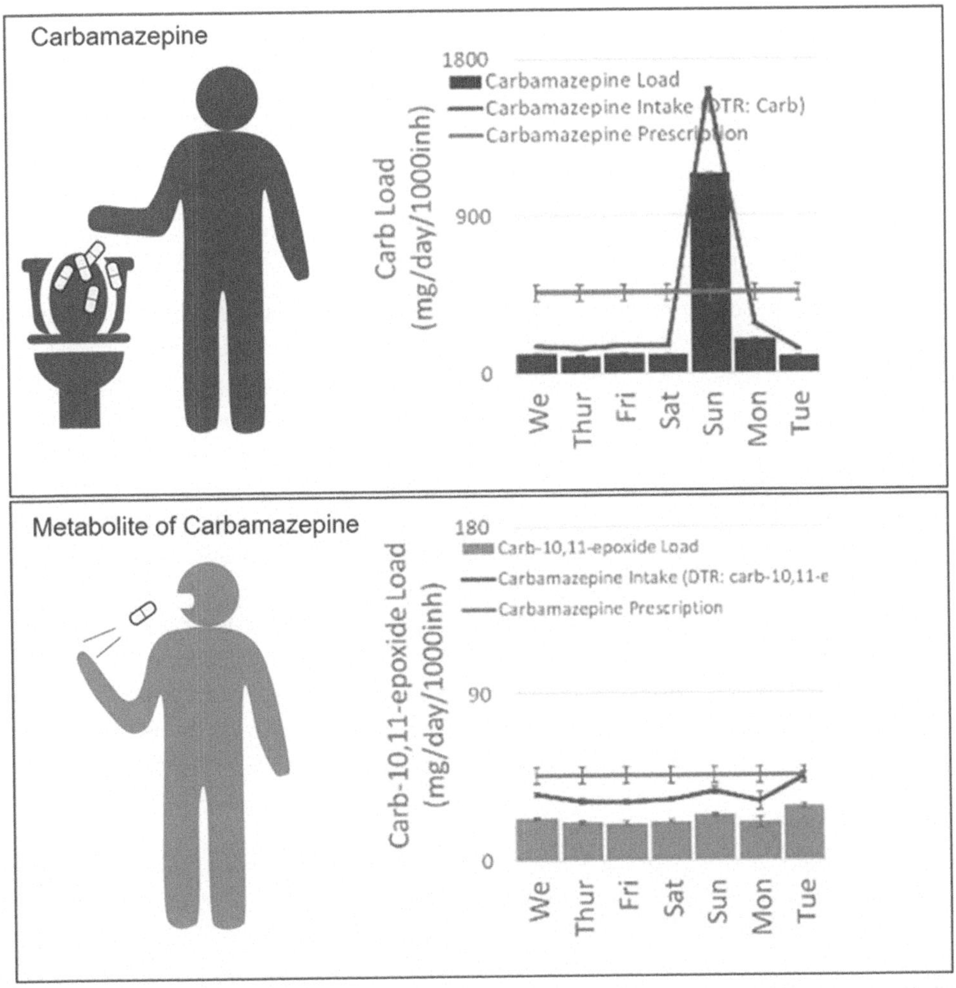

Figure 3 | Comparison of carbamazepine loads and intake calculated using carbamazepine-1-,11-epoxide and prescribed over a 7-day sampling week (adapted from Wade *et al.* 2022).

Presence of chemicals in faeces and urine

Chemical residues are excreted through urine and stool and can be detected in wastewater, usually reported as a percentage of the target residue (Baker *et al.* 2014). Understanding of human pharmacokinetics including excretion rates are therefore essential to back-calculate the approximate total exposure or consumption of chemicals (Zhang *et al.* 2019).

A comparison between the parent compounds and their metabolites for WBE found a higher certainty using metabolic markers (Baker *et al.* 2014), while other studies went further to suggest that only metabolic markers should be used in calculations where possible (Kasprzyk-Hordern *et al.* 2021).

An extensive list of excretion rates of biomarkers for lifestyle (e.g., illicit drugs, alcohol), exposure (e.g., pesticides), health (e.g., pharmaceuticals), and population (e.g., endogenous compounds) is available from Gracia-Lor *et al.* (2017). These estimates are from studies published after 2000, and primarily after 2010. However, some metabolites only have urine or faecal excretion percentage reported, and rarely both. This means some studies estimating exposure have only used excretion rates from urine in back-calculations (van Nuijs *et al.* 2015), likely underestimating exposure since often only phase I metabolites of chemicals are excreted in urine. A more accurate approach would be to include both phase I and II metabolites, which can be found in urine and faeces (Daughton 2018; Escolà Casas *et al.* 2021).

Further research is needed to understand excretion data for metabolites, which are often less complete than for parent compounds and varies depending on individual and medical conditions (Escolà Casas *et al.* 2021). Unfortunately, further pharmacokinetic studies on excretion rates are not always feasible, due to high costs, time requirements, and ethical considerations (Gracia-Lor *et al.* 2017). Alternative approaches include *in vitro* studies using liver enzymes (Mardal *et al.* 2016) and computer modelling (Reid *et al.* 2014). However, such data provide information on metabolic products and general insight into metabolism, but not quantitative information on excretion rates that could be used for WBE. Clinical proteomics have also led to the identification of new biomarkers. However, due to the time and expense required for clinical trials, very few are being confirmed clinically (Nkuipou-Kenfack *et al.* 2017; Rice & Kasprzyk-Hordern 2019). Nevertheless, the identification of human proteins in wastewater offers an opportunity for community-wide monitoring of diseases (Carrascal *et al.* 2020), and further studies on key biomarkers including metabolites as well as other endogenously formed entities such as proteins are needed. Emphasis should be put on improving understanding of pharmacokinetics to help quantify excretion factors for stool and urine, particularly where population variation is likely.

Impact of wastewater on chemical detection

Chemical stability in wastewater varies depending on the original compound and the conditions in wastewater. The stability of illicit drug biomarkers has been reviewed elsewhere (McCall *et al.* 2016a). A study investigating 43 pharmaceuticals showed that the majority remained stable in a pressurised sewer of 7.6 km, with a concentration error of $\pm 10\%$ for most pharmaceuticals, with some at ~60% (Jelic *et al.* 2015). Temperature impacts biomarker stability, with one study showing that half-lives of biomarkers in colder temperatures were significantly longer than in warmer temperatures (Senta *et al.* 2014). McCall *et al.* (2016b) also found that temperature, pH, hydraulics, residence times, sorption processes, and biodegradation rates affect drug stability in sewers. Biofilms may also influence the stability of illicit drug biomarkers (McCall *et al.* 2016a; Li *et al.* 2018) and particulate matter in sewage has been shown to affect results, where a large proportion of the analyte may partition onto particulate matter (Baker *et al.* 2012). A study from Tang *et al.* (2020) found phthalate metabolites in wastewater were in much higher concentrations than predicted from urinary excretion, therefore whether metabolites are formed from parent compounds within wastewater environments should be investigated.

Further studies are required to estimate correction factors for metabolites that are to be used in WBE, particularly through understanding the conditions in wastewater that may influence interpretations, including the amount and type of sewage biofilm and different environmental conditions (e.g., temperature, pH and biodegradation) (McCall *et al.* 2016a).

Sampling methodology and storage requirements for chemicals

Analysis of chemicals and drugs in wastewater is most accurately conducted through 24-h flow-proportional composite sampling. While both time- and flow-proportional sampling is commonly applied, flow-proportional instead of time-proportional samples (e.g., every 20 min), have been found to provide an accurate indication of peak input of chemicals and biomarkers excreted into wastewater (Ort *et al.* 2010; Petrie *et al.* 2017). Grab or spot sampling does not allow for full analyte capture due to common diurnal variabilities in wastewater composition linked with variable human excretion patterns as well as external factors, like rainfall. To preserve chemicals for accurate analysis, storage time and temperature are crucial.

Samples need to be transported to laboratories refrigerated (if over a short period of time, e.g., a few hours) or frozen (for overnight transport). If chilled, they must be immediately frozen on arrival to prevent analyte degradation. Acidification can improve the stability of chemicals, by preventing some drugs from partitioning to the water fraction (Baker & Kasprzyk-Hordern 2011).

Passive sampling has been investigated (Moschet *et al.* 2015; Townsend *et al.* 2018; Rimayi *et al.* 2019). A comparison of passive and active sampling undertaken by Petrie *et al.* (2016), found the concentration of chemicals captured from passive sampling was within a factor of 2 of the known concentration. However, passive samplers might not achieve the same resolution as composite samples (Jones *et al.* 2015), but may nevertheless be used when analyte detection alone is a suitable measure for public health purposes. Further research is needed before it can be used as a standard alternative sampling method where there is insufficient infrastructure for composite sampling.

To preserve chemicals, understanding storage time and temperature for chemical stability is crucial. Chemicals are likely to degrade under repeated freeze–thawing cycles and over long-term storage. In a study by Fedorova *et al.* (2014), influent wastewater samples were spiked with 124 pharmaceuticals and personal care products to determine the impact of different storage conditions on analyte detection. Storage conditions at 4 °C resulted in greater amounts of analyte being detected compared to samples kept at −18 °C, with 15 analytes undetected after a freeze/thaw cycle. An understanding of the stability of target compounds and biomarkers, and appropriately designed standards for storage is necessary, to ensure consistency and comparability among studies.

Laboratory analysis methodology for chemicals

Chemical markers are present in wastewater at low concentrations, spanning from sub-ppt to ppm levels. Ultra-performance liquid chromatography is the method of choice for the separation of structurally wide-ranging groups of analytes with variable polarities, but gas chromatography can be also utilised for more volatile targets, also with common derivatisation steps utilised for more polar analytes.

In order to enable analysis at trace ppt levels, analyte enrichment is required. Solid phase extraction with wide-ranging materials is widely utilised to both clean-up and enrich liquid samples. QuEChERS method as well as pressurised liquid extraction or accelerated solvent extraction are also suitable for the extraction of analytes from solid samples such as suspended particulate matter in wastewater (Nieto *et al.* 2010; Kachhawaha *et al.* 2017; Pang *et al.* 2017). High-throughput methods are increasingly being used and within the context of WBE can provide more insight into multiple targets. Triple quadrupole mass spectrometry is usually applied to undertake multi-residue targeted analysis of >100 markers with one analytical run and is highly useful for large catchment studies (Proctor *et al.* 2019). High-resolution mass spectrometry can also allow untargeted screening for novel chemicals, with subsequent data mining allowing the potential to expand and develop analytes of interest (Lopardo *et al.* 2019b).

Further research on high-resolution mass spectrometry approaches is required for data mining and identifying further steps in sensing approaches. High-resolution analysis approaches would enable detection of an expanded biomarker suite, and allow multilevel estimation of public exposure to stressors and resulting effects (e.g., disease status).

Further studies are also necessary on chemical population markers. Population normalisation is necessary for all aspects of WBE, where inputs into the sewer may vary over the course of the day or be diluted by stormwater. Estimation of population may also provide insights into population mobility and travel. Metabolic markers for population normalisation are being explored, such as metabolites for caffeine or nicotine (Senta *et al.* 2015), or prescription pharmaceuticals such as atenolol (Lai *et al.* 2011). Further application of these biomarkers in WBE studies, and more information on how these products are metabolised, are necessary to use these to quantitatively estimate population.

Comparison of chemical WBE data

Several studies have compared prescription and WBE data (Carballa *et al.* 2008; Lai *et al.* 2011; Verlicchi *et al.* 2014; Escolà Casas *et al.* 2021). Data on illicit drug market trends and confiscated drugs have been used to sense-check WBE illicit drug data (Kankaanpää *et al.* 2016; Krizman-Matasic *et al.* 2019; Gonzalez *et al.* 2020). Considerations for evaluating WBE for pharmaceutical consumption include ensuring prescription data are at an appropriate temporal and spatial scale, where national averaged data are insufficient for comparison (Carballa *et al.* 2008; Lai *et al.* 2011; Verlicchi *et al.* 2014). An example of a study with a higher temporal resolution is Escolà Casas *et al.* (2021), which compared pharmaceutical detection via WBE against a monthly aggregated pharmaceutical dataset. Such datasets need to be explored and compared against WBE data to ensure data are correct, identify and explain discrepancies, and account for any variation.

Summary and suggestions for further research

Chemicals can also act as indicators of human health. However, these are known to undergo changes in the body and/or the sewer environment, affecting the choice of chemical markers for surveillance, collection and storage of samples, and data interpretation.

- Further research is needed to understand excretion data for metabolites and endogenously formed entities, such as proteins. Improved understanding of pharmacokinetics is necessary to help find excretion factors in stool and urine, and how these may vary due to treatment received or age group.
- There is a lack of information on biomarker stability, particularly for metabolites. Further studies should investigate conditions in the sewer which may impact interpretations of WBE data, including sewage biofilm and different environmental conditions.
- Improved understanding of passive sampling could support WBE in areas where composite sampling is not feasible. Further understanding of the stability of various biomarkers is necessary to design appropriate standards, to ensure consistency and comparability among studies.
- Expansion of the biomarker suite coupled with high-resolution mass spectrometry approaches could allow multilevel estimation of public exposure and resulting effects. Exploration and application of population level biomarkers in WBE studies are necessary to have an improved estimated of population.
- WBE data need to be compared to appropriate datasets at an appropriate temporal and spatial scale to identify and explain discrepancies and account for any variation.

ADDITIONAL CONSIDERATIONS FOR IMPLEMENTATION

Integration into public health monitoring

Looking forward, WBE can be used for strategic and urgent public health surveillance needs, and to support environmental surveillance (Figure 4). Consideration should be given to the detection approach necessary for each area of interest and the subsequent response to an increase in infection or positive detection.

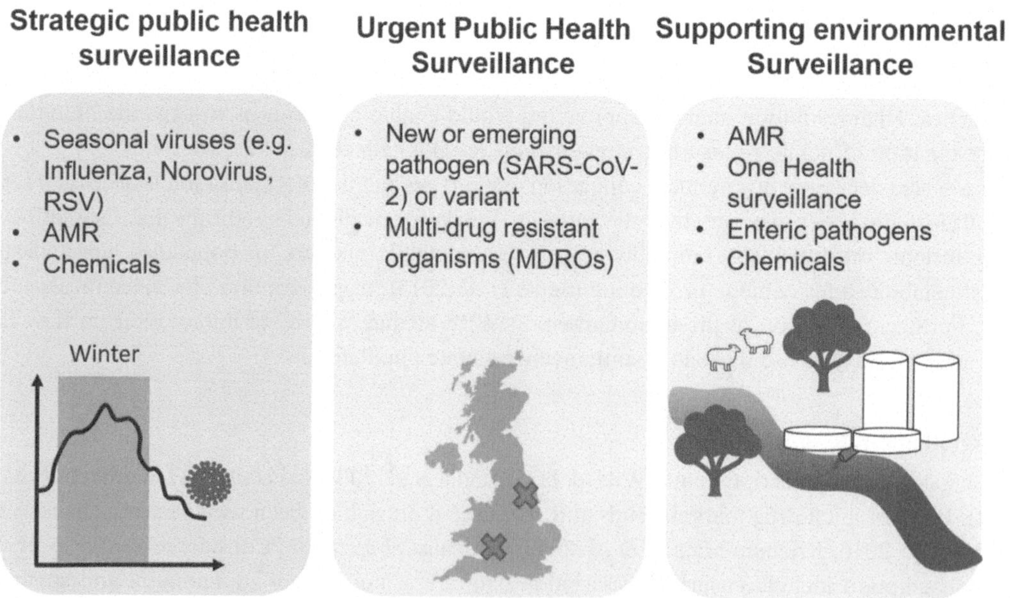

Figure 4 | WBE could be used to support public health surveillance through strategic public health surveillance, urgent public health surveillance, and by supporting environmental surveillance.

To be useful for public health monitoring, WBE needs to be integrated and add value to other public health initiatives, including clinical case reports, self-reporting systems, or social media tracking systems like FluDetector (Lampos 2016) or ZoeApp (Menni *et al.* 2021). Current epidemiological systems typically do not integrate clinical and environmental monitoring together. However, such an approach is necessary to ensure both groups are openly communicating, sharing insights, and harmonising approaches. Communication and collaboration with clinical and public health teams are also necessary for ensuring the data collected are useful in informing practice (e.g., identification of clinically relevant resistance within a community would need to be communicated to veterinary and clinical practitioners). Likewise, people working in healthcare might share with WBE teams requirements for novel targets. Collaboration with organisations collecting animal and environmental data could also be beneficial for developing a One Health surveillance framework.

Further research is needed between people working in WBE and those in healthcare settings to develop improved ways for public health users to receive these data and the best communication tools. Further discussions across clinical and environmental sectors is crucial for the design of WBE to ensure this is integrated into existing monitoring and provides a practical application. In addition, communication across the general public warrants further consideration. Publicly available dashboards during the COVID-19 pandemic showing daily case rates should be considered for additional WBE applications, allowing the public to have transparent access to information, raise awareness of health trends and provide information to make informed decisions. An interactive website or centralised dashboard providing clear, accessible summarised/non-identifiable data, integrating environmental, wastewater, and clinical data would be a potential way to communicate this information.

Having an overall understanding and ability to communicate uncertainties is also important. Within the UK, sewage networks are regionally diverse and can be publicly and privately owned, leading to differences in the way sampling is delivered (Wade *et al.* 2022). Some of the challenges may be especially evident in smaller rural communities, which may have hundreds of small wastewater treatment sites (Farkas *et al.* 2020a). Being able to communicate this uncertainty to relevant clinical stakeholders, the media and the general public is necessary to ensure data transparency and guide informed interpretations and responses.

Ethical implications

Despite wastewater being considered an ethical way of collecting data, further consideration and engagement of social scientists and the public is required to assess the ethics of WBE in applications beyond the pandemic. It is widely assumed that WBE uses aggregate measurements and cannot be used to identify individual persons. However, concerns have been raised over the privacy of data obtained from small populations or marginalised communities (McClary-Gutierrez *et al.* 2021), as well as over how the data will be used. Individual consent for the collection of wastewater samples for SARS-CoV-2 analysis has not been deemed necessary (Kwiatkowska *et al.* 2021). While data should be made publicly available and transparent, it is necessary to navigate how best to ethically and legally handle data, as well as communicate this information. Further research is required to gain community feedback and support open access information and prevent disagreements and stigmatisation (Polo *et al.* 2020). Regions or marginalised communities may be associated with negative health, and this could have a negative impact on well-being, local economies, and potentially insurance policies (Wang *et al.* 2020; Harris *et al.* 2021). These discussions are particularly urgent as the technology and applications for WBE advance.

CONCLUSIONS

Research on WBE has proven it is a potentially useful tool in monitoring the presence of health markers, providing a theoretically constant stream of information on the health status of whole communities. Proposed uses include tracking viral pathogens, AMR, and chemicals (i.e., drug and pharmaceutical metabolites, markers of human disease or pollution exposure). However, each application of wastewater has its own challenges and research needs, including varying disease dynamics, non-human reservoirs, instability of markers in sewers and storage and interpretation for public health. Research into these areas would allow WBE to be developed as a useful tool for public health to supplement existing surveillance. However, WBE methods and protocols have been developed in high-income countries with improved sanitation. With greater globalisation, emerging threats to health can occur anywhere, and spread around the world rapidly. International cooperation and coordination is necessary to monitor for future threats, and if WBE is to play a role in this, comparable datasets across communities and countries are needed. In particular, we should consider and support the development and validation of these techniques for lower income settings to ensure improved global health surveillance.

FUNDING

The United Kingdom Government (Department of Health and Social Care) fund the sampling, testing, and data analysis of wastewater in England. The Welsh Government (Technical Advisory Cell) funds the sampling, testing, and data analysis of wastewater in Wales. The following funding sources are acknowledged by the contributing authors: 'National COVID-19 Wastewater Epidemiology Surveillance programme (NE/V010441/1) UKRI-NERC COVID Response (D.W.G., D.L.J., B.K.-H.); 'Building an Early Warning System for community-wide infectious disease spread: SARS-CoV-2 tracking in Africa via environment fingerprinting' (EP/V028499/1) GCRF/Newton Fund (B.K.-H.).

DATA AVAILABILITY STATEMENT

All relevant data are included in the paper or its Supplementary Information.

CONFLICT OF INTEREST

The authors declare there is no conflict.

REFERENCES

Aarestrup, F. M. & Woolhouse, M. E. 2020 Using sewage for surveillance of antimicrobial resistance. *Science* **367** (6478), 630–632.

Adriaenssens, E. M., Farkas, K., Harrison, C., Jones, D. L., Allison, H. E. & McCarthy, A. J. 2018 Viromic analysis of wastewater input to a river catchment reveals a diverse assemblage of RNA viruses. *mSystems*. https://doi.org/10.1128/mSystems.00025-18.

Adriaenssens, E. M., Farkas, K., McDonald, J. E., Jones, D. L., Allison, H. E. & McCarthy, A. J. 2021 Tracing the fate of wastewater viruses reveals catchment-scale virome diversity and connectivity. *Water Research* **203**, 117568. https://doi.org/10.1016/j.watres.2021.117568.

Agga, G. E., Arthur, T. M., Durso, L. M., Harhay, D. M. & Schmidt, J. W. 2015 Antimicrobial-resistant bacterial populations and antimicrobial resistance genes obtained from environments impacted by livestock and municipal waste. *PLoS One* **10** (7), e0132586.

Ahmed, W., Bivins, A., Bertsch, P. M., Bibby, K., Gyawali, P., Sherchan, S. P., Simpson, S. L., Thomas, K. V., Verhagen, R., Kitajima, M. & Mueller, J. F. 2021 Intraday variability of indicator and pathogenic viruses in 1-h and 24-h composite wastewater samples: implications for wastewater-based epidemiology. *Environmental Research* **193**, 110531.

Al-Duroobi, H., Moghadam, S. V., Phan, D. C., Jafarzadeh, A., Matta, A. & Kapoor, V. 2021 Wastewater surveillance of SARS-CoV-2 corroborates heightened community infection during the initial peak of COVID-19 in Bexar County, Texas. *FEMS Microbes* **2**. https://doi.org/10.1093/femsmc/xtab015.

Allard, A., Albinsson, B. & Wadell, G. 1992 Detection of adenoviruses in stools from healthy persons and patients with diarrhea by two-step polymerase chain reaction. *Journal of Medical Virology* **37** (2), 149–157.

Amahmid, O., El Guarmri, Y., Rakibi, Y., Ouizat, S., Yazidi, M., Razoki, B., Kaid Rassou, K., Asmama, S., Bouhoum, K. & Belghyti, D. 2021 Occurrence of SARS-CoV-2 in excreta, sewage, and environment; epidemiological significance and potential risks. *International Journal of Environmental Health Research* **31**, 1–14.

Aquino de Carvalho, N., Stachler, E. N., Cimabue, N. & Bibby, K. 2017 Evaluation of phi6 persistence and suitability as an enveloped virus surrogate. *Environmental Science & Technology* **51** (15), 8692–8700.

Atkinson, B., Hearn, P., Afrough, B., Lumley, S., Carter, D., Aarons, E. J., Simpson, A. J., Brooks, T. J. & Hewson, R. 2016 Detection of Zika virus in semen. *Emerging Infectious Diseases* **22** (5), 940.

Avgeris, M., Adamopoulos, P. G., Galani, A., Xagorari, M., Gourgiotis, D., Trougakos, I. P., Voulgaris, N., Dimopoulos, M.-A., Thomaidis, N. S. & Scorilas, A. 2021 Novel Nested-Seq approach for SARS-CoV-2 real-time epidemiology and in-depth mutational profiling in wastewater. *International Journal of Molecular Sciences* **22**, 8498.

Baker, D. R. & Kasprzyk-Hordern, B. 2011 Multi-residue analysis of drugs of abuse in wastewater and surface water by solid-phase extraction and liquid chromatography–positive electrospray ionisation tandem mass spectrometry. *Journal of Chromatography a* **1218** (12), 1620–1631.

Baker, D. R., Očenášková, V., Kvicalova, M. & Kasprzyk-Hordern, B. 2012 Drugs of abuse in wastewater and suspended particulate matter – further developments in sewage epidemiology. *Environment International* **48**, 28–38.

Baker, D. R., Barron, L. & Kasprzyk-Hordern, B. 2014 Illicit and pharmaceutical drug consumption estimated via wastewater analysis. Part A: chemical analysis and drug use estimates. *Science of the Total Environment* **487**, 629–641.

Baltrusaitis, K., Santillana, M., Crawley, A. W., Chunara, R., Smolinski, M. & Brownstein, J. S. 2017 Determinants of participants' follow-up and characterization of representativeness in flu near you, a participatory disease surveillance system. *JMIR Public Health and Surveillance* **3** (2), e7304.

Baquero, F., Martínez, J. L. & Cantón, R. 2008 Antibiotics and antibiotic resistance in water environments. *Current Opinion in Biotechnology* **19** (3), 260–265.

Barbosa, M. R. F., Garcia, S. C., de Castro Bruni, A., Machado, F. S., de Oliveira, R. X., Dropa, M., da Costa, A. C., Leal, E., Brandao, C. J., da Silva, R. L. O., Iko, B. Y., Kondo, V. K. M., de Araujo, R. S., da Silveira, V. B., de Andrade, T. M., Nunes, D. R., Janini, L. M. R., Braconi,

C. T., Maricato, J. T. & Sato, M. I. Z. 2022 One-year surveillance of SARS-CoV-2 in wastewater from vulnerable urban communities in metropolitan Sao Paulo, Brazil. *Journal of Water and Health* **20** (2), 471–490.

Baz-Lomba, J. A., Salvatore, S., Gracia-Lor, E., Bade, R., Castiglioni, S., Castrignanò, E., Causanilles, A., Hernandez, F., Kasprzyk-Hordern, B., Kinyua, J. & McCall, A. K. 2016 Comparison of pharmaceutical, illicit drug, alcohol, nicotine and caffeine levels in wastewater with sale, seizure and consumption data for 8 European cities. *BMC Public Health* **16** (1), 1–11.

Bibby, K., Fischer, R. J., Casson, L. W., Stachler, E., Haas, C. N. & Munster, V. J. 2015 Persistence of Ebola virus in sterilized wastewater. *Environmental Science & Technology Letters* **2** (9), 245–249.

Biello, D. 2007 A stash in every flush. *Scientific American* **297** (6), 28–30.

Bivins, A. & Bibby, K. 2021 Wastewater surveillance during mass COVID-19 vaccination on a college campus. *Environmental Science & Technology Letters* **8**, 792–798. https://doi.org/10.1021/acs.estlett.1c00519.

Bivins, A., Kaya, D., Bibby, K., Simpson, S. L., Bustin, S. A., Shanks, O. C. & Ahmed, W. 2021 Variability in RT-qPCR assay parameters indicates unreliable SARS-CoV-2 RNA quantification for wastewater surveillance. *Water Research* **203**, 117516. https://doi.org/10.1016/j.watres.2021.117516.

Blaak, H., Kemper, M. A., de Man, H., van Leuken, J. P., Schijven, J. F., van Passel, M. W., Schmitt, H. & de Roda Husman, A. M. 2021 Nationwide surveillance reveals frequent detection of carbapenemase-producing Enterobacterales in Dutch municipal wastewater. *Science of The Total Environment* **776**, 145925.

Bofill-Mas, S., Albinana-Gimenez, N., Clemente-Casares, P., Hundesa, A., Rodriguez-Manzano, J., Allard, A., Calvo, M. & Girones, R. 2006 Quantification and stability of human adenoviruses and polyomavirus JCPyV in wastewater matrices. *Applied and Environmental Microbiology* **72** (12), 7894–7896.

Brown, C. & Fry, C. 2020 Framework of actions to contain carbapenemase-producing Enterobacterales. Freeman & S. Hopkins, eds. PHE Publications, London. https://assets.publishing.service.gov.uk/government/uploads/system/uploads/attachment_data/file/923385/Framework_of_actions_to_contain_CPE.pdf.

Brown, M. R., Wade, M. J., McIntyre-Nolan, S., Bassano, I., Denise, H., Bass, D., Bentley, J., Bunce, J. T., Grimsley, J., Hart, A. & Hoffmann, T. 2021 *JBC and Defra: Wastewater Monitoring of SARS-CoV-2 Variants in England: Demonstration Case Study for Bristol (December 2020 to March 2021), 8 April 2021.*

Buelow, E., Bayjanov, J. R., Majoor, E., Willems, R. J., Bonten, M. J., Schmitt, H. & van Schaik, W. 2018 Limited influence of hospital wastewater on the microbiome and resistome of wastewater in a community sewerage system. *FEMS Microbiology Ecology* **94** (7), fiy087.

Buelow, E., Rico, A., Gaschet, M., Lourenço, J., Kennedy, S. P., Wiest, L., Ploy, M. C. & Dagot, C. 2020 Hospital discharges in urban sanitation systems: long-term monitoring of wastewater resistome and microbiota in relationship to their eco-exposome. *Water Research X* **7**, 100045.

Burnham, J. P., Olsen, M. A. & Kollef, M. H. 2019 Re-estimating annual deaths due to multidrug-resistant organism infections. *Infection Control & Hospital Epidemiology* **40** (1), 112–113.

Bustin, S., Benes, V., Garson, J. A., Hellemans, J., Huggett, J., Kubista, M., Mueller, R., Nolan, T., Pfaffl, M., Shipley, G. L., Vandesompele, J. & Wittwer, C. T. 2009 The MIQE guidelines: minimum information for publication of quantitative real-time PCR experiments. *Clinical Chemistry* **55**, 611–622. https://doi.org/10.1373/clinchem.2008.112797.

Carballa, M., Omil, F. & Lema, J. M. 2008 Comparison of predicted and measured concentrations of selected pharmaceuticals, fragrances and hormones in Spanish sewage. *Chemosphere* **72** (8), 1118–1123.

Carr, V. R., Witherden, E. A., Lee, S., Shoaie, S., Mullany, P., Proctor, G. B., Gomez-Cabrero, D. & Moyes, D. L. 2020 Abundance and diversity of resistomes differ between healthy human oral cavities and gut. *Nature Communications* **11** (1), 1–10.

Carrascal, M., Abian, J., Ginebreda, A. & Barceló, D. 2020 Discovery of large molecules as new biomarkers in wastewater using environmental proteomics and suitable polymer probes. *Science of the Total Environment* **747**, 141145.

Cassemiro, K. M. S. D. M., Burlandy, F. M., Barbosa, M. R., Chen, Q., Jorba, J., Hachich, E. M., Sato, M. I., Burns, C. C. & da Silva, E. E. 2016 Molecular and phenotypic characterization of a highly evolved type 2 vaccine-derived poliovirus isolated from seawater in Brazil, 2014. *PLoS One* **11** (3), e0152251.

Castiglioni, S., Bijlsma, L., Covaci, A., Emke, E., Hernández, F., Reid, M., Ort, C., Thomas, K. V., Van Nuijs, A. L., De Voogt, P. & Zuccato, E. 2013 Evaluation of uncertainties associated with the determination of community drug use through the measurement of sewage drug biomarkers. *Environmental Science & Technology* **47** (3), 1452–1460.

Castrignanò, E., Yang, Z., Feil, E. J., Bade, R., Castiglioni, S., Causanilles, A., Gracia-Lor, E., Hernandez, F., Plósz, B. G., Ramin, P. & Rousis, N. I. 2020 Enantiomeric profiling of quinolones and quinolones resistance gene qnrS in European wastewaters. *Water Research* **175**, 115653.

Caucci, S., Karkman, A., Cacace, D., Rybicki, M., Timpel, P., Voolaid, V., Gurke, R., Virta, M. & Berendonk, T. U. 2016 Seasonality of antibiotic prescriptions for outpatients and resistance genes in sewers and wastewater treatment plant outflow. *FEMS Microbiology Ecology* **92** (5), fiw060.

Cavalheiro, N. D. P., Santos, A. C. D. O., Melo, C. E., Morimitsu, S. R. & Barone, A. A. 2008 Hepatitis C virus detection in the semen of infected patients. *Brazilian Journal of Infectious Diseases* **12**, 358–361.

Cavany, S., Bivins, A., Wu, Z., North, D., Bibby, K. & Perkins, T. A. 2022 Inferring SARS-CoV-2 RNA shedding into wastewater relative to time of infection. *Epidemiol. Infect.* **7**, 150, e21. doi:10.1017/S0950268821002752.

CDC 2022 *Wastewater Surveillance Testing Methods*. Available from: https://www.cdc.gov/healthywater/surveillance/wastewater-surveillance/testing-methods.html.

Chau, K., Barker, L., Budgell, E., Vihta, K. D., Sims, N., Kasprzyk-Hordern, B., Harriss, E., Crook, D., Read, D., Walker, S. & Stoesser, N. 2021 Systematic review of wastewater surveillance of antimicrobial resistance in human populations. *Environment International* **162**, 107171.

Chen, C., Zhao, B., Yang, X. & Li, Y. 2011 Role of two-way airflow owing to temperature difference in severe acute respiratory syndrome transmission: revisiting the largest nosocomial severe acute respiratory syndrome outbreak in Hong Kong. *Journal of the Royal Society Interface* **8** (58), 699–710.

Cheng, G., Ning, J., Ahmed, S., Huang, J., Ullah, R., An, B., Hao, H., Dai, M., Huang, L., Wang, X. & Yuan, Z. 2019 Selection and dissemination of antimicrobial resistance in agri-food production. *Antimicrobial Resistance & Infection Control* **8** (1), 1–13.

Cheung, K. S., Hung, I. F., Chan, P. P., Lung, K. C., Tso, E., Liu, R., Ng, Y. Y., Chu, M. Y., Chung, T. W., Tam, A. R. & Yip, C. C. 2020 Gastrointestinal manifestations of SARS-CoV-2 infection and virus load in fecal samples from a Hong Kong cohort: systematic review and meta-analysis. *Gastroenterology* **159** (1), 81–95.

Chitambar, S. D., Joshi, M. S., Sreenivasan, M. A. & Arankalle, V. A. 2001 Fecal shedding of hepatitis A virus in Indian patients with hepatitis A and in experimentally infected Rhesus monkey. *Hepatology Research* **19** (3), 237–246.

Choi, P. M., Tscharke, B. J., Donner, E., O'Brien, J. W., Grant, S. C., Kaserzon, S. L., Mackie, R., O'Malley, E., Crosbie, N. D., Thomas, K. V. & Mueller, J. F. 2018 Wastewater-based epidemiology biomarkers: past, present and future. *TrAC Trends in Analytical Chemistry* **105**, 453–469.

Collignon, P., Beggs, J. J., Walsh, T. R., Gandra, S. & Laxminarayan, R. 2018 Anthropological and socioeconomic factors contributing to global antimicrobial resistance: a univariate and multivariable analysis. *The Lancet Planetary Health* **2** (9), e398–e405.

Corchis-Scott, R., Geng, Q., Seth, R., Ray, R., Beg, M., Biswas, N., Charron, L., Drouillard, K. D., D'Souza, R., Heath, D. D., Houser, C., Lawal, F., McGinlay, J., Menard, S. L., Porter, L. A., Rawlings, D., Scholl, M. L., Siu, K., Tong, Y., Weisener, C. G. & McKay, R. 2021 Averting an outbreak of SARS-CoV-2 in a University residence hall through wastewater surveillance. *Microbiology Spectrum* **9**, e0079221. https://doi.org/10.1128/Spectrum.00792-21.

Crank, K., Chen, W., Bivins, A., Lowry, S. & Bibby, K. 2022 Contribution of SARS-CoV-2 RNA shedding routes to RNA loads in wastewater. *Science of the Total Environment* **806**, 150376.

Crits-Christoph, A., Kantor, R. S., Olm, M. R., Whitney, O. N., Al-Shayeb, B., Lou, Y. C., Flamholz, A., Kennedy, L. C., Greenwald, H., Hinkle, A. & Hetzel, J. 2021 Genome sequencing of sewage detects regionally prevalent SARS-CoV-2 variants. *MBio* **12** (1), e02703–e02720.

Dantas, G. & Sommer, M. O. 2012 Context matters – the complex interplay between resistome genotypes and resistance phenotypes. *Current Opinion in Microbiology* **15** (5), 577–582.

Daughton, C. G. 2018 Monitoring wastewater for assessing community health: Sewage Chemical-Information Mining (SCIM). *Science of The Total Environment* **619**, 748–764.

De Crom, S. C. M., Obihara, C. C., de Moor, R. A., Veldkamp, E. J. M., van Furth, A. M. & Rossen, J. W. A. 2013 Prospective comparison of the detection rates of human enterovirus and parechovirus RT-qPCR and viral culture in different pediatric specimens. *Journal of Clinical Virology* **58** (2), 449–454.

Dharmadhikari, T., Rajput, V., Yadav, R., Boaragaonkar, R., Patil, D., Kale, S., Kamble, S. P., Dastager, S. G. & Dharne, M. S. 2021 High throughput sequencing based direct detection of SARS-CoV-2 fragments in wastewater of Pune, West India. *Science of the Total Environment* **807** (Pt 3), 151038.

Escolà Casas, M. E., Schröter, N. S., Zammit, I., Castaño-Trias, M., Rodriguez-Mozaz, S., Gago-Ferrero, P. & Corominas, L. 2021 Showcasing the potential of wastewater-based epidemiology to track pharmaceuticals consumption in cities: comparison against prescription data collected at fine spatial resolution. *Environment International* **150**, 106404.

Elder, F. C., Proctor, K., Barden, R., Gaze, W. H., Snape, J., Feil, E. J. & Kasprzyk-Hordern, B. 2021 Spatiotemporal profiling of antibiotics and resistance genes in a river catchment: human population as the main driver of antibiotic and antibiotic resistance gene presence in the environment. *Water Research* **203**, 117533.

Elliott, J., Whitaker, M., Bodinier, B., Eales, O., Riley, S., Ward, H., Cooke, G., Darzi, A., Chadeau-Hyam, M. & Elliott, P. 2021 Predictive symptoms for COVID-19 in the community: REACT-1 study of over 1 million people. *PLoS Medicine* **18** (9), e1003777.

European Monitoring Centre for Drugs, Drug Addiction and Europeiska centrumet för kontroll av narkotika och narkotikamissbruk 2003 *Annual Report 2011: The State of the Drugs Problem in Europe*. Publications Office of the European Union.

Falman, J. C., Fagnant-Sperati, C. S., Kossik, A. L., Boyle, D. S. & Meschke, J. S. 2019 Evaluation of secondary concentration methods for poliovirus detection in wastewater. *Food and Environmental Virology* **11** (1), 20–31.

Falzone, L., Musso, N., Gattuso, G., Bongiorno, D., Palermo, C. I., Scalia, G., Libra, M. & Stefani, S. 2020 Sensitivity assessment of droplet digital PCR for SARS-CoV-2 detection. *International Journal of Molecular Medicine* **46** (3), 957–964.

Farkas, K., Hassard, F., McDonald, J. E., Malham, S. K. & Jones, D. L. 2017 Evaluation of molecular methods for the detection and quantification of pathogen-derived nucleic acids in sediment. *Frontiers in Microbiology* **8**, 53. https://doi.org/10.3389/fmicb.2017.00053.

Farkas, K., Marshall, M., Cooper, D., McDonald, J. E., Malham, S. K., Peters, D. E., Maloney, J. D. & Jones, D. L. 2018 Seasonal and diurnal surveillance of treated and untreated wastewater for human enteric viruses. *Environmental Science and Pollution Research* **25** (33), 33391–33401.

Farkas, K., Hillary, L. S., Malham, S. K., McDonald, J. E. & Jones, D. L. 2020a Wastewater and public health: the potential of wastewater surveillance for monitoring COVID-19. *Current Opinion in Environmental Science & Health* **17**, 14–20.

Farkas, K., Mannion, F., Hillary, L. S., Malham, S. K. & Walker, D. I. 2020b Emerging technologies for the rapid detection of enteric viruses in the aquatic environment. *Current Opinion in Environmental Science & Health* **16**, 1–6. https://doi.org/10.1016/j.coesh.2020.01.007.

Farkas, K., Hillary, L. S., Thorpe, J., Walker, D. I., Lowther, J. A., McDonald, J. E., Malham, S. K. & Jones, D. L. 2021 Concentration and quantification of SARS-CoV-2 RNA in wastewater using polyethylene glycol-based concentration and qRT-PCR. *Methods and Protocols* **4** (1), 17.

Fattore, E., Davoli, E., Castiglioni, S., Bosetti, C., Depaolini, A. R., Marzona, I., Zuccato, E. & Fanelli, R. 2016 Wastewater-based epidemiological evaluation of the effect of air pollution on short-acting beta-agonist consumption for acute asthma treatment. *Environmental Research* **150**, 106–111.

Fedorova, G., Golovko, O., Randak, T. & Grabic, R. 2014 Storage effect on the analysis of pharmaceuticals and personal care products in wastewater. *Chemosphere* **111**, 55–60.

Feng, J., Li, B., Jiang, X., Yang, Y., Wells, G. F., Zhang, T. & Li, X. 2018 Antibiotic resistome in a large-scale healthy human gut microbiota deciphered by metagenomic and network analyses. *Environmental Microbiology* **20** (1), 355–368.

Fernandez-Cassi, X., Timoneda, N., Martinez-Puchol, S., Rusinol, M., Rodriguez-Manzano, J., Figuerola, N., Bofill-Mas, S., Abril, J. F. & Girones, R. 2018 Metagenomics for the study of viruses in urban sewage as a tool for public health surveillance. *Sci. Total Env.* **618**, 870–880.

Flach, C. F., Hutinel, M., Razavi, M., Åhrén, C. & Larsson, D. J. 2021 Monitoring of hospital sewage shows both promise and limitations as an early-warning system for carbapenemase-producing Enterobacterales in a low-prevalence setting. *Water Research* **200**, 117261.

FluSurvey 2017 *FluSurvey UKHSA Flusurvey: Home.* (accessed 24 November 2021).

Fout, G. S., Brinkman, N. E., Cashdollar, J. L., Griffin, S. M., McMinn, B. R., Rhodes, E. R., Varughese, E. A., Karim, M. R., Grimm, A. C., Spencer, S. K. & Borchardt, M. A. 2010 Method 1615: measurement of norovirus and enterovirus occurrence in water by culture and RT-qPCR. *Journal of Visualized Experiments* **16** (107), e52646.

Fox, J. D., Briggs, M., Ward, P. A. & Tedder, R. S. 1990 Human herpesvirus 6 in salivary glands. *The Lancet* **336** (8715), 590–593.

Fumian, T. M., Fioretti, J. M., Lun, J. H., Dos Santos, I. A. L., White, P. A. & Miagostovich, M. P. 2019 Detection of norovirus epidemic genotypes in raw sewage using next generation sequencing. *Environment International* **123**, 282–291.

Galani, A., Aalizadeh, R., Kostakis, M., Markou, A., Alygizakis, N., Lytras, T., Adamopoulos, P. G., Peccia, J., Thompson, D. C., Kontou, A. & Karagiannidis, A. 2022 SARS-CoV-2 wastewater surveillance data can predict hospitalizations and ICU admissions. *Science of The Total Environment* **804**, 150151.

Ganesh, A. & Lin, J. 2013 Waterborne human pathogenic viruses of public health concern. *International Journal of Environmental Health Research* **23** (6), 544–564.

Garner, E., Davis, B., Milligan, E., Blair, M. F., Keenum, I., Maile-Moskowitz, A., Pan, J., Gnegy, M., Liguiori, K., Gupta, S., Prussin II, A. J., Marr, L. C., Heath, L. S., Vikesland, P. J., Zhang, L. & Pruden, A. 2021 Next generation sequencing approaches to evaluate water and wastewater quality. *Water Research* **194**, 116907.

Gillings, M. R., Gaze, W. H., Pruden, A., Smalla, K., Tiedje, J. M. & Zhu, Y. G. 2015 Using the class 1 integron-integrase gene as a proxy for anthropogenic pollution. *ISME Journal* **9** (6), 1269–1279.

Gonzalez, R., Curtis, K., Bivins, A., Bibby, K., Weir, M. H., Yetka, K., Thompson, H., Keeling, D., Mitchell, J. & Gonzalez, D. 2020 COVID-19 surveillance in Southeastern Virginia using wastewater-based epidemiology. *Water Research* **186**, 116296.

González-Mariño, I., Baz-Lomba, J. A., Alygizakis, N. A., Andrés-Costa, M. J., Bade, R., Bannwarth, A., Barron, L. P., Been, F., Benaglia, L., Berset, J. D. & Bijlsma, L. 2020 Spatio-temporal assessment of illicit drug use at large scale: evidence from 7 years of international wastewater monitoring. *Addiction* **115** (1), 109–120.

Gracia-Lor, E., Castiglioni, S., Bade, R., Been, F., Castrignanò, E., Covaci, A., González-Mariño, I., Hapeshi, E., Kasprzyk-Hordern, B., Kinyua, J. & Lai, F. Y. 2017 Measuring biomarkers in wastewater as a new source of epidemiological information: current state and future perspectives. *Environment International* **99**, 131–150.

Guajardo-Leiva, S., Chnaiderman, J., Gaggero, A. & Díez, B. 2020 Metagenomic insights into the sewage RNA virosphere of a large city. *Viruses* **12**, 1050.

Guillois, Y., Abravanel, F., Miura, T., Pavio, N., Vaillant, V., Lhomme, S., Le Guyader, F. S., Rose, N., Le Saux, J. C., King, L. A. & Izopet, J. 2016 High proportion of asymptomatic infections in an outbreak of hepatitis E associated with a spit-roasted piglet, France, 2013. *Clinical Infectious Diseases* **62** (3), 351–357.

Gullberg, E., Cao, S., Berg, O. G., Ilbäck, C., Sandegren, L., Hughes, D. & Andersson, D. I. 2011 Selection of resistant bacteria at very low antibiotic concentrations. *PLoS Pathogens* **7** (7), e1002158.

Gupta, C. L., Tiwari, R. K. & Cytryn, E. 2020a Platforms for elucidating antibiotic resistance in single genomes and complex metagenomes. *Environment International* **138**, 105667.

Gupta, S., Parker, J., Smits, S., Underwood, J. & Dolwani, S. 2020b Persistent viral shedding of SARS-CoV-2 in faeces–a rapid review. *Colorectal Disease* **22** (6), 611–620.

Habtewold, J., McCarthy, D., McBean, E., Law, I., Goodridge, L., Habash, M. & Murphy, H. M. 2022 Passive sampling, a practical method for wastewater-based surveillance of SARS-CoV-2. *Environmental Research* **204**, 112058.

Haramoto, E., Kitajima, M., Hata, A., Torrey, J. R., Masago, Y., Sano, D. & Katayama, H. 2018 A review on recent progress in the detection methods and prevalence of human enteric viruses in water. *Water Research* **135**, 168–186. https://doi.org/10.1016/j.watres.2018.02.004.

Harris, S., Morris, C., Morris, D., Cormican, M. & Cummins, E. 2014 Antimicrobial resistant Escherichia coli in the municipal wastewater system: effect of hospital effluent and environmental fate. *Science of the Total Environment* **468**, 1078–1085.

Harris, T. F., Yelowitz, A. & Courtemanche, C. 2021 Did COVID-19 change life insurance offerings? *Journal of Risk and Insurance* **88** (4), 831–861.

Hart, O. E. & Halden, R. U. 2020 Simulated 2017 nationwide sampling at 13,940 major US sewage treatment plants to assess seasonal population bias in wastewater-based epidemiology. *Science of the Total Environment* **727**, 138406.

Harman, C., Reid, M. & Thomas, K. V. 2011 In situ calibration of a passive sampling device for selected illicit drugs and their metabolites in wastewater, and subsequent year-long assessment of community drug usage. *Environ. Sci. Technol.* **45** (13), 5676–5682.

Hassoun-Kheir, N., Stabholtz, Y., Kreft, J. U., De La Cruz, R., Romalde, J. L., Nesme, J., Sørensen, S. J., Smets, B. F., Graham, D. & Paul, M. 2020 Comparison of antibiotic-resistant bacteria and antibiotic resistance genes abundance in hospital and community wastewater: a systematic review. *Science of the Total Environment* **743**, 140804.

Hassoun-Kheir, N., Stabholz, Y., Kreft, J., de la Cruz, R., Dechensne, A., Smets, B. F., Romalde, J., Lema, A., Balboa, S., Garcia-Riestra, C., Torres-Sangiao, E., Neuberger, A., Graham, D., Quintela-Baluja, M., Stekel, D. J., Graham, J., Pruden, A., Nesme, J. & Paul, M. 2021 EMBRACE-WATERS statement: Recommendations for reporting of studies on antimicrobial resistance in wastewater and related aquatic environments. *One Health* **13**, 100339.

Hata, A., Katayama, H. & Furumai, H. 2015 Organic substances interfere with reverse transcription-quantitative PCR-based virus detection in water samples. *Applied and Environmental Microbiology* **81** (5), 1585–1593.

Hata, A., Kitajima, M., Haramoto, E., Lee, S., Ihara, M., Gerba, C. P. & Tanaka, H. 2018 Next-generation amplicon sequencing identifies genetically diverse human astroviruses, including recombinant strains, in environmental waters. *Scientific Reports* **8**, 1–9.

Heijnen, L. & Medema, G. 2011 Surveillance of influenza A and the pandemic influenza A (H1N1) 2009 in sewage and surface water in the Netherlands. *Journal of Water and Health* **9** (3), 434–442.

Hendriksen, R. S., Munk, P., Njage, P., Van Bunnik, B., McNally, L., Lukjancenko, O., Röder, T., Nieuwenhuijse, D., Pedersen, S. K., Kjeldgaard, J. & Kaas, R. S. 2019 Global monitoring of antimicrobial resistance based on metagenomics analyses of urban sewage. *Nature Communications* **10** (1), 1–1.

Herold, M., d'Hérouël, A. F., May, P., Delogu, F., Wienecke-Baldacchino, A., Tapp, J., Walczak, C., Wilmes, P., Cauchie, H. M., Fournier, G. & Ogorzaly, L. 2021 Genome sequencing of SARS-CoV-2 allows monitoring of variants of concern through wastewater. *Water* **13** (21), 3018.

Hillary, L. S., Farkas, K., Maher, K. H., Lucaci, A., Thorpe, J., Distaso, M. A., Gaze, W. H., Paterson, S., Burke, T., Connor, T. R., McDonald, J. E., Malham, S. K. & Jones, D. L. 2021 Monitoring SARS-CoV-2 in municipal wastewater to evaluate the success of lockdown measures for controlling COVID-19 in the UK. *Water Research* **200**, 117214.

Hiller, C. X., Hübner, U., Fajnorova, S., Schwartz, T. & Drewes, J. E. 2019 Antibiotic microbial resistance (AMR) removal efficiencies by conventional and advanced wastewater treatment processes: a review. *Science of the Total Environment* **685**, 596–608.

Hoffmann, T. & Alsing, J. 2021 Faecal shedding models for SARS-CoV-2 RNA amongst hospitalised patients and implications for wastewater-based epidemiology. *medRxiv*.

Holmes, E. C., Goldstein, S. A., Rasmussen, A. L., Robertson, D. L., Crits-Christoph, A., Wertheim, J. O., Anthony, S. J., Barclay, W. S., Boni, M. F., Doherty, P. C., Farrar, J., Geoghegan, J. L., Jiang, X., Beibowitz, J. L., Neil, S. J. D., Skern, T., Weiss, S. R., Worobey, M., Andersen, K. G., Garry, R. F. & Rambaut, A. 2021 *Cell* **184** (19), 4848–4856.

Hovi, T., Shulman, L. M., Van Der Avoort, H., Deshpande, J., Roivainen, M. & De Gourville, E. M. 2012 Role of environmental poliovirus surveillance in global polio eradication and beyond. *Epidemiology & Infection* **140** (1), 1–13.

Hu, Y., Lu, S., Song, Z., Wang, W., Hao, P., Li, J., Zhang, X., Yen, H. L., Shi, B., Li, T. & Guan, W. 2013a Association between adverse clinical outcome in human disease caused by novel influenza A H7N9 virus and sustained viral shedding and emergence of antiviral resistance. *The Lancet* **381** (9885), 2273–2279.

Hu, Y., Yang, X., Qin, J., Lu, N., Cheng, G., Wu, N., Pan, Y., Li, J., Zhu, L., Wang, X. & Meng, Z. 2013b Metagenome-wide analysis of antibiotic resistance genes in a large cohort of human gut microbiota. *Nature Communications* **4** (1), 1–7.

Huijbers, P. M., Larsson, D. J. & Flach, C. F. 2020 Surveillance of antibiotic resistant Escherichia coli in human populations through urban wastewater in ten European countries. *Environmental Pollution* **261**, 114200.

Huizer, M., ter Laak, T. L., de Voogt, P. & van Wezel, A. P. 2021 Wastewater-based epidemiology for illicit drugs: A critical review on global data. *Water Research* **207**, 117789.

Hutinel, M., Huijbers, P. M. C., Fick, J., Åhrén, C., Larsson, D. G. J. & Flach, C. F. 2019 Population-level surveillance of antibiotic resistance in Escherichia coli through sewage analysis. *Eurosurveillance* **24** (37), 1800497.

Iaconelli, M., Valdazo-González, B., Equestre, M., Ciccaglione, A. R., Marcantonio, C., Della Libera, S. & La Rosa, G. 2017 Molecular characterization of human adenoviruses in urban wastewaters using next generation and Sanger sequencing. *Water Research* **121**, 240–247.

Islam, M. A., Islam, M., Hasan, R., Hossain, M. I., Nabi, A., Rahman, M., Goessens, W. H. F., Endtz, H. P., Boehm, A. B. & Faruque, S. M. 2017 Environmental spread of New Delhi metallo-β-lactamase-1-producing multidrug-resistant bacteria in Dhaka, Bangladesh. *Applied Environmental Microbiology* **83**, e00793-17.

Izquierdo-Lara, R., Elsinga, G., Heijnen, L., Munnink, B. B. O., Schapendonk, C. M., Nieuwenhuijse, D., Kon, M., Lu, L., Aarestrup, F. M., Lycett, S. & Medema, G. 2021 Monitoring SARS-CoV-2 circulation and diversity through community wastewater sequencing, the Netherlands and Belgium. *Emerging Infectious Diseases* **27** (5), 1405.

Jahne, M. A., Brinkman, N. E., Keely, S. P., Zimmerman, B. D., Wheaton, E. A. & Garland, J. L. 2020 Droplet digital PCR quantification of norovirus and adenovirus in decentralized wastewater and greywater collections: implications for onsite reuse. *Water Research* **169**, 115213. https://doi.org/https://doi.org/10.1016/j.watres.2019.115213.

Jartti, T., Lehtinen, P., Vanto, T., Hartiala, J., Vuorinen, T., Mäkelä, M. J. & Ruuskanen, O. 2006 Evaluation of the efficacy of prednisolone in early wheezing induced by rhinovirus or respiratory syncytial virus. *The Pediatric Infectious Disease Journal* **25** (6), 482–488.

Jelic, A., Rodriguez-Mozaz, S., Barceló, D. & Gutierrez, O. 2015 Impact of in-sewer transformation on 43 pharmaceuticals in a pressurized sewer under anaerobic conditions. *Water Research* **68**, 98–108.

Jeulin, H., Salmon, A., Bordigoni, P. & Venard, V. 2011 Diagnostic value of quantitative PCR for adenovirus detection in stool samples as compared with antigen detection and cell culture in haematopoietic stem cell transplant recipients. *Clinical Microbiology and Infection* **17** (11), 1674–1680.

Jiao, Y. N., Zhou, Z. C., Chen, T., Wei, Y. Y., Zheng, J., Gao, R. X. & Chen, H. 2018 Biomarkers of antibiotic resistance genes during seasonal changes in wastewater treatment systems. *Environmental Pollution* **234**, 79–87.

Jmii, H., Gharbi-Khelifi, H., Assaoudi, R. & Aouni, M. 2021 Detection of SARS-CoV-2 in the sewerage system in Tunisia: a promising tool to confront COVID-19 pandemic. *Future Virology* **17**, 751–759.

Jones, L., Ronan, J., McHugh, B., McGovern, E. & Regan, F. 2015 Emerging priority substances in the aquatic environment: a role for passive sampling in supporting WFD monitoring and compliance. *Analytical Methods* **7** (19), 7976–7984.

Jones, D. L., Baluja, M. Q., Graham, D. W., Corbishley, A., McDonald, J. E., Malham, S. K., Hillary, L. S., Connor, T. R., Gaze, W. H., Moura, I. B. & Wilcox, M. H. 2020 Shedding of SARS-CoV-2 in feces and urine and its potential role in person-to-person transmission and the environment-based spread of COVID-19. *Science of the Total Environment* **749**, 141364.

Jong, M. C., Harwood, C. R., Blackburn, A., Snape, J. R. & Graham, D. W. 2020 Impact of redox conditions on antibiotic resistance conjugative gene transfer frequency and plasmid fate in wastewater ecosystems. *Environmental Science & Technology* **54** (23), 14984–14993.

Jorgensen, J. H. & Turnidge, J. D. 2015 Susceptibility test methods: dilution and disk diffusion methods. *In*: J. H. Jorgensen, K. C. Carroll, G. Funke, M. A. Pfaller, M. L. Landry, S. S. Richter & D. W. Warnock, eds. Manual of Clinical Microbiology, 11th edn. American Society for Microbiology Press, pp. 1253-1273.

Kachhawaha, A. S., Nagarnaik, P. M., Jadhav, M., Pudale, A., Labhasetwar, P. K. & Banerjee, K. 2017 Optimization of a modified QuEChERS methods for multiresidue analysis of pharmaceuticals and personal care products in sewage and surface water by LC-MS/MS. *Journal of AOAC International* **100** (3), 592–597.

Kankaanpää, A., Ariniemi, K., Heinonen, M., Kuoppasalmi, K. & Gunnar, T. 2016 Current trends in Finnish drug abuse: wastewater based epidemiology combined with other national indicators. *Science of the Total Environment* **568**, 864–874.

Kaplan, J. B. 2010 Biofilm dispersal. *Journal of Dental Research* **89** (3), 205–218.

Kaplan, E. H., Wang, D., Wang, M., Malik, A. A., Zulli, A. & Peccia, J. 2020 Aligning SARS-CoV-2 indicators via an epidemic model: application to hospital admissions and RNA detection in sewage sludge. *Health Care Manag. Sci* **24** (2), 320 –329.

Kariuki, S. M., Selhorst, P., Norman, J., Cohen, K., Rebe, K., Williamson, C. & Dorfman, J. R. 2020 Detectable HIV-1 in semen in individuals with very low blood viral loads. *Virology Journal* **17** (1), 1–5.

Karthikeyan, S., Levy, J. I., De Hoff, P., Humphrey, G., Birmingham, A., Jepsen, K., Farmer, S., Tubb, H. M., Valles, T., Tribelhorn, C. E. & Tsai, R. 2021 Wastewater sequencing uncovers early, cryptic SARS-CoV-2 variant transmission. *medRxiv.*

Kasprzyk-Hordern, B., Proctor, K., Jagadeesan, K., Watkins, S., Standerwick, R., Barden, R. & Barnett, J. 2021 Diagnosing down-the-drain disposal of unused pharmaceuticals at a river catchment level: unrecognized sources of environmental contamination that require nontechnological solutions. *Environmental Science & Technology* **55** (17), 11657–11666.

Kaye, K. S., Gupta, V., Mulgirigama, A., Joshi, A. V., Scangarella-Oman, N. E., Yu, K., Ye, G. & Mitrani-Gold, F. S. 2021 Antimicrobial resistance trends in urine Escherichia coli isolates from adult and adolescent females in the United States from 2011–2019: rising ESBL strains and impact on patient management. *Clinical Infectious Diseases* **73** (11), 1992–1999.

Kevill, J. L., Pellett, C., Farkas, K., Brown, M. R., Bassano, I., Denise, H., McDonald, J. E., Malham, S. K., Porter, J., Warren, J. & Evens, N. P. 2022 A comparison of precipitation and filtration-based SARS-CoV-2 recovery methods and the influence of temperature, turbidity, and surfactant load in urban wastewater. *Science of the Total Environment* **808**, 151916.

Kirby, A. E., Walters, M. S., Jennings, W. C., Fugitt, R., LaCross, N., Mattioli, M., Marsh, Z. A., Roberts, V. A., Mercante, J. W., Yoder, J. & Hill, V. R. 2021 Using wastewater surveillance data to support the COVID-19 response – United States, 2020–2021. *Morbidity and Mortality Weekly Report* **70** (36), 1242.

Knapp, C. W., Engemann, C. A., Hanson, M. L., Keen, P. L., Hall, K. J. & Graham, D. W. 2008 Indirect evidence of transposon-mediated selection of antibiotic resistance genes in aquatic systems at low-level oxytetracycline exposures. *Environmental Science & Technology* **42** (14), 5348–5353.

Kolahchi, Z., De Domenico, M., Uddin, L. Q., Cauda, V., Grossmann, I., Lacasa, L., Grancini, G., Mahmoudi, M. & Rezaei, N. 2021 COVID-19 and Its global economic impact. *Advances in Experimental Medicine and Biology* **1318**, 825–837.

Kotwal, G. & Cannon, J. L. 2014 Environmental persistence and transfer of enteric viruses. *Current Opinion in Virology* **4**, 37–43.

Krizman-Matasic, I., Senta, I., Kostanjevecki, P., Ahel, M. & Terzic, S. 2019 Long-term monitoring of drug consumption patterns in a large-sized European city using wastewater-based epidemiology: comparison of two sampling schemes for the assessment of multiannual trends. *Science of the Total Environment* **647**, 474–485.

Kwiatkowska, R., Ruhaak, A., Kasprzyk-Hordern, B., Hassard, F., Lundy, L., Di Cesare, M., Hickman, M. & Singer, A. 2021 *Wastewater-based Epidemiology and Group Privacy: the Elephant in the Sewer?* doi:10.31219/osf.io/q8rcd.

Lago, P. M., Gary Jr, H. E., Pérez, L. S., Cáceres, V., Olivera, J. B., Puentes, R. P., Corredor, M. B., Jímenez, P., Pallansch, M. A. & Cruz, R. G. 2003 Poliovirus detection in wastewater and stools following an immunization campaign in Havana, Cuba. *International Journal of Epidemiology* **32** (5), 772–777.

Lai, F. Y., Ort, C., Gartner, C., Carter, S., Prichard, J., Kirkbride, P., Bruno, R., Hall, W., Eaglesham, G. & Mueller, J. F. 2011 Refining the estimation of illicit drug consumptions from wastewater analysis: co-analysis of prescription pharmaceuticals and uncertainty assessment. *Water Research* **45** (15), 4437–4448.

Lamba, M., Graham, D. W. & Ahammad, S. Z. 2017 Hospital wastewater releases of carbapenem-resistance pathogens and genes in urban India. *Environmental Science & Technology* **51** (23), 13906–13912.

Lamba, M., Gupta, S., Shukla, R., Graham, D. W., Sreekrishnan, T. R. & Ahammad, S. Z. 2018 Carbapenem resistance exposures via wastewaters across New Delhi. *Environment International* **119**, 302–308.

Lampejo, T. 2020 Influenza and antiviral resistance: an overview. *European Journal of Clinical Microbiology & Infectious Diseases* **39** (7), 1201–1208.

Lampos, V. 2016 Flu Detector: Estimating influenza-like illness rates from online user-generated content. arXiv preprint arXiv:1612.03494.

Lampos, V., Miller, A. C., Crossan, S. & Stefansen, C. 2015 Advances in nowcasting influenza-like illness rates using search query logs. *Scientific Reports* **5** (1), 1–10.

La Rosa, G., Pourshaban, M., Iaconelli, M. & Muscillo, M. 2010 Quantitative real-time PCR of enteric viruses in influent and effluent samples from wastewater treatment plants in Italy. *Annali dell'Istituto superiore di sanita* **46**, 266–273.

Larsen, J., Raisen, C. L., Ba, X., Sadgrove, N. J., Padilla-González, G. F., Simmonds, M. S., Loncaric, I., Kerschner, H., Apfalter, P., Hartl, R. & Deplano, A. 2022 Emergence of methicillin resistance predates the clinical use of antibiotics. *Nature* **602**, 135–141.

Leonard, A. F., Zhang, L., Balfour, A. J., Garside, R., Hawkey, P. M., Murray, A. K., Ukoumunne, O. C. & Gaze, W. H. 2018 Exposure to and colonisation by antibiotic-resistant E. coli in UK coastal water users: environmental surveillance, exposure assessment, and epidemiological study (Beach Bum Survey). *Environment International* **114**, 326–333.

Li, J., Gao, J., Thai, P. K., Sun, X., Mueller, J. F., Yuan, Z. & Jiang, G. 2018 Stability of illicit drugs as biomarkers in sewers: from lab to reality. *Environmental Science & Technology* **52** (3), 1561–1570.

Li, D., Reid, C. J., Kudinha, T., Jarocki, V. M. & Djordjevic, S. P. 2020 Genomic analysis of trimethoprim-resistant extraintestinal pathogenic Escherichia coli and recurrent urinary tract infections. *Microbial Genomics* **6** (12). https://doi.org/10.1099/mgen.0.000475.

Li, L., Nesme, J., Quintela-Baluja, M., Balboa, S., Hashsham, S., Williams, M. R., Yu, Z., Sørensen, S. J., Graham, D. W., Romalde, J. L., Dechesne, A. & Smets, B. F. 2021 Extended-spectrum β-lactamase and carbapenemase genes are substantially and sequentially reduced during conveyance and treatment of urban sewage. *Environmental Science & Technology* **55** (9), 5939–5949.

Li, X., Kulandaivelu, J., Guo, Y., Zhang, S., Shi, J., O'Brien, J., Arora, S., Kumar, M., Sherchan, S. P., Honda, R., Jackson, G., Luby, S. P. & Jiang, G. 2022 SARS-CoV-2 shedding sources in wastewater and implications for wastewater-based epidemiology. *Journal of Hazardous Materials* **432**, 128667.

Lion, T., Kosulin, K., Landlinger, C., Rauch, M., Preuner, S., Jugovic, D., Pötschger, U., Lawitschka, A., Peters, C., Fritsch, G. & Matthes-Martin, S. 2010 Monitoring of adenovirus load in stool by real-time PCR permits early detection of impending invasive infection in patients after allogeneic stem cell transplantation. *Leukemia* **24** (4), 706–714.

Locarnini, S. 2004 Movecular virology of hepatitis B virus. *Semin. Liver Dis.* **24**, 3–10.

Lopardo, L., Petrie, B., Proctor, K., Youdan, J., Barden, R. & Kasprzyk-Hordern, B. 2019a Estimation of community-wide exposure to bisphenol A via water fingerprinting. *Environment International* **125**, 1–8.

Lopardo, L., Rydevik, A. & Kasprzyk-Hordern, B. 2019b A new analytical framework for multi-residue analysis of chemically diverse endocrine disruptors in complex environmental matrices utilising ultra-performance liquid chromatography coupled with high-resolution tandem quadrupole time-of-flight mass spectrometry. *Analytical and Bioanalytical Chemistry* **411** (3), 689–704.

Ludden, C., Reuter, S., Judge, K., Gouliouris, T., Blane, B., Coll, F., Naydenova, P., Hunt, M., Tracey, A., Hopkins, K. L. & Brown, N. M. 2017 Sharing of carbapenemase-encoding plasmids between Enterobacteriaceae in UK sewage uncovered by MinION sequencing. *Microbial Genomics* **3** (7). https://doi.org/10.1099/mgen.0.000114.

Mancini, P., Bonanno Ferraro, G., Iaconelli, M., Suffredini, E., Valdazo-González, B., Della Libera, S., Divizia, M. & La Rosa, G. 2019 Molecular characterization of human Sapovirus in untreated sewage in Italy by amplicon-based Sanger and next-generation sequencing. *Journal of Applied Microbiology* **126** (1), 324–331.

Marano, R. B., Fernandes, T., Manaia, C. M., Nunes, O., Morrison, D., Berendonk, T. U., Kreuzinger, N., Telson, T., Corno, G., Fatta-Kassinos, D. & Merlin, C. 2020 A global multinational survey of cefotaxime-resistant coliforms in urban wastewater treatment plants. *Environment International* **144**, 106035.

Mardal, M., Miserez, B., Bade, R., Portolés, T., Bischoff, M., Hernández, F. & Meyer, M. R. 2016 3-Fluorophenmetrazine, a fluorinated analogue of phenmetrazine: studies on in vivo metabolism in rat and human, in vitro metabolism in human CYP isoenzymes and microbial biotransformation in Pseudomonas Putida and wastewater using GC and LC coupled to (HR)-MS techniques. *Journal of Pharmaceutical and Biomedical Analysis* **128**, 485–495.

Martínez, J. L. 2008 Antibiotics and antibiotic resistance genes in natural environments. *Science* **321** (5887), 365–367.

Martinez, J. L. 2009 Environmental pollution by antibiotics and by antibiotic resistance determinants. *Environmental Pollution* **157** (11), 2893–2902.

Martínez, J. L., Coque, T. M. & Baquero, F. 2015 What is a resistance gene? Ranking risk in resistomes. *Nature Reviews Microbiology* **13** (2), 116–123.

Matheson, C. D., Gurney, C., Esau, N. & Lehto, R. 2010 Assessing PCR inhibition from humic substances. *The Open Enzyme Inhibition Journal* **3** (1), 38–45.

Maunula, L., Klemola, P., Kauppinen, A., Söderberg, K., Nguyen, T., Pitkänen, T., Kaijalainen, S., Simonen, M. L., Miettinen, I. T., Lappalainen, M. & Laine, J. 2009 Enteric viruses in a large waterborne outbreak of acute gastroenteritis in Finland. *Food and Environmental Virology* **1** (1), 31–36.

McCall, A. K., Bade, R., Kinyua, J., Lai, F. Y., Thai, P. K., Covaci, A., Bijlsma, L., Van Nuijs, A. L. & Ort, C. 2016a Critical review on the stability of illicit drugs in sewers and wastewater samples. *Water Research* **88**, 933–947.

McCall, A. K., Scheidegger, A., Madry, M. M., Steuer, A. E., Weissbrodt, D. G., Vanrolleghem, P. A., Kraemer, T., Morgenroth, E. & Ort, C. 2016b Influence of different sewer biofilms on transformation rates of drugs. *Environmental Science & Technology* **50** (24), 13351–13360.

McCall, C., Wu, H., O'Brien, E. & Xagoraraki, I. 2021 Assessment of enteric viruses during a hepatitis outbreak in Detroit MI using wastewater surveillance and metagenomic analysis. *Journal of Applied Microbiology* **131** (3), 1539–1554.

McCann, C. M., Christgen, B., Roberts, J. A., Su, J. Q., Arnold, K. E., Gray, N. D., Zhu, Y. G. & Graham, D. W. 2019 Understanding drivers of antibiotic resistance genes in High Arctic soil ecosystems. *Environment International* **125**, 497–504.

McClary-Gutierrez, J. S., Mattioli, M. C., Marcenac, P., Silverman, A. I., Boehm, A. B., Bibby, K., Balliet, M., de los Reyes, F. L., Gerrity, D., Griffith, J. F., Holden, P. A., Katehis, D., Kester, G., LaCross, N., Lipp, E. K., Meiman, J., Noble, R. T., Brossard, D. & McLellan, S. L. 2021 SARS-CoV-2 wastewater surveillance for public health action. *Emerging Infectious Diseases* **27** (9), 1–8. https://doi.org/10.3201/eid2709.210753.

McMinn, B. R., Korajkic, A., Kelleher, J., Herrmann, M. P., Pemberton, A. C., Ahmed, W., Villegas, E. N. & Oshima, K. 2021 Development of a large volume concentration method for recovery of coronavirus from wastewater. *Science of The Total Environment* **774**, 145727.

Medema, G., Heijnen, L., Elsinga, G., Italiaander, R. & Brouwer, A. 2020 Presence of SARS-Coronavirus-2 RNA in sewage and correlation with reported COVID-19 prevalence in the early stage of the epidemic in the Netherlands. *Environmental Science & Technology Letters* **7** (7), 511–516.

Medina, W. R., Eramo, M., Tu, A., Fahrenfeld, M. & L, N. 2020 Sewer biofilm microbiome and antibiotic resistance genes as function of pipe material, source of microbes, and disinfection: field and laboratory studies. *Environmental Science: Water Research and Technology* **2020** (6), 2122–2137.

Meir-Gruber, L., Manor, Y., Gefen-Halevi, S., Hindiyeh, M. Y., Mileguir, F., Azar, R., Smollan, G., Belausov, N., Rahav, G., Shamiss, A. & Mendelson, E. 2016 Population screening using sewage reveals pan-resistant bacteria in hospital and community samples. *PloS one* **11** (10), e0164873.

Melchor-Martínez, E. M., Jiménez-Rodríguez, M. G., Martínez-Ruiz, M., Peña-Benavides, S. A., Iqbal, H. M., Parra-Saldívar, R. & Sosa-Hernández, J. E. 2021 Antidepressants surveillance in wastewater: overview extraction and detection. *Case Studies in Chemical and Environmental Engineering* **3**, 100074.

Mendelson, M. & Matsoso, M. P. 2015 The World Health Organization global action plan for antimicrobial resistance. *SAMJ: South African Medical Journal* **105** (5), 325–325.

Menni, C., Klaser, K., May, A., Polidori, L., Capdevila, J., Louca, P., Sudre, C. H., Nguyen, L. H., Drew, D. A., Merino, J. & Hu, C. 2021 Vaccine side-effects and SARS-CoV-2 infection after vaccination in users of the COVID symptom study app in the UK: a prospective observational study. *The Lancet Infectious Diseases* **21** (7), 939–949.

Minodier, L., Charrel, R. N., Ceccaldi, P. E., Van Der Werf, S., Blanchon, T., Hanslik, T. & Falchi, A. 2015 Prevalence of gastrointestinal symptoms in patients with influenza, clinical significance, and pathophysiology of human influenza viruses in faecal samples: what do we know? *Virology Journal* **12** (1), 1–9.

Modjarrad, K., Giersing, B., Kaslow, D. C., Smith, P. G. & Moorthy, V. S. 2016 WHO consultation on respiratory syncytial virus vaccine development report from a World Health Organization meeting held on 23–24 March 2015. *Vaccine* **34** (2), 190–197.

Morse, S. S., Mazet, J. A., Woolhouse, M., Parrish, C. R., Carroll, D., Karesh, W. B., Zambrana-Torrelio, C., Lipkin, W. I. & Daszak, P. 2012 Prediction and prevention of the next pandemic zoonosis. *The Lancet* **380** (9857), 1956–1965.

Moschet, C., Vermeirssen, E. L., Singer, H., Stamm, C. & Hollender, J. 2015 Evaluation of in-situ calibration of Chemcatcher passive samplers for 322 micropollutants in agricultural and urban affected rivers. *Water Research* **71**, 306–317.

Mughini-Gras, L., Dorado-García, A., van Duijkeren, E., van den Bunt, G., Dierikx, C. M., Bonten, M. J., Bootsma, M. C., Schmitt, H., Hald, T., Evers, E. G. & de Koeijer, A. 2019 Attributable sources of community-acquired carriage of Escherichia coli containing β-lactam antibiotic resistance genes: a population-based modelling study. *The Lancet Planetary Health* **3** (8), e357–e369.

Muirhead, A., Zhu, K., Brown, J., Basu, M., Brinton, M. A., Costa, F., Hayat, M. J. & Stauber, C. E. 2020 Zika Virus RNA persistence in sewage. *Environmental Science & Technology Letters* **7** (9), 659–664.

Murray, A. K., Zhang, L., Yin, X., Zhang, T., Buckling, A., Snape, J. & Gaze, W. H. 2018 Novel insights into selection for antibiotic resistance in complex microbial communities. *MBio* **9** (4), e00969-18.

Naughton, C. C., Roman, F. A., Alvarado, A. G. F., Tariqi, A. Q., Deeming, M. A., Bibby, K., Bivins, A., Rose, J. B., Medema, G., Ahmed, W. & Katsivelis, P. 2021 Show us the data: Global COVID-19 wastewater monitoring efforts, equity, and gaps. *medRxiv*.

Niederman, J. C., Miller, G., Pearson, H. A., Pagano, J. S. & Dowaliby, J. M. 1976 Infectious mononucleosis: Epstein–Barr-Virus shedding in saliva and the oropharynx. *New England Journal of Medicine* **294** (25), 1355–1359.

Nieto, A., Borrull, F., Pocurull, E. & Marce, R. M. 2010 Pressurized liquid extraction: a useful technique to extract pharmaceuticals and personal-care products from sewage sludge. *TrAC Trends in Analytical Chemistry* **29** (7), 752–764.

Nkuipou-Kenfack, E., Zuerbig, P. & Mischak, H. 2017 The long path towards implementation of clinical proteomics: exemplified based on CKD273. *PROTEOMICS–Clinical Applications* **11** (5–6), 1600104.

O'Brien, J. W., Choi, P. M., Li, J., Thai, P. K., Jiang, G., Tscharke, B. J., Mueller, J. F. & Thomas, K. V. 2019 Evaluating the stability of three oxidative stress biomarkers under sewer conditions and potential impact for use in wastewater-based epidemiology. *Water Research* **166**, 115068.

Ogorzaly, L., Walczak, C., Galloux, M., Etienne, S., Gassilloud, B. & Cauchie, H.-M. 2015 Human adenovirus diversity in water samples using a next-generation amplicon sequencing approach. *Food and Environmental Virology* **7**, 112–121. https://doi.org/10.1007/s12560-015-9194-4.

Ort, C., Lawrence, M. G., Rieckermann, J. & Joss, A. 2010 Sampling for pharmaceuticals and personal care products (PPCPs) and illicit drugs in wastewater systems: are your conclusions valid? A critical review. *Environmental Science & Technology* **44** (16), 6024–6035.

Ort, C., Van Nuijs, A. L., Berset, J. D., Bijlsma, L., Castiglioni, S., Covaci, A., de Voogt, P., Emke, E., Fatta-Kassinos, D., Griffiths, P. & Hernández, F. 2014 Spatial differences and temporal changes in illicit drug use in Europe quantified by wastewater analysis. *Addiction* **109** (8), 1338–1352.

Paget, J., Spreeuwenberg, P., Charu, V., Taylor, R. J., Iuliano, A. D., Bresee, J., Simonsen, L. & Viboud, C. 2019 Global mortality associated with seasonal influenza epidemics: new burden estimates and predictors from the GLaMOR project. *Journal of Global Health* **9** (2), 020421.

Pal, C., Bengtsson-Palme, J., Kristiansson, E. & Larsson, D. J. 2016 The structure and diversity of human, animal and environmental resistomes. *Microbiome* **4** (1), 1–15.

Pallares-Vega, R., Leal, L. H., Fletcher, B. N., Vias-Torres, E., van Loosdrecht, M. C., Weissbrodt, D. G. & Schmitt, H. 2021 Annual dynamics of antimicrobials and resistance determinants in flocculent and aerobic granular sludge treatment systems. *Water Research* **190**, 116752.

Pandey, D., Verma, S., Verma, P., Mahanty, B., Dutta, K., Daverey, A. & Arunachalam, K. 2021 SARS-CoV-2 in wastewater: challenges for developing countries. *International Journal of Hygiene and Environmental Health* **231**, 113634.

Pang, L., Yang, P., Ge, L., Du, J. & Zhang, H. 2017 Accelerated solvent extraction combined with solid phase extraction for the determination of organophosphate esters from sewage sludge compost by UHPLC-MS/MS. *Analytical and Bioanalytical Chemistry* **409** (5), 1435–1440.

Papapetropoulou, M. & Vantarakis, A. C. 1998 Detection of adenovirus outbreak at a municipal swimming pool by nested PCR amplification. *Journal of Infection* **36** (1), 101–103.

Park, S. K., Lee, C. W., Park, D. I., Woo, H. Y., Cheong, H. S., Shin, H. C., Ahn, K., Kwon, M. J. & Joo, E. J. 2021 Detection of SARS-CoV-2 in fecal samples from patients with asymptomatic and mild COVID-19 in Korea. *Clinical Gastroenterology and Hepatology* **19** (7), 1387–1394.

Pärnänen, K. M., Narciso-da-Rocha, C., Kneis, D., Berendonk, T. U., Cacace, D., Do, T. T., Elpers, C., Fatta-Kassinos, D., Henriques, I., Jaeger, T. & Karkman, A. 2019 Antibiotic resistance in European wastewater treatment plants mirrors the pattern of clinical antibiotic resistance prevalence. *Science Advances* **5** (3), eaau9124.

Peccia, J., Zulli, A., Brackney, D. E., Grubaugh, N. D., Kaplan, E. H., Casanovas-Massana, A., Ko, A. I., Malik, A. A., Wang, D., Wang, M. & Warren, J. L. 2020 Measurement of SARS-CoV-2 RNA in wastewater tracks community infection dynamics. *Nature Biotechnology* **38** (10), 1164–1167.

Petrie, B., Youdan, J., Barden, R. & Kasprzyk-Hordern, B. 2016 Multi-residue analysis of 90 721 emerging contaminants in liquid and solid environmental matrices by ultra-high-performance 722 liquid chromatography tandem mass spectrometry. *J. Chromatogr. A* **1431**, 64–78.

Petrie, B., Proctor, K., Youdan, J., Barden, R. & Kasprzyk-Hordern, B. 2017 Critical evaluation of monitoring strategy for the multi-residue determination of 90 chiral and achiral micropollutants in effluent wastewater. *Science of the Total Environment* **579**, 569–578.

Piddock, L. J. 2016 Assess drug-resistance phenotypes, not just genotypes. *Nature Microbiology* **1** (8), 1–2.

Plowright, R. K., Reaser, J. K., Locke, H., Woodley, S. J., Patz, J. A., Becker, D. J., Oppler, G., Hudson, P. J. & Tabor, G. M. 2021 Land use-induced spillover: a call to action to safeguard environmental, animal, and human health. *The Lancet Planetary Health* **5** (4), E237–E245.

Poirel, L., Kämpfer, P. & Nordmann, P. 2002 Chromosome-encoded Ambler class A β-lactamase of Kluyvera georgiana, a probable progenitor of a subgroup of CTX-M extended-spectrum *β*-lactamases. *Antimicrobial Agents and Chemotherapy* **46** (12), 4038–4040.

Pollock, J., Muwonge, A., Hutchings, M. R., Mainda, G., Bronsvoort, B. M., Gally, D. L. & Corbishley, A. 2020 Resistance to change: AMR gene dynamics on a commercial pig farm with high antimicrobial usage. *Scientific Reports* **10** (1), 1–10.

Polo, D., Quintela-Baluja, M., Corbishley, A., Jones, D. L., Singer, A. C., Graham, D. W. & Romalde, J. L. 2020 Making waves: wastewater-based epidemiology for COVID-19–approaches and challenges for surveillance and prediction. *Water Research* **186**, 116404.

Poulsen, C. S., Kaas, R. S., Aarestrup, F. M. & Pamp, S. J. 2021 Standard sample storage conditions have an impact on inferred microbiome composition and antimicrobial resistance patterns. *Microbiology Spectrum* **9** (2), e01387–21.

Pouwels, K. B., House, T., Robotham, J. V., Birrell, P., Gelman, A. B., Bowers, N., Boreham, I., Thomas, H., Lewis, J., Bell, I. & Bell, J. I. 2020 Community prevalence of SARS-CoV-2 in England: results from the ONS coronavirus infection survey pilot. *MedRxiv*.

Prestinaci, F., Pezzotti, P. & Pantosti, A. 2015 Antimicrobial resistance: a global multifaceted phenomenon. *Pathogens and Global Health* **109** (7), 309–318.

Proctor, K., Petrie, B., Barden, R., Arnot, T. & Kasprzyk-Hordern, B. 2019 Multi-residue ultra-performance liquid chromatography coupled with tandem mass spectrometry method for comprehensive multi-class anthropogenic compounds of emerging concern analysis in a catchment-based exposure-driven study. *Analytical and Bioanalytical Chemistry* **411** (27), 7061–7086.

Pruden, A., Vikesland, P. J., Davis, B. C. & de Roda Husman, A. M. 2021 Seizing the moment: now is the time for integrated global surveillance of antimicrobial resistance in wastewater environments. *Current Opinion in Microbiology* **64**, 91–99.

Quintela-Baluja, M., Abouelnaga, M., Romalde, J., Su, J., Yu, Y., Gomez-Lopez, M., Smets, B., Zhu, Y. & Graham, D. W. 2019 Spatial ecology of a wastewater network defines the antibiotic resistance genes in downstream receiving waters. *Water Research* **162**, 347–357.

Quintela-Baluja, M., Frigon, D., Abouelnaga, M., Jobling, K., Romalde, J. L., Lopez, M. G. & Graham, D. W. 2021 Dynamics of integron structures across a wastewater network–Implications to resistance gene transfer. *Water Research* **206**, 117720.

Rajko-Nenow, P., Waters, A., Keaveney, S., Flannery, J., Tuite, G., Coughlan, S., O'Flaherty, V. & Doré, W. 2013 Norovirus genotypes present in oysters and in effluent from a wastewater treatment plant during the seasonal peak of infections in Ireland in 2010. *Applied and Environmental Microbiology* **79** (8), 2578–2587.

Randazzo, W., Truchado, P., Cuevas-Ferrando, E., Simón, P., Allende, A. & Sánchez, G. 2020 SARS-CoV-2 RNA in wastewater anticipated COVID-19 occurrence in a low prevalence area. *Water Research* **181**, 115942.

Rani, A., Ranjan, R., Metwally, A., Brennan, D. C., Finn, P. W. & Perkins, D. L. 2020 Abundance of multidrug resistance efflux pumps in the urinary metagenome of kidney transplant patients. *Biomed Res Int* **2020**, 5421269.

Recht, J., Schuenemann, V. J. & Sánchez-Villagra, M. R. 2020 Host diversity and origin of zoonoses: the ancient and the new. *Animals* **10** (9), 1672.

Reid, M. J., Baz-Lomba, J. A., Ryu, Y. & Thomas, K. V. 2014 Using biomarkers in wastewater to monitor community drug use: a conceptual approach for dealing with new psychoactive substances. *Science of the Total Environment* **487**, 651–658.

Rice, J. & Kasprzyk-Hordern, B. 2019 A new paradigm in public health assessment: water fingerprinting for protein markers of public health using mass spectrometry. *TrAC Trends in Analytical Chemistry* **119**, 115621.

Rice, J., Kannan, A. M., Castrignanò, E., Jagadeesan, K. & Kasprzyk-Hordern, B. 2020 Wastewater-based epidemiology combined with local prescription analysis as a tool for temporal monitoring of drugs trends-A UK perspective. *Science of the Total Environment* **735**, 139433.

Rimayi, C., Chimuka, L., Gravell, A., Fones, G. R. & Mills, G. A. 2019 Use of the Chemcatcher® passive sampler and time-of-flight mass spectrometry to screen for emerging pollutants in rivers in Gauteng Province of South Africa. *Environmental Monitoring and Assessment* **191** (6), 1–20.

Rios, G., Lacoux, C., Leclercq, V., Diamant, A., Lebrigand, K., Lazuka, A., Soyeux, E., Lacroix, S., Fassy, J., Couesnon, A. & Thiery, R. 2021 Monitoring SARS-CoV-2 variants alterations in nice neighborhoods by wastewater nanopore sequencing. *The Lancet Regional Health-Europe* **10**, 100202.

Rose, G., Shaw, A. G., Sim, K., Wooldridge, D. J., Li, M. S., Gharbia, S., Misra, R. & Kroll, J. S. 2017 Antibiotic resistance potential of the healthy preterm infant gut microbiome. *PeerJ* **5**, e2928.

Rousis, N. I., Gracia-Lor, E., Reid, M. J., Baz-Lomba, J. A., Ryu, Y., Zuccato, E., Thomas, K. V. & Castiglioni, S. 2020 Assessment of human exposure to selected pesticides in Norway by wastewater analysis. *Science of The Total Environment* **723**, 138132.

Ruotsalainen, M., Hyvärinen, M. K., Piippo-Savolainen, E. & Korppi, M. 2013 Adolescent asthma after rhinovirus and respiratory syncytial virus bronchiolitis. *Pediatric Pulmonology* **48** (7), 633–639.

Sabrià, A., Pintó, R. M., Bosch, A., Bartolomé, R., Cornejo, T., Torner, N., Martínez, A., de Simón, M., Domínguez, A. & Guix, S. 2016 Norovirus shedding among food and healthcare workers exposed to the virus in outbreak settings. *Journal of Clinical Virology* **82**, 119–125.

Santos, V. S., Gurgel, R. Q., Cuevas, L. E. & Martins-Filho, P. R. 2020 Prolonged fecal shedding of SARS-CoV-2 in pediatric patients. A quantitative evidence synthesis. *Journal of Pediatric Gastroenterology and Nutrition* **71** (2), 150–152.

Sattar, S. A. & Westwood, J. C. 1977 Isolation of apparently wild strains of poliovirus type 1 from sewage in the Ottawa area. *Canadian Medical Association Journal* **116** (1), 25.

Schang, C., Crosbie, N. D., Nolan, M., Poon, R., Wang, M., Jex, A., John, N., Baker, L., Scales, P., Schmidt, J. & Thorley, B. R. 2021 Passive sampling of SARS-CoV-2 for wastewater surveillance. *Environmental Science & Technology* **55** (15), 10432–10441.

Senta, I., Krizman, I., Ahel, M. & Terzic, S. 2014 Assessment of stability of drug biomarkers in municipal wastewater as a factor influencing the estimation of drug consumption using sewage epidemiology. *Science of the Total Environment* **487**, 659–665.

Senta, I., Gracia-Lor, E., Borsotti, A., Zuccato, E. & Castiglioni, S. 2015 Wastewater analysis to monitor use of caffeine and nicotine and evaluation of their metabolites as biomarkers for population size assessment. *Water Research* **74**, 23–33.

Senta, I., Rodriguez-Mozaz, S., Corominas, L. & Petrovic, M. 2020 Wastewater-based epidemiology to assess human exposure to personal care and household products–A review of biomarkers, analytical methods, and applications. *Trends in Environmental Analytical Chemistry* **28**, e00103.

Shang, Y., Li, H. & Zhang, R. 2021 Effects of pandemic outbreak on economies: evidence from business history context. *Front. Public Health.* **9**, 632043 doi: 10.3389/fpubh.2021.632043.

Sharp, P. M. & Hahn, B. H. 2011 Origins of HIV and the AIDS pandemic. *Cold Spring Harbor Perspectives in Medicine* **1** (1), a006841.

Sidhu, J. P., Ahmed, W., Palmer, A., Smith, K., Hodgers, L. & Toze, S. 2017 Optimization of sampling strategy to determine pathogen removal efficacy of activated sludge treatment plant. *Environmental Science and Pollution Research* **24** (23), 19001–19010.

Sims, N. & Kasprzyk-Hordern, B. 2020 Future perspectives of wastewater-based epidemiology: monitoring infectious disease spread and resistance to the community level. *Environment International* **139**, 105689.

Singer, A. C., Järhult, J. D., Grabic, R., Khan, G. A., Fedorova, G., Fick, J., Lindberg, R. H., Bowes, M. J., Olsen, B. & Söderström, H. 2013 Compliance to oseltamivir among two populations in Oxfordshire, United Kingdom affected by influenza A (H1N1) pdm09, November 2009–a wastewater epidemiology study. *PLoS One* **8** (4), e60221.

Sun, Y., Shen, Y. X., Liang, P., Zhou, J., Yang, Y. & Huang, X. 2016 Multiple antibiotic resistance genes distribution in ten large-scale membrane bioreactors for municipal wastewater treatment. *Bioresource Technology* **222**, 100–106.

Swift, C. L., Isanovic, M., Velez, K. E. C. & Norman, R. S. 2021 Community-level SARS-CoV-2 sequence diversity revealed by wastewater sampling. *Science of the Total Environment* **801**, 149691.

Tacconelli, E., De Angelis, G., Cataldo, M. A., Mantengoli, E., Spanu, T., Pan, A., Corti, G., Radice, A., Stolzuoli, L., Antinori, S. & Paradisi, F. 2009 Antibiotic usage and risk of colonization and infection with antibiotic-resistant bacteria: a hospital population-based study. *Antimicrobial Agents and Chemotherapy* **53** (10), 4264–4269.

Tang, J. W., Lam, T. T., Zaraket, H., Lipkin, W. I., Drews, S. J., Hatchette, T. F., Heraud, J. M., Koopmans, M. P., Abraham, A. M., Baraket, A. & Bialasiewicz, S. 2017 Global epidemiology of non-influenza RNA respiratory viruses: data gaps and a growing need for surveillance. *The Lancet Infectious Diseases* **17** (10), e320–e326.

Tang, S., He, C., Thai, P., Vijayasarathy, S., Mackie, R., Toms, L. M. L., Thompson, K., Hobson, P., Tscharke, B., O'Brien, J. W. & Mueller, J. F. 2020 Concentrations of phthalate metabolites in Australian urine samples and their contribution to the per capita loads in wastewater. *Environment International* **137**, 105534.

Tao, Z., Chen, P., Cui, N., Lin, X., Ji, F., Liu, Y., Xiong, P., Zhang, L., Xu, Q. & Song, Y. 2020 Detection of enteroviruses in urban sewage by next generation sequencing and its application in environmental surveillance. *Science of the Total Environment* **728**, 138818.

Taylor, S. C., Lapperriere, G. & Germain, H. 2017 Droplet digital PCR versus qPCR for gene expression analysis with low abundant targets: from variable nonsense to publication quality data. *Scientific Reports* **7**, 2409.

Thomaidis, N. S., Gago-Ferrero, P., Ort, C., Maragou, N. C., Alygizakis, N. A., Borova, V. L. & Dasenaki, M. E. 2016 Reflection of socioeconomic changes in wastewater: licit and illicit drug use patterns. *Environmental Science & Technology* **50** (18), 10065–10072.

Thomas, K. V. & Reid, M. J. 2011 What else can the analysis of sewage for urinary biomarkers reveal about communities? *Science of the Total Environment* **45, 18**, 7611–7612.

Thomas, K. V., Bijlsma, L., Castiglioni, S., Covaci, A., Emke, E., Grabic, R., Hernández, F., Karolak, S., Kasprzyk-Hordern, B., Lindberg, R. H. & de Alda, M. L. 2012 Comparing illicit drug use in 19 European cities through sewage analysis. *Science of the Total Environment* **432**, 432–439.

Townsend, I., Jones, L., Broom, M., Gravell, A., Schumacher, M., Fones, G. R., Greenwood, R. & Mills, G. A. 2018 Calibration and application of the Chemcatcher® passive sampler for monitoring acidic herbicides in the river Exe, UK catchment. *Environmental Science and Pollution Research* **25** (25), 25130–25142.

Tsimtsiou, Z., Tsiligianni, I., Papaioannou, A., Gougourelas, D., Kolokas, K., Gkizlis, V., Chatzimanolis, E. & Domeyer, P. J. 2021 Understanding what people think and feel about adult vaccinations and the associated barriers in Greece: development and validation of the attitude towards adult vaccination (ATAVAC) scale. *Health & Social Care in the Community* **29** (3), 818–828.

UKHSA 2021 *English Surveillance Programme for Antimicrobial Utilisation and Resistance (ESPAUR) Report 2020 to 2021 English Surveillance Programme for Antimicrobial Utilisation and Resistance (ESPAUR) Report 2020 to 2021.* (publishing.service.gov.uk) (accessed 24 November 2021).

University of California 2022 *COVID19Poops Dashboard.* Available from: 502 https://www.covid19wbec.org/covidpoops19. (accessed 18 January 2021).

USEPA 2001 *Method 1602: Male-Specific (F+) and Somatic Coliphage in Water by Single Agar Layer (SAL) Procedure. EPA Document 821-R-01-029 April.*

van Binnendijk, R. S., van den Hof, S., van den Kerkhof, H., Kohl, R. H., Woonink, F., Berbers, G. A., Conyn-van Spaendonck, M. A. & Kimman, T. G. 2003 Evaluation of serological and virological tests in the diagnosis of clinical and subclinical measles virus infections during an outbreak of measles in The Netherlands. *The Journal of Infectious Diseases* **188** (6), 898–903.

van Doorn, A. S., Meijer, B., Frampton, C. M., Barclay, M. L. & de Boer, N. K. 2020 Systematic review with meta-analysis: SARS-CoV-2 stool testing and the potential for faecal-oral transmission. *Alimentary Pharmacology & Therapeutics* **52** (8), 1276–1288.

van Nuijs, A. L., Covaci, A., Beyers, H., Bervoets, L., Blust, R., Verpooten, G., Neels, H. & Jorens, P. G. 2015 Do concentrations of pharmaceuticals in sewage reflect prescription figures? *Environmental Science and Pollution Research* **22** (12), 9110–9118.

Van Poelvoorde, L. A., Saelens, X., Thomas, I. & Roosens, N. H. 2020 Next-generation sequencing: an eye-opener for the surveillance of antiviral resistance in influenza. *Trends in Biotechnology* **38** (4), 360–367.

Verlicchi, P., Al Aukidy, M., Jelic, A., Petrović, M. & Barceló, D. 2014 Comparison of measured and predicted concentrations of selected pharmaceuticals in wastewater and surface water: a case study of a catchment area in the Po Valley (Italy). *Science of the Total Environment* **470**, 844–854.

Vincent-Hubert, F., Morga, B., Renault, T. & Le Guyader, F. S. 2017 Adsorption of norovirus and ostreid herpesvirus type 1 to polymer membranes for the development of passive samplers. *Journal of Applied Microbiology* **122** (4), 1039–1047.

Volkmann, H., Schwartz, T., Kirchen, S., Stofer, C. & Obst, U. 2007 Evaluation of inhibition and cross-reaction effects on real-time PCR applied to the total DNA of wastewater samples for the quantification of bacterial antibiotic resistance genes and taxon-specific targets. *Molecular and Cellular Probes* **21** (2), 125–133.

von Linstow, M., Eugen-Olsen, J., Koch, A., Winther, T. N., Westh, H. & Hogh, B. 2006 Excretion patterns of human metapneumovirus and respiratory syncytial virus among young children. *European Journal of Medical Research* **11** (8), 329.

Wade, M., Jones, D., Singer, A., Hart, A., Corbishley, A., Spence, C., Morvan, M., Zhang, C., Pollock, M., Hoffmann, T. & Singleton, P. 2020 *Wastewater COVID-19 Monitoring in the UK: Summary for SAGE–19/11/20.* Joint Biosecurity Centre.

Wade, M. J., Jacomo, A. L., Armenise, E., Brown, M. R., Bunce, J. T., Cameron, G. J., Fang, Z., Farkas, K., Gilpin, D. F., Graham, D. W. & Grimsley, J. M. 2022 Understanding and managing uncertainty and variability for wastewater monitoring beyond the pandemic: lessons learned from the United Kingdom national COVID-19 surveillance programmes. *Journal of Hazardous Materials* **424**, 127456.

Wagner, M., Lampos, V., Cox, I. J. & Pebody, R. 2018 The added value of online user-generated content in traditional methods for influenza surveillance. *Sci. Rep.* **8**, 13963. https://doi.org/10.1038/s41598-018-32029-6.

Wang, F., Qiao, M., Su, J., Zhou, X. & Zhu, Y. 2014 High throughput profiling of antibiotic resistance genes in urban park soils with reclaimed water irrigation. *Environ. Sci. Technol.* **48** (16), 9079–9085.

Wang, Y., Zhang, D., Wang, X. & Fu, Q. 2020 How does COVID-19 affect China's insurance market? *Emerging Markets Finance and Trade* **56** (10), 2350–2362.

Ward, R. L. & Ashley, C. S. 1980 Effects of wastewater sludge and its detergents on the stability of rotavirus. *Applied and Environmental Microbiology* **39** (6), 1154–1158.

WEF 2021 Antimicrobial Resistance and Water: The risks and costs for economies and societies. World Economic Forum Briefing Paper August 2021.

Weiss, A., Jellingsø, M. & Sommer, M. O. A. 2020 Spatial and temporal dynamics of SARS-CoV-2 in COVID-19 patients: a systematic review and meta-analysis. *EBioMedicine* **58**, 102916.

Whaley, S. G., Berkow, E. L., Rybak, J. M., Nishimoto, A. T., Barker, K. S. & Rogers, P. D. 2017 Azole antifungal resistance in Candida albicans and emerging non-albicans Candida species. *Frontiers in Microbiology* **7**, 2173.

White, L., Hopkins, K. L., Meunier, D., Perry, C. L., Pike, R., Wilkinson, P., Pickup, R. W., Cheesbrough, J. & Woodford, N. 2016 Carbapenemase-producing Enterobacteriaceae in hospital wastewater: a reservoir that may be unrelated to clinical isolates. *Journal of Hospital Infection* **93** (2), 145–151.

Wigginton, K. R., Ye, Y. & Ellenberg, R. M. 2015 Emerging investigators series: the source and fate of pandemic viruses in the urban water cycle. *Environmental Science: Water Research & Technology* **1** (6), 735–746.

Woolhouse, M. E., Howey, R., Gaunt, E., Reilly, L., Chase-Topping, M. & Savill, N. 2008 Temporal trends in the discovery of human viruses. *Proceedings of the Royal Society B: Biological Sciences* **275** (1647), 2111–2115.

Woolhouse, M., Ward, M., Van Bunnik, B. & Farrar, J. 2015 Antimicrobial resistance in humans, livestock and the wider environment. *Philosophical Transactions of the Royal Society B: Biological Sciences* **370** (1670), 20140083.

World Health Organization 2017 *Prioritization of Pathogens to Guide Discovery, Research and Development of new Antibiotics for Drug-Resistant Bacterial Infections, Including Tuberculosis (No. WHO/EMP/IAU/2017.12).* World Health Organization.

World Health Organization 2021a *Global Antimicrobial Resistance and use Surveillance System (GLASS) Report: 2021.*

World Health Organization 2021b *WHO Integrated Global Surveillance on ESBL-Producing E. Coli Using a 'One Health' Approach: Implementation and Opportunities.*

Wyres, K. L., Hawkey, J., Mirčeta, M., Judd, L. M., Wick, R. R., Gorrie, C. L., Pratt, N. F., Garlick, J. S., Watson, K. M., Pilcher, D. V. & McGloughlin, S. A. 2021 Genomic surveillance of antimicrobial resistant bacterial colonisation and infection in intensive care patients. *BMC Infectious Diseases* **21** (1), 1–11.

Xu, Y., Li, X., Zhu, B., Liang, H., Fang, C., Gong, Y., Guo, Q., Sun, X., Zhao, D., Shen, J. & Zhang, H. 2020 Characteristics of pediatric SARS-CoV-2 infection and potential evidence for persistent fecal viral shedding. *Nature Medicine* **26** (4), 502–505.

Ye, Y., Ellenberg, R. M., Graham, K. E. & Wigginton, K. R. 2016 Survivability, partitioning, and recovery of enveloped viruses in untreated municipal wastewater. *Environmental Science & Technology* **50** (10), 5077–5085.

Zeng, J., Pan, Y., Yang, J., Hou, M., Zeng, Z. & Xiong, W. 2019 Metagenomic insights into the distribution of antibiotic resistome between the gut-associated environments and the pristine environments. *Environment International* **126**, 346–354.

Zhang, N., Lu, M., Duan, X., Liu, C. C. & Wang, H. 2019 In situ calibration of direct analysis in real time-mass spectrometry for direct quantification: urine excretion rate index creatinine as an example. *Talanta* **201**, 134–142.

Zhang, A. N., Gaston, J. M., Dai, C. L., Zhao, S., Poyet, M., Groussin, M., Yin, X., Li, L. G., van Loosdrecht, M., Topp, E. & Gillings, M. R. 2021a An omics-based framework for assessing the health risk of antimicrobial resistance genes. *Nature Communications* **12** (1), 1–11.

Zhang, Y., Cen, M., Hu, M., Du, L., Hu, W., Kim, J. J. & Dai, N. 2021b Prevalence and persistent shedding of fecal SARS-CoV-2 RNA in patients with COVID-19 infection: a systematic review and meta-analysis. *Clinical and Translational Gastroenterology* **12** (4), e00343.

Zhao, Y., Xia, Q., Yin, Y. & Wang, Z. 2016 Comparison of droplet digital PCR and quantitative PCR assays for quantitative detection of Xanthomonas citri Subsp. Citri. *PLoS One* **11** (7), e0159004.

Zheng, Y., Xu, H., Yang, M., Zeng, Y., Chen, H., Liu, R., Li, Q., Zhang, N. & Wang, D. 2020. Epidemiological characteristics and clinical features of 32 critical and 67 noncritical cases of COVID-19 in Chengdu. *Journal of Clinical Virology* **127**, 104366. https://doi.org/10.1016/j.jcv.2020.104366.

Zhu, Z., Liu, Y., Xu, L., Guan, W., Zhang, X., Qi, T., Shi, B., Song, Z., Liu, X., Wan, Y. & Tian, D. 2015 Extra-pulmonary viral shedding in H7N9 Avian Influenza patients. *Journal of Clinical Virology* **69**, 30–32.

Zuccato, E., Chiabrando, C., Castiglioni, S., Calamari, D., Bagnati, R., Schiarea, S. & Fanelli, R. 2005 Cocaine in surface waters: a new evidence-based tool to monitor community drug abuse. *Environmental Health* **4** (1), 1–7.

First received 22 January 2022; accepted in revised form 28 July 2022. Available online 12 August 2022